Environmental Modeling and Health Risk Analysis (ACTS/RISK)

T0190301

Environmental Modeling and Health Risk Analysis (ACTS/RISK)

Mustafa M. Aral

 Springer

Prof. Mustafa M. Aral
Georgia Institute of Technology
School of Civil and Environmental Engineering
790 Atlantic Dr. NW, Mason Bldg.
Atlanta Georgia 30332-0355
USA
mustafa.aral@ce.gatech.edu

Additional material to this book can be downloaded from http://extra.springer.com

ISBN 978-94-007-9219-7   ISBN 978-90-481-8608-2 (eBook)
DOI 10.1007/978-90-481-8608-2
Springer Dordrecht Heidelberg London New York

© Springer Science+Business Media B.V. 2010
Softcover re-print of the Hardcover 1st edition 2010
No part of this work may be reproduced, stored in a retrieval system, or transmitted in any form or by any
means, electronic, mechanical, photocopying, microfilming, recording or otherwise, without written
permission from the Publisher, with the exception of any material supplied specifically for the purpose
of being entered and executed on a computer system, for exclusive use by the purchaser of the work.

Cover illustration: Environmental exposure theme superimposed on the Vitruvian Man by Leonardo
Da Vinci. Image created by Mustafa M. Aral.

Printed on acid-free paper

Springer is part of Springer Science+Business Media (www.springer.com)

Preface

Environmental Modeling and Health Risk Analysis (ACTS/RISK)

The purpose of this book is to provide the reader with an integrated perspective on several fields. First, it discusses the fields of environmental modeling in general and multimedia (the term "multimedia" is used throughout the text to indicate that environmental transformation and transport processes are discussed in association with three environmental media: air, groundwater and surface water pathways) environmental transformation and transport processes in particular; it also provides a detailed description of numerous mechanistic models that are used in these fields. Second, this book presents a review of the topics of exposure and health risk analysis. The Analytical Contaminant Transport Analysis System (ACTS) and Health RISK Analysis (RISK) software tools are an integral part of the book and provide computational platforms for all the models discussed herein. The most recent versions of these two software tools can be downloaded from the publisher's web site. The author recommends registering the software on the web download page so that users can receive updates about newer versions of the software.

This book is intended to support instruction in environmental quality modeling in surface water, air and groundwater pathways that are linked to exposure and health risk analysis. The book is based on the author's many years of experience in field applications as well as in classroom teaching on these topics. As such, it should serve as a valuable tool and reference for practicing professionals as well as for graduate and undergraduate students. It is currently used as a textbook in the School of Civil and Environmental Engineering at the Georgia Institute of Technology in a senior-level undergraduate class that frequently includes graduate students.

Studies on environmental quality modeling can be traced back to G.I. Taylor's seminal work on diffusion processes in 1921. Since then, the scientific field of the analysis of advection, diffusion and dispersion processes has experienced considerable progress with the introduction of many innovative concepts, principles and applications. Now, in what may be identified as the field of air and water quality modeling, there are numerous models which make use of these principles in providing solutions to complex problems. There are a number of excellent textbooks which are available on air, groundwater and surface water quality modeling, and they are cited throughout the book. However, this book differs from others in that its first purpose is to provide an integrated view of basic principles of environmental quality modeling in these seemingly different media, as well as a comprehensive review of the analytical models that are available in these fields. The reader will recognize that the basic principles and modeling tools described in each chapter for air, groundwater and surface water pathways are very similar, at least in mathematical form. This is because both air and water are fluids, so the transport and transformation processes in each medium are governed by similar processes that mathematically follow the same principles. The author hopes that practicing professionals and students who are interested in these topics will find this integrated approach useful.

During the past decade, exposure and health risk analysis has also become an important and inseparable part of environmental assessment. This is primarily because we, as scientists and engineers, are no longer only interested in environmental characterization, remediation and management, but we are also interested in health effects or ecosystem hazards associated with pollutants which are present in the environment or released into the environment. Similarly, numerous models for exposure and health risk analysis have been developed in the literature as well, and it is the second purpose of this book to provide an integrated view of these topics and link them to environmental transformation and transport models.

The models discussed in this book have been coded for easy access and use in ACTS and RISK. These two software tools have been developed as WINDOWS^{TM} based applications to provide professionals in environmental engineering and environmental health with a compact resource for the analytical methods discussed in this text. These models can be used to evaluate the transport and transformation of contaminants in multimedia environments (air, surface water, soil and groundwater) as well as to perform exposure and health risk analysis. The multimedia transport and transformation models included in this software and reviewed in this book are state-of-the-art analytic tools that can be used in the analysis of steady state and time dependent contaminant transformation and transport processes. For the analysis of cases that may involve uncertainty in input parameters, Monte Carlo methods have been developed and are dynamically linked with all pathway models included in the ACTS and RISK software. In the Monte Carlo analysis mode, all or a selected subset of input parameters of a particular model may be characterized in terms of statistical distributions provided in the software, allowing statistical distributions of contaminant concentrations or exposure risk to be evaluated at a particular exposure point at a particular point in time.

Currently, the total number of environmental transformation and transport and exposure models that are included in the ACTS and RISK software exceeds 300 (when all subcategory models for each pathway are considered). These models may be used to evaluate and understand how chemical and pathway specific properties of the media impact the transformation and transport and the overall exposure and health risk assessment processes. In addition to serving as a documentation of the technical background of the models used in the ACTS and RISK software, the book also serves as the reference document for these software tools.

Author

Dr. Mustafa M. Aral is a professor and the director of the Multimedia Environmental Simulations Laboratory, a research center at the School of Civil and Environmental Engineering, Georgia Institute of Technology, Atlanta, Georgia, USA. He received his B.S. degree in civil engineering from Middle East Technical University, Ankara, Turkey, and M.S. and Ph.D. degrees from the School of Civil and Environmental Engineering, Georgia Institute of Technology. He specializes in the fields of environmental modeling, water resources engineering, mathematical, numerical and optimization methods, and exposure and health risk analysis. He is a registered professional engineer in the State of Georgia with more than 35 years of research and consulting experience in environmental quality and quantity modeling, environmental fluid mechanics, optimal management of engineered and environmental systems, site characterization and health risk assessment studies. Dr. Aral has authored and co-authored more than 300 technical publications and reports. He is the editor of the International Journal on Water Quality, Exposure and Health published by Springer publishers and he is also an associate editor of several other technical journals in water resources and environmental engineering. Dr. Aral has received numerous technical awards in publication excellence and research development categories from several state and federal agencies, and professional organizations. He is the principal author and developer of the ACTS and RISK software that is used in this book.

Acknowledgements

Writing a book on the complex subject of this book required significant amount of dedication and time, which most of us give unselfishly to achieve the lifelong objectives dictated by a career choice we made long ago. We also know that everything in life is a zero-sum game, i.e. some things must give way to accomplish others. Over the years, in the course of this project, several things had to give way and this may be the proper place to acknowledge those, who stood on the sidelines or who in one way or another contributed to the successful completion of this book. Foremost among those is my family, whose kindness and support should be mentioned first. Faculty and research associates of the author, who were instrumental in the completion of this project are too many to include here without the risk of omitting some of them in error. However, without any order of priority, I would like to thank many scientists from the Agency for Toxic Substances and Disease Registry (ATSDR) at the Department of Health and Human Services (DHHS), USA, who supported various activities of our research program at the Georgia Institute of Technology over the years. Finally, I must also acknowledge the numerous graduate students who contributed to the overall effort at different stages.

This book is intended to serve as a comprehensive resource for the advancements and contributions made in environmental transformation and transport modeling in general and the associated health risk assessment topics in particular. The selection of the title, "Environmental Modeling and Health Risk Analysis" stems from an ambitious objective of providing a review of air, surface water and groundwater quality modeling topics and linking these models with exposure and health risk analysis. Given this broad objective, one quickly recognizes that these topics cannot be covered at an elementary level. Although every effort has been made to provide the reader with basic information that is necessary to evaluate and use the models described in this book, it is expected that the reader will be familiar with environmental transformation and transport modeling principles, statistical methods and general modeling concepts at the level of an undergraduate course on these topics. This preliminary knowledge is necessary for the reader to make full use of the information and modeling tools provided in this book.

In this information age, what good would a book on multimedia environmental modeling and health risk analysis be if it is not accompanied by a software tool that provides the reader with a user-friendly interface to access the models? The extended title of the book, "ACTS and RISK" reflects the second ambitious goal of this project. ACTS and RISK are two accompanying software tools that I and my students developed in a WINDOWS-XP^{TM} based environment. In the ACTS software tool the analytical solutions of the contaminant transformation and transport models described in the book are included in a user friendly platform which makes all of these models readily available for use in the analysis of complex problems that may be described for each pathway. In RISK, the exposure models described in the book are presented again in a user-friendly platform for the user to evaluate exposure of populations or individuals to contaminants by linking the environmental model simulation results to multi-pathway exposure routes that may exist in the environment. The ACTS and RISK software are linked through transferable data structures and can be used in an integrated manner first to evaluate environmental transformation and transport of contaminants in multimedia environmental pathways and then to link the results of this analysis to health risk evaluation. Thus, in addition to being a reference text book for environmental modeling, this book also serves as the primary resource for the ACTS and RISK software tools which can be used in practical applications as demonstrated in the included exercises. Furthermore, both software include a stochastic analysis tool (Monte Carlo Analysis), that extends the application of all models included in ACTS and RISK to the stochastic analysis mode. The development of this comprehensive software was a very challenging task and could not have been completed without the contributions of several researchers at Georgia Tech and technical personnel at ATSDR. I would like to again acknowledge the contributions of several students and research engineers involved in the development of this software. Without their efforts, this project could not have been completed at its present level. A list of these research faculty and former students are included in the "contributors list" in Appendix 3. Finally, last but not least, the contributions of numerous undergraduate students should be acknowledged, who helped with the debugging of the software application over the years in a classroom environment.

Atlanta, Georgia, USA   Mustafa M. Aral
February 2010

neither of these models were considering harmonious ways of combining these two variables. A review and combination of the better parts of the earlier management philosophies revealed the concept of Sustainable Environmental Management as the resolution of the conflict between these two variables. The basic philosophy behind this approach was outlined in the Brundtland World Commission on Environment and Development report (Colby 1990). In this approach the environment and economics are considered to be parts of a mutually supporting ecosystem. Long-term issues and long-term solutions became a key consideration for this model.

In this evolution, it is not very difficult to anticipate the next step if one asks the right questions. The proper questions to ask may be: "Can there be a global or uniform environmental policy and a management model?" "Based on their characteristics, should different issues, different regions and different applications have unique environmental management strategies?"; and "Are we really worried about proper environmental management strategies for the sake of the environment and economics, or are there other reasons?" It seems that there is a more important reason behind this evolution that led us to the concept of sustainable environmental management. We now realize that one of the main purposes of our search for a proper environmental management model is the protection of populations from adverse environmental stressors which may lead to detrimental health effects. These effects, which are an outcome of the selected environmental management strategy, may be economic or environmental in nature, or directly related to health effects. When we include the concept of "health effects" in the overall picture and emphasize and recognize its importance, the policies, principles and methods we work with will change considerably. In the earlier management models that were discussed above the "health effects" issue was not forgotten, but it was not emphasized as the primary policy issue. In the earlier management models health effects appeared mostly as a concept, as an issue to worry about, measure, document and possibly correct. Again, in the environmental management models summarized above, the emphasis on health effects appears to be more pronounced when the management model emphasizes environmental concerns rather than economic ones. When we realize the importance and the depth of the "health effects" concept, we will quickly abandon the philosophy of the Sustainable Environmental Management, mainly because it still reflects a two-dimensional perspective of a three-dimensional problem (Aral 2005).

Now we should expand the preliminary premise that is stated at the beginning of this chapter. The premise that considers a multitude of present-day environmental issues can be restated as: All human interventions to natural environments, our demand for built environments and natural or forced disasters will sooner or later be associated with health issues.

Based on this premise, it is clear that all environmental intrusion will have health effect implications imbedded in them. This is apparent and repeatedly acknowledged in most current studies on environmental management. Accordingly, the next stage of environmental management model we work with may be identified as Environmental Management for Sustainable Populations. Here the term environment implies built and/or natural environments. In this management model, the goal will be the long-term harmonious management of economic resources and environmental preservation, for the health, safety and prosperity of sustainable populations. Policy decisions that will be made in this phase will now explicitly include a very complex element, i.e. the dynamic and also very delicate "population" or "human" element. When populations are explicitly included in the overall management framework, social policy, ethics and health issues assume a very important role in the management strategy. It can be anticipated that in order to identify and resolve the problems of this management style, scientists from the fields of social sciences, public policy, health sciences, basic sciences, and also engineering need to work more closely together than they have in the past. To establish this working environment more barriers need to be broken, new rules need to be established, and more importantly, a common language has to be introduced. Technological, scientific and holistic advances made in each field need to be translated into this common language and put to use for the ultimate goal of maintaining sustainable populations. In this approach economic incentives and environmental constraints have to be considered harmoniously, with the main emphasis placed on the protection and preservation of human health and sustainability of populations.

As expected, this management model will require the collaboration of various disciplines in the overall framework. When scientists from diverse backgrounds are involved in an applied or theoretical problem, the first issue that needs to be addressed is the difference in technical language used by the team members and the implied meaning and importance of the terms, as well as the expectations for the input data requirements of a specific problem and the expectations for the outcome of the individual and team effort. For these issues to be resolved in a harmonious way, members of the team should spend considerable effort on learning the terminology and expectations for each other's scientific fields and the limitations or boundaries of knowledge that a team member may bring to the group. In the following chapters of this book our goal is to define this common language, and provide an understanding of data requirements and the uncertainty in input data as well as in outcomes. We strive to do this as much as possible from an environmental modeling perspective, without creating a new language or principles of our own.

In this book, the discussion of this topic starts with a discussion of environmental transformation and transport concepts. Toxic perturbations introduced by humans on the present-day earth have raised fundamental questions about our understanding of various processes in environmental, geochemical and biological cycles, and in the transformation of the toxic substances in multimedia environments. More and more, scientists are recognizing that the environment must be considered as a whole, and scientific and regulatory approaches alike must take into account the complex interactions between multimedia and inter-media pathways to understand the propagation of these toxic perturbations in the environment. These observations have imposed new demands on environmental and health scientists for understanding the interactions between these cycles and their effect on the environment and ultimately on human health.

We must also distinguish the difference between the two synonymous terms that are used routinely in this field, namely *contamination* and *pollution*. These two terms appear frequently in the technical literature and also in the common language but may be used in different contexts when transformation and transport processes are considered as opposed to environmental health concerns. Contamination is commonly associated with the presence of an alien substance in the environment. The adverse effects of this alien substance are not implied. Pollution is commonly associated with adverse ecological or health effects. Contamination that is present in the environment at low concentrations and thus does not cause adverse environmental or health effects, should not be confused with pollution. This definition conforms to the observation that there are naturally occurring contaminants in the environment and most of them do not cause health hazards at low concentration levels. It is when these contaminant levels exceed a certain threshold and cause health effects that they are classified as environmental pollution. That is the case with arsenic, which exists in most soils around the world as a contaminant. However, the contaminant levels of arsenic observed in the delta of Bangladesh elevates it to a pollution level with significant health effects outcome (Meharg 2005).

Contaminants released into the environment are distributed among environmental media such as air, water, soil and vegetation as a result of complex physical, chemical and biological processes. Thus, environmental pollution by contaminants is a multimedia and multi-pathway migration problem, and environmental assessment, exposure risk assessment and the design of appropriate environmental remediation and exposure evaluation methods require that we carefully consider the transport, transformation, and accumulation of pollutants in the environment as a whole. Methods proposed to evaluate environmental or exposure characterization in this envirosphere must consider all pathways and the interactions between these pathways. In the scientific literature, the multimedia approach to environmental and exposure analysis is identified as Total Environmental Characterization (TEC) or Total Exposure Analysis (TEA) (Fig. 1.1). Applied or theoretical research activities

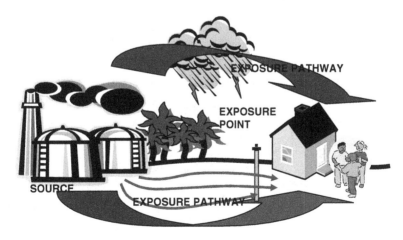

Fig. 1.1 Multimedia environmental exposure pathways

within this field are closely linked with research activities in the subsurface, surface and atmospheric sciences and in the toxicology, epidemiology and health fields. Given this complex picture, the analysis of these problems should follow a complex systems approach. That is, events should not be examined in isolation but should use an integrated multidisciplinary approach instead.

In health effects studies, the environment we are concerned with can be divided into the ambient and inner environments. The ambient environment is the environment where a human body resides or functions while the inner environment is the vital organs of the human body (Moeller 1997). The interaction between the two environments occurs primarily through three contact surfaces: the skin; the gastrointestinal tract (GI); and, the membrane lining of the lungs. Thus, as environmental health scientists, we are primarily concerned with dermal, ingestion and inhalation exposure pathways to environmental contaminants. The skin protects the human body from exposure to contaminants through contact. The gastrointestinal tract protects the body from ingested contaminants. The lungs protect the body from inhaled contaminants. In a human body, there are also secondary protection organs such as the kidneys and liver which filter and extract contaminants absorbed into the body. There are also secondary extraction mechanisms such as vomiting, coughing and diarrhea which discard the contaminants through mechanical reaction to the presence of contaminants. The interaction of the inner and ambient environments is a complex process which involves numerous uncertainties (Fig. 1.2). Moreover,

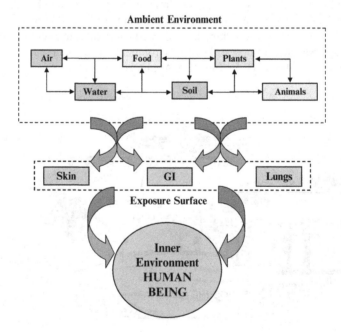

Fig. 1.2 Environment and potential exposure routes

we should also acknowledge that the ambient environment and the inner environment are in a constant state of change. The ambient environment changes due to variability in source loadings as well as in the effect of transformation, transport and natural processes in the environment. The inner environment changes due to aging or the effect of other physiologic changes during our lifespan. These add to the complexity of exposure analysis and health risk assessment methodologies.

Starting with the ambient environment, the presence of contaminants in air, groundwater and surface water pathways has raised significant public concern associated with risk to human health and ecological destruction. In response to these concerns and for purposes of environmental conservation, regulatory agencies and researchers have been actively involved in developing exposure and risk assessment methods that rely on environmental pathway analysis to quantify potential health and ecosystem hazards associated with environmental contaminants.

The purpose of human exposure and risk assessment studies is to identify and quantify past, present and potential future exposures to toxicants that may cause health effects, and qualitatively and quantitatively describe the risk associated with such exposure. Human exposure and risk assessment studies primarily include the following steps:

    i. Identification and evaluation of sources of toxicants (type, amount, duration and geographic location of release);
    ii. Determination of concentration levels of toxicants in the environment (air, water, soil, plants animals and food);
    iii. Identification of pathways and routes of exposure;
    iv. Identification of point of contact of concentrations;
    v. Determination of intensity, duration and frequency of exposure;
    vi. Determination of dose resulting from exposure;
    vii. Determination of health effects of exposure–dose;
    viii. Estimation of number of persons exposed; and,
    ix. Identification of high-risk groups (groups that are highly exposed or that are more susceptible to adverse effects).

Thus, to prevent adverse effects of environmental pollution on humans it is important to know the following: the environmental migration and transformation of source concentrations; the exposure levels; the health effects of contaminants; and the exposed populations and population subgroups that may be at a higher risk for adverse health effects. Using this information, scientists and managers may implement policies and measures to reduce toxicant exposure and risk to exposure.

In exposure and risk assessment studies, activities and processes identified in Fig. 1.3 constitute the main components of the overall study. In a risk evaluation process, as indicated in Fig. 1.3, one must link contaminant levels at source locations to risk of affecting human health at other locations. The sequences of studies that are needed to ascertain this link are identified in the boxes at the center of Fig. 1.3. First contaminant sources must be linked to exposure through multi-pathway environmental transformation and transport models. Next, the duration

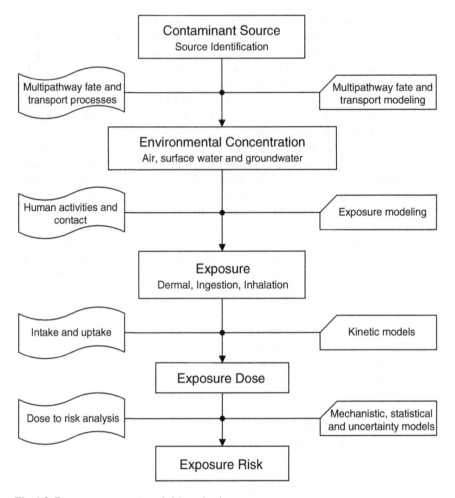

Fig. 1.3 Exposure assessment and risk evaluation process

and level of exposure must be determined. This leads to exposure risk, which can finally be linked to risk of adverse health effects. The boxes to the left of the central path identify the processes involved in moving from the previous step to the next, such as transformation and transport processes that link contaminant source concentrations to environmental or point-of-contact exposure concentrations. The boxes to the right of the central path give the mechanistic models that are used in the analysis of the processes on the left. Thus contaminant source identification, multimedia transformation and transport processes, human or ecological exposure to the contaminants along these pathways and the toxicological impact of the contaminants represent a comprehensive sequence of studies that are required for the completion of an environmental health risk assessment study.

The release of contaminants into the environment may be accidental or controlled based on our present-day understanding of acceptable environmental discharges and associated hazard levels. We have to accept the fact that a zero contaminant concentration release from industrial, municipal and domestic facilities or other potential contaminant sources is not attainable in today's society. This is a fact we have to live with.

In most cases, contaminant source concentrations and the duration of contaminant release from these facilities or other contaminated sites can be estimated or measured. If this data is not available, then historical contamination events at a site and the reconstruction of the contamination history in the environment can be attempted. This is a task for environmental engineers and in this process the tools used are environmental models. The next step in a health risk assessment, environmental assessment or ecological risk assessment study is the evaluation of potential migration and transformation of these contaminants in multimedia environments such as air, water and soil. The evaluation starts with the identification of the entry flux of a toxic substance from its source to the environment. The subsequent transport and diffusion fluxes must balance the source flux with adjustments made for chemical and biological reactions and sinks, which are lumped under the term "environmental transformation and transport processes." Thus, the fundamental principle of environmental assessment and management has its roots in the conservation of mass principle. Clearly there is no point in formulating an assessment or management strategy without considering the complete mass balance of contaminants in the envirosphere, which includes all potential pathways for the analysis of migration of contaminants. The most important pathways to consider in this continuum are the air, soil, groundwater and surface water pathways. Methods used in the analysis of transformation and transport processes in these pathways constitute the main theme of this book although animal and plant pathways are also an important part of this analysis. As defined earlier, we will identify the multimedia approach to environmental exposure characterization as Total Environmental Characterization (TEC) and the analysis required to perform this characterization as Total Exposure Analysis (TEA).

In order to address public concerns on long-term adverse effects of contaminants released to the environment, or to conduct studies that consider environmental remediation alternatives, scientists and engineers use transformation and transport models, risk assessment models or other environmental management tools. These tools are based on mathematical concepts. Based on this background information, the purpose of this book is threefold: (i) to provide the reader with an overview of the basic principles involved in environmental quality, exposure and risk analysis modeling, and to familiarize the reader with the terminology and language of the multidisciplinary fields involved in health risk assessment studies (see Appendices 1 and 2); (ii) to provide the reader with a concise summary of the most important and commonly used mathematical models that are available in the literature on environmental quality, exposure and risk analysis modeling; and, (iii) to provide the reader with a user-friendly computational platform through which these models can be accessed and used in site-specific applications (Anderson et al. 2007).

Pollution sources in the environment are most commonly linked to the production and handling of hazardous materials. The relative importance of various categories of potential contamination sources is shown in Fig. 1.4a. As seen in Fig. 1.4a, the most important contributors to environmental pollution are industrial disposal and storage operations. The cost of cleanup of these contaminated sites is very high as shown in Fig. 1.4b. Thus, similar to the idea behind preventative medicine, it may be important to pay attention to source and pathway control measures in environmental management as a first resort before contaminants are introduced and extensive pollution of the environment occurs.

In the context of the involvement of various disciplines in environmental management, we will start with a review of the functions and terminology used in different fields that comprise environmental health. It is recommended that the reader should supplement this review with the reading material referenced in each section. In addition to the description of the subfields covered in each section, the

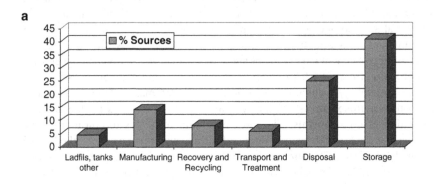

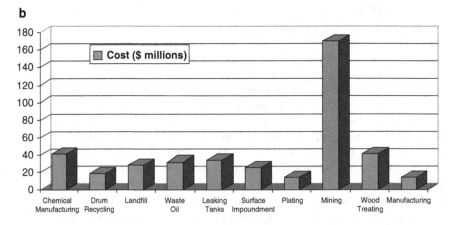

Fig. 1.4 (**a**) Distribution of contamination sources based on production and handling, (**b**) estimated cost of clean-up operations

acronyms and abbreviations used in these subfields are given in Appendix 1. Definitions of important terms used in these subfields are provided in Appendix 2. These two appendices will introduce the reader to the terminology used in the broad environmental health field. It is important for the reader to familiarize themselves with these definitions since they will be used extensively in the following chapters.

1.1 Environmental Processes

Pollutants originating from various natural or manmade sources may reach the human exposure point through a single or a combination of complex environmental pathways. To determine exposure levels to these contaminants, the best approach would be the measurement of ambient concentrations in the environment at the exposure point. However, more often than not, this is not possible since most of the time the outcome of exposure, which leads to health effects, is recognized long after the exposure has occurred. In such cases, to determine exposure levels at contamination sites, the other alternative would be the measurement of biologic markers of dose to arrive at human contact level concentrations if such markers are available for a specific pollutant. The biomarker approach does not give us information on which pathways or sources may be the cause of exposure. However, for well established pollutants such as "lead," the concentrations of lead in the blood of an exposed person would be a good indicator of the exposure level. Another alternative for obtaining this information would be the modeling of transformation and transport processes in the environment to arrive at human contact point concentrations of environmental contaminants.

There are a number of processes which affect the behavior and existence of contaminants or natural substances in the environment. These processes, which may be of chemical, biological or physical nature are commonly identified as transformation and transport processes. It is important for the reader to have a general understanding of these processes, both in terms of the physical, biological and chemical processes involved and also the mathematical representation of these processes, before a detailed discussion is attempted. Thus, the reader should first familiarize themselves with various concepts used in the environmental transformation and transport modeling field. The list of processes and their definitions given in Appendix 2 provides a concise summary of the environmental processes that are discussed in this book. The mathematical treatment of these processes will be given in more detail in the following chapters based on the terminology used in each of the three pathways considered and emphasized in this book.

At different rates, the sum or a subset of the processes described in Appendix 2 will affect the transformation and transport of pollutants in the environment. A schematic representation of these processes is illustrated in Fig. 1.5. In this book we will focus on the mathematical definitions and the analytical solutions of the mathematical models used in defining the transformation and transport

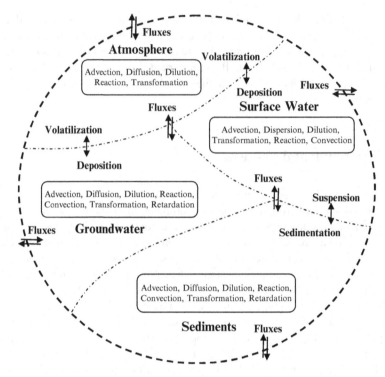

Fig. 1.5 Transfer and transformation of pollutants in the environment

processes in air, surface water and groundwater pathways. The purpose is to quantify the effect of these processes on the migration of contaminants in the three pathways.

Environmental quality parameters are measured in mass or concentration units such as milligrams, milligrams per liter and moles per liter (mg, mg L^{-1}, mol L^{-1}). Thus, the key concept in defining environmental processes will be the mass balance principle. Since mass balance is based on an accounting principle of the components of a system, one first has to define a domain for the system in which the mass will be balanced. This domain is identified as the "control volume" (CV). The size of the control volume is problem-dependent but needs to be defined clearly so that we know if an entity is inside or outside its boundaries. We treat the control volume as a bulk volume and assign the properties of the variables we define in it as bulk variables. Thus, the size of the CV should yield parameter values that will be representative of the overall region that is under consideration. If the size of the CV selected and thus the bulk parameter values assigned to the system within this CV is not representative of the overall region of analysis the selection of a series of smaller CV sizes will be needed to represent the parameters of the region. A representative control volume can then be selected with the criteria that the parameters that will be used in defining the process under study may be represented in terms of average values of the parameter within the control volume without adversely affecting or misrepresenting the behavior of the process in question. The averaging process for a property P can be given as shown in Eq. (1.1) (Fig. 1.6). Here the property P is averaged over an appropriate control volume $\mathcal{V}^*$ such that this property can be used in terms of its average value without adversely affecting the outcome of the process. On the other hand, the volume $\mathcal{V}^*$ must be small enough in comparison with the overall solution domain so that it can be treated as a point. This volume thus defined is also identified as "Representative Elementary Volume" (REV) in some literature (Bear 1972). Note that the symbol $\mathcal{V}^*$ is used for volume in this book to distinguish the use of the symbol V which is used for velocity in later chapters.

$$P_{ave}\left(x_i, y_i, z_i\right) = \frac{1}{\mathcal{V}^*} \int_{\mathcal{V}^*} \left(P\left(x_i, y_i, z_i\right) - P\left(x, y, z\right)\right) dx\,dy\,dz \qquad (1.1)$$

Further insight into the selection of the appropriate size of the REV can be provided if, for example, we identify the parameter in question as the porosity of the soil. Porosity of soil medium is identified as the ratio of the volume of voids in a REV to the bulk volume of the REV. In Fig. 1.6 the granular skeleton of the soil is represented with solid blocks, while the space between the solid grains indicates the volume of voids. In Fig. 1.6, possible choices for the REV are indicated by the choice of circles with increasing radius. If one chooses the smallest circle the REV will be composed of the void region only and the porosity of the medium will be calculated as one. If this smallest circle falls on the solid granular region, then the REV will be composed of the solid region and the porosity will be calculated as zero. Obviously both of these estimates are not a good representative of the porosity of the soil media under study. In order to represent the porosity of this soil appropriately, the REV size must be increased. Similar to this lower limit definition, an upper limit to the size of the REV also exists if one considers non-homogeneous

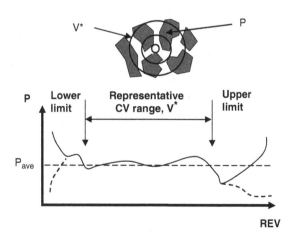

Fig. 1.6 Control volume size for property P

soil layers where soils of different porosities are involved. In that case the size of the REV should not exceed the size of the homogeneous porosity region of the soil. Otherwise the averaging process will include properties of different soils which may not be desired. There are significant uncertainties associated with the definition and the size of REV. At one level, the REV may not exist; at another level, even if it exists, it may be very difficult to determine its size. For those cases statistical theory has also been used in the literature to define this volume where the REV is treated as a random function and the REV is replaced by a volume with a certain scale of analysis (Dagan 1986).

To study flow field of fluids in a continuum followed by transformation and transport analysis, one must also define the "scale" of the problem of interest. As shown in Fig. 1.7, again using the soil medium analogy, the appropriate scales of analysis may range from molecular scale, to micro scale (pore space), to macro scale (the soil medium) and finally to mega scale (the aquifer medium). The scales indicated in Fig. 1.7 are all appropriate scales of analysis given the type of analysis of interest. In the field of chemistry and physics molecular scales may be more appropriate. Whereas, in environmental applications macro or mega scales will be more appropriate. Suggested ranges of scales in this process may be given as: (i) the laboratory scale ($10^{-1} \sim 10^{0}$ m – molecular and micro scale); (ii) the local scale ($10^{1} \sim 10^{2}$ m – micro and macro scale); and, (iii) the regional scale ($10^{3} \sim 10^{5}$ m – mega scale). Variability of parameter values is also associated with the scale of the analysis selected. The expected parameter variability is indicated in the lower part of Fig. 1.7.

Parameters used in subsurface analysis, such as porosity, intrinsic permeability etc. are relatively well defined. For these parameters scale issues may arise when the field representation of them is necessary to solve a problem in a non-homogeneous domain. Other parameters such as longitudinal and transverse hydrodynamic diffusion coefficients, decay of organic matter and other chemicals, heat transfer to atmosphere through water surface, erosion and deposition of sediments in natural

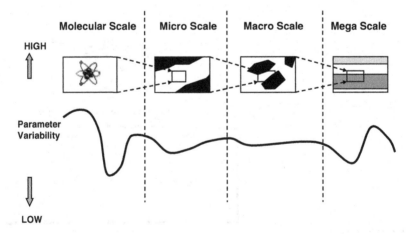

Fig. 1.7 Scale of application and parameter variability

environments have been studied by many researchers, with significant variations in the appropriate mathematical description of the phenomena involved. Keeping these variations in mind, the complexity of the mathematical model adopted in a study should be associated with the availability of data and the simplifications that can be introduced to the physical system without introducing significant errors in representing the system. This simplification is also associated with the selection of the appropriate scale for the model. For a mathematical model to properly represent the system modeled, the selection of the scale and the mathematical model complexity issues may render the modeling effort almost an art rather than a science.

The accounting of mass within the control volume requires knowledge of input and output fluxes across the boundaries, the transport characteristics within the control volume and across the boundaries, as well as information on sources, sinks and accumulation within the control volume (Fig. 1.8) (Clark 1996; Schnoor 1996). This model can be described by the following equation for a representative control volume:

$$\sum \text{Mass flux in} - \sum \text{Mass flux out} \pm \sum (\text{sources/sinks})$$
$$= \text{Rate of change of mass within CV} \tag{1.2}$$

In terms of the terminology of environmental processes (Appendix 2), fluxes in and out of a REV can be defined in terms of advection, diffusion, dispersion, convection, conduction and radiation processes. One should recognize that the terms used in the previous sentence to identify "fluxes" refer to significantly different environmental processes (Appendix 2). The source and sink terms can be defined in terms of deposition, adsorption, decay and numerous other chemical and biological reaction processes. The overall accounting process is then associated with the rate of change of mass within the control volume. If the system is at a steady state, the rate of change of mass within a control volume is zero and Eq. (1.2) simplifies to:

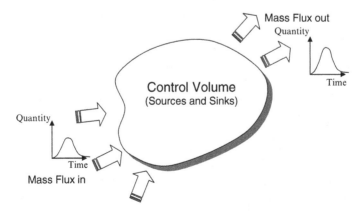

Fig. 1.8 Mass balance and control volume approach to environmental modeling

$$\sum \text{Mass flux in} - \sum \text{Mass flux out} \pm \sum (\text{sources/sinks}) = 0 \qquad (1.3)$$

Most of the models that are used in contaminant transformation and transport simulations, which are based on the conservation of mass principle, can be studied as deterministic models. That is, these models will yield one expected outcome at a spatial point and time, based on a given set of initial and boundary conditions and a set of parameters used in defining the process. In this approach one assumes that there is no uncertainty in conceptualization, data, model structure or the scale selected. It is well established in the literature that there are numerous uncertainties in each phase of the modeling effort, which may lead to the predictive uncertainty. To address uncertainty issues, models may also be used in a probabilistic sense, yielding not only the expected outcome, but also the variance of that expected outcome. In the probabilistic analysis of the models reviewed in this book, the Monte Carlo approach will be adopted to address uncertainty issues. There are also more recent approaches that can be categorized as non-probabilistic analysis or possibilistic analysis. The possibilistic approach has demonstrated that it can be employed in addressing uncertainty in environmental or health risk modeling where the uncertainty is heuristic. The possibilistic approaches include the Fuzzy systems approach (Kosko 1997; Kentel and Aral 2004; Kentel and Aral 2005), will not be covered in this book but the reader is referred to the above references since this type of analysis is important health risk analysis.

1.2   Environmental Modeling Concepts

A review of the modeling field indicates that several environmental models with varied degrees of complexity and different simulation objectives are available in the literature. One problem with most of these models is that it is often very difficult to implement them. These difficulties are due in part to the inaccessibility of the computer codes used in the solution, and in part to the problem-oriented design employed in the development of these models and codes. Thus some of these models are either never used or used by few users who have access to their computational platforms.

Models and model building is at the core of environmental management studies and significant time and effort must be spent to make proper decisions to appropriately represent the system being modeled. Several authors have discussed extensively the importance of models and model building in their books on scientific methods (Rosenbluth and Wiener 1945; Bloschl and Sivapalan 1995; Schnoor 1996). The following statement can be considered to be a consensus:

> No substantial part of the universe is so simple that it can be grasped and controlled without abstraction. Abstraction consists in replacing the part of the universe under consideration by a model of similar but simpler structure. Models . . . are thus a central necessity of scientific procedures.

Thus, a scientific model can be defined as an abstraction of some real system, an abstraction that can be used for prediction and management purposes. The purpose of a scientific model is to enable the analyst to determine how one or more changes in various aspects of the modeled system may affect other aspects of the system or the system as a whole. Because models are not a precise and complete depiction of the real system, they need to be presented and analyzed in a computational environment which should include an analysis of uncertainty. Uncertainty analysis may take the form of sensitivity analysis, or for more complicated applications, statistical uncertainty analysis may be utilized. We should also emphasize the difference between two commonly used terms in modeling "uncertainty" and "variability." As expected they refer to two distinct concepts:

Uncertainty is a measure of the knowledge of the magnitude of a parameter. Uncertainty can be reduced by research, i.e., the parameter value can be refined through further experimentation or further data collection.

Variability is a measure of the heterogeneity of a parameter or the inherent variability in a chemical property. Variance cannot be reduced by further research, but a model can be developed such that it would mimic the variability of the parameter used in the model.

There are many advantages to the use of mathematical models. According to (Fishman 1996), these advantages are:

    i. Enable investigators to organize their theoretical beliefs and observations about a system and to deduce the logical implications of this organization;

    ii. Lead to improved system understanding;

    iii. Bring into perspective the need for detail and relevance;

    iv. Expedite the analysis;

    v. Provide a framework for testing the desirability of system modifications;

    vi. Allow for easier manipulation than the system itself permits;

    vii. Permit control over more sources of variation than direct study of a system would allow; and,

    viii. Analysis is generally less costly than observing the system.

On the other hand, there are at least three reservations one should always bear in mind while constructing and using a model (Rubinstein 1981). First, there is no guarantee that the time and effort devoted to modeling will return useful results and satisfactory benefits. Occasional failures are expected to occur because of limited resources allocated to modeling. More often, however, failure results when the investigator relies too much on the method and not enough on ingenuity in constructing the model. The proper balance between the two is the key to success in modeling. The second reservation concerns the tendency of the investigator to treat his or her mathematical description of the problem as the best representation of the reality. One should be open minded in understanding the limitations of the proposed model. The third reservation concerns the use of the model outside the predictive range of the model developed. When working with a model, care must be given to ensure that the analysis remains within the valid representation range of the model. These are important concepts of concern when working with models.

It is well known that model design, almost by definition, is a pragmatic process. The simulation objectives determine the basic form, usability, and generality of the model proposed. Further, an investigation of the various environmental models, approved by U.S. Nuclear Regulatory Commission Regulatory Guide (Till and Meyer 1983), which focuses on their usability and applicability in predicting the transport of effluents in a surface water environment following an accidental spill, clearly indicates the necessity of the availability of user-friendly and well documented computer models. In this book, a review of the most common environmental models used in environmental health risk assessment studies is provided for the groundwater, air and surface water pathways, along with a user-friendly software interface to implement them and to facilitate their use.

Environmental transformation and transport models are built for the following purposes: (i) to evaluate the transformation and transport of contaminants in the environment by quantifying physical, chemical and biological processes that affect migration; (ii) to evaluate dynamic point-of-contact concentration levels that may have occurred in the past, are occurring presently or will occur in the future; and, (iii) to evaluate the outcome of different scenarios under various loading or management action alternatives. Since determination of exposure concentrations to toxicants constitutes the first step in health risk assessment, and direct field measurements may not be always available environmental modeling is becoming more and more of a permanent part of environmental health risk assessment studies.

Among the models that are available for environmental modeling, the first category of models may be identified as empirical models. In these models the description of cause-and-effect relationships is based on observational data sets with minimum analytic understanding of how the system works based on the relationships developed through the analysis of the data. These models are tied to empirical constants obtained from field or experimental data which may become the source of considerable uncertainty in applications.

The other category of models may be identified as mechanistic models. When we express the cause-and-effect relationships for a certain process or a system in terms of mathematical equations (differential or algebraic), the resulting models are identified as mechanistic (deterministic). Mechanistic models, in principle, reflect our understanding of how the system works, and they are based on certain accounting principles such as conservation of mass, energy or momentum. The complexity of these models depends on the level of detail for a process in a specific model or the dimensionality of the model developed.

Model accuracy and reliability are two of the more important aspects of modeling, which should not be overlooked. If a model is to be accepted as a reliable predictive tool, the numerical error bounds generated in computation should be within acceptable limits, and the model should be calibrated regionally or locally using available data. Proceeding in this direction, much of the recent work done in environmental quality modeling has been oriented towards improving models and incorporating better numerical solution techniques, the accuracy of which by far surpasses the availability and accuracy of the field parameter data that have to be used with such models. Scarcity of the field data, especially in air, groundwater and surface water quality modeling, is well known to researchers and engineers working in this field. Currently there is some disagreement among researchers as to whether higher priority should be placed on still further developments in model sophistication or on parameter prediction to improve accuracy.

A very simplistic model may use a very crude definition of a physical process, with few parameters to define the process. A very complex model may use a very detailed definition of a physical process, with a significant increase in parameters that is used to define the process. Naturally, improved sophistication of models is associated with an increase in the number of model parameters. Since it is likely that many of the additional parameters included in the model would be defined only in qualitative terms or with lesser accuracy, a relatively more sophisticated model can be less reliable than a simpler version. On the other hand, some systems and some physical phenomena are so complex in nature that there is often little reason to believe that good simulations are possible with simplified representations. In such cases, the need for more detailed and realistic models should be clear. A simple and crude example can be found in the case of effluent transport models for a river system. Given our current understanding and knowledge of turbulence characteristics, secondary currents, roughness concepts and sediment transport characteristics of natural rivers, it may be overly ambitious to develop a three-dimensional effluent transport model for a river network system just because it is possible numerically. Going to the other extreme, if in order to simplify such a model, that is, in order to reduce the model's dependence on complex field parameters, if one ignores the diffusive transport terms while keeping the convective transport terms in the analysis, the reliability of the model becomes questionable, at least for certain problem types such as accidental spills of pollutants or daily cyclic variation of spills, as is the case in sewage output. Thus, it is not necessarily true that models become more accurate as more complex definitions are used to define the model's processes. Inaccuracies may also result from the increase in the number of parameters associated with the detailed definition of a process or system. As observed in many applications, the likelihood of accurately defining these parameters is very low, resulting in an inherent loss of accuracy for complex models. On the other hand simplifying models has pitfalls as indicated in the example above. Thus, in developing models, the optimum solution is between these two extremes. In an attempt to achieve this balanced goal, an effort is made in this book to introduce the reader to one-, two- and three-dimensional screening level models and analytical solutions to these models, which, in most cases, provide sufficient detail for understanding the bounds of the problem at hand at a screening level.

Evaluation of advection and dispersion of effluents in natural or manmade environments is a complex phenomenon, especially if an effort is made to cover all aspects of their evaluation. In an industrialized society, a great variety of pollutants may get mixed into groundwater, surface waters or air. Dissolved matters such as chemicals, radioactive materials and salt, solid matters such as sediments, and temperature gradients introduced by power plants can be cited as a few of the sources of environmental pollution. Different models are needed to describe the transport characteristics of different pollutants. Thus, in environmental model building, the decision or selection of the contaminant type is the first step which needs to be addressed. A conservative chemical behaves differently than a non-conservative chemical. The stage of effluent transport is another variable that needs to be considered, since mathematical models describing initial mixing zones are considerably different than mathematical models that are used to evaluate conditions for well mixed zones. In building an environmental model, the third variable to consider is the choice of model dimensions. Given the present knowledge in numerical and analytical methods, it is usually tempting to develop a three-dimensional model, with the assumption that the parameters needed in implementing such a model are readily available. Thus, determination of the dimensionality of physical and kinematic parameters is the third complexity encountered in modeling transformation and transport of pollutants in natural or manmade environments. Within this set of available choices and options the best approach to modeling is very difficult to identify. That is why modeling is considered to be both a science and art in the current literature.

In the course of time, a number of deterministic, empirical or stochastic models have been proposed to predict mass transport in multipathway environments such as air, groundwater and surface water. Contaminant transformation and transport models, as they are treated in this book, fall under the category of mechanistic models. These models are generic models which may be used in the analysis of a wide range of conditions and site specific applications. Mechanistic models may also be used in a statistical sense, in which case one or more of the parameters will be defined in terms of probability density functions. This approach would yield the outcome in terms of statistical (probability) distributions. This mode of analysis, i.e. Monte Carlo analysis, will be used extensively in this book. Stochastic models seek to identify the probability of the occurrence of a given outcome based on probabilistic variations that are introduced to the model. They may be used to identify the variability in output based on variability in input parameters or variability of the boundary conditions of the problem analyzed.

The environmental modeling field has its own terminology and associated definitions. A review of the important terms used in this field is given in Appendix 2. In addition to the definitions of the terminology given in Appendix 2, the acronyms and abbreviations given in Appendix 1 are commonly used in the environmental modeling literature. It is important for the reader to familiarize themselves with their definitions.

1.3 Environmental Toxicology

Chemicals on earth are plenty and diverse. In addition to their presence, the chemical industry worldwide manufactures and markets thousands of new synthetic chemicals each year. Thus, it is safe to say that we are constantly being exposed to natural or synthetic chemicals in our ambient environment. The task of environmental health scientists is to ensure that the public health is not adversely affected by exposure to these chemicals. Exposure aspects of environmental health effect studies are commonly considered in epidemiologic investigations. Adversity or other measures of the effect of exposure is studied in the field of toxicology. Scientists who conduct laboratory studies on animals to understand, quantify and estimate health effects of a wide range of toxic substances, are referred to as toxicologists. Their work traditionally consists of quantifying the effects of one toxicant on a single or multiple animal species. However, this perspective of defining the work of a toxicologist or the toxicology field would be too restrictive since it is obvious that understanding the health effects of exposure is more complex than one chemical and one organism link. In our ambient environment, we are exposed to multiple toxicants at various doses during various exposure durations. Thus, the link among toxicants, exposure and health effects is much more complex than the data that can be extracted in a laboratory study. Nowadays, it would be more proper to identify the professional activities of toxicologists in a broader perspective. Under this umbrella, the work of toxicologists maybe grouped in three categories (Williams et al. 2000):

i. Descriptive toxicology: In this group, scientists' work primarily focuses on the toxicity testing of chemicals. The studies performed in this category are designed to generate toxicity information that can be used to identify the various organ toxicities that the test agent is capable of inducing under a wide range of exposure conditions.

ii. Research/mechanistic toxicology: Under this category, scientists study the toxicant in more detail for the purpose of gaining an understanding of how the toxicant initiates those biochemical or physiological changes within the cell or tissue that result in toxicity. Thus, the goal here is to understand the chain of biologic or biochemical events a toxicant triggers in a cell to create a toxic outcome.

iii. Environmental/applied toxicology: The studies described in the two categories above are conducted in a laboratory setting. In the applied toxicology category the scientist's focus is on the chemicals in the ambient environment. The purpose of the studies under this category is the use of descriptive and mechanistic toxicology results to identify some measure of safe dose of the toxicant through risk assessment methods.

While the laboratory studies are of significant importance, the evaluation of the combined effects of toxicants in the ambient environment on populations is much more complex. In the environmental toxicology area not only should the exposure to mixed toxicants be considered but their effects on multiple species must also be observed to assess the impacts of the toxicants. To complicate the overall picture further, one has to realize that the effect of toxicants on biologic entities is not always direct. The indirect effect, such as the release of sulfur dioxide into the atmosphere, which results in acid rain, has far more devastating impact on populations than one toxicant coming in contact with one organism. In this sense, the environmental toxicology field is much more complicated, and it is quite possible that accurate prediction of the effect of toxicants on populations is unlikely to be achieved in the near future. However, using epidemiologic studies and risk assessment methods, it is possible to define some measure of safe dose and/or some pathway that may link toxicants in the environment to adverse effects on populations. This information is very useful in managing environmental health concerns.

The toxicology field is a quite diverse field of science. It is important for the professionals working in the environmental health field to familiarize themselves with various concepts and methods employed in this field to be able to understand the outcomes of toxicology studies and use them in environmental health analysis. For this purpose the following references are recommended, (Sullivan and Krieger 1987; Ottobani 1991; Ballantine and Sullivan 1992; Ballantine et al. 1993; Eaton and Klassen 1996; Moeller 1997; Williams et al. 2000). The list of terms that are used in the toxicology field and their definitions are given in Appendix 2. It is important for the reader to familiarize themselves with the terminology used in the toxicology field as a starting point.

1.4   Exposure Analysis

Exposure can be defined as the contact of a chemical, physical or biological agent in the ambient environment with the exposure surface of an organism. In exposure assessment, the goal is to identify potentially hazardous and toxic chemicals, the frequency and duration of exposure to these chemicals and the routes of exposure to populations. This information or data forms the basis of an exposure study. This data may be obtained through actual monitoring of the contaminated environment, through mathematical modeling, or through scientific estimates based on data for similar events or observations made elsewhere. In total exposure characterization, all potential exposure pathways are evaluated to identify the total quantity of chemicals that are potentially internalized by exposed populations. The three principal exposure pathways that are commonly considered in these studies are exposure through inhalation, ingestion and dermal contact. Each of these exposure pathways has their own sub-pathways. For example, ingestion exposure may occur through ingestion of tap water, ingestion of water while swimming, ingestion of food etc. Similarly, dermal exposure may occur through contact with water while swimming or showering or through coming into contact with contaminated soil, dust or vapor. There are two approaches that may be used in evaluation of exposure. EPA regulations state that the exposure assessment should be based on the criteria of "Maximum Exposed Individual" (MEI). This implies that the MEI, on a daily basis, breathes contaminated air, ingests and is in dermal contact with contaminated soil, consumes contaminated water, fish, beef and dairy products etc. In this approach, all of these sources would contain the upper-bound estimates from the source. The assumptions in the case of MEI are: (i) the exposure occurs at the same location; and, (ii) the exposed person resides and works at this location for an entire lifetime of 70 years. The other approach is based on the "Reasonably Maximally

Exposed Individual" (RMI) concept. In this case, only ten percent of the food intake is produced at the exposure point and exposure duration is considered to be 30 years.

Similar to the transport processes, environmental modeling and the toxicology fields, the exposure analysis field also has its specific terminology and associated definitions. The terms included in Appendix 2 contain definitions for the most commonly used exposure assessment terms. A more comprehensive list may also be found in (Hogan 2000; USEPA 2005) or other USEPA exposure assessment manuals. Thus, following the format used earlier, a review of the important definitions used in this field is included in Appendix 2.

The utility of the software tools introduced in this text may become important in the exposure characterization stage of the risk analysis paradigm. If the environmental exposure data is available, i.e. after the exposure damage has occurred, the health risks from environmental chemicals can be estimated using standard risk assessment methodologies. However, in most health risk assessment studies, the purpose is to estimate health risk based on uncertainty in exposure data and in some cases (preferably) prior to the occurrence of the exposure or ecological damage. Thus, a typical exposure and ecological risk assessment process involves a sequence of computations and/or measurements to provide information on health and ecological risks for individuals or populations from multimedia environmental pathways. Steps involved in an exposure and ecologic impact assessment process may be identified as follows:

    i. Measurement or estimation of historical contaminant source levels and source locations in the environment;

    ii. Calculation of migration of the source contamination through relevant environmental pathways to provide information on exposure duration and human contact concentration levels; and,

    iii. Calculation of internalized concentrations through inhalation, ingestion and dermal contact pathways.

In the second step of the above list environmental modeling tools are used. A substantial number of these modeling tools are included in the ACTS computational platform provided with this book. The third step of the above list includes the use of models that are included in the RISK computational platform which is also provided with this book. These software tools will allow the user to conduct the necessary analysis in deterministic or probabilistic modes.

1.5 Environmental Risk Analysis

Origins of the risk assessment field can be traced back to 3200 BC during which time the Asipu, a group of priests in the Tigris–Euphrates valley, established a methodology that included hazard identification, generation of alternatives to avoid hazards, data collection to document hazards including signs from gods, and report creation to document findings. The next study that may be linked to the risk assessment field is seen around 1200 AD when King Edward II had to deal with the air pollution (smoke) in London and the associated health problems. In 1285, King Edward II established a commission to study the problem. In 1298 the commission called for voluntary reductions in the use of soft coal. In 1307 a royal proclamation banned the use of soft coal. This proclamation was followed by the formation of a second commission to study why it was not being followed. This example of a regulatory effort may be the first signs of what lay ahead. In the environmental risk assessment field, it is relatively easy to identify a hazard and establish a rule to prevent exposure. However, the implementation of the rule to its fullest extent may be more difficult to achieve.

Other more sophisticated studies later appear in radiation control. For example, one of the first comprehensive risk assessment studies was completed in 1975, when cancer deaths due to a nuclear core meltdown were investigated (USNRC 1978). In environmental applications, the purpose and objectives of risk analysis have evolved into is a scientific framework to make informed decisions which have the potential to produce the best management strategies in minimizing health risks and maximizing environmental conservation. Environmental risk analysis or risk based environmental simulation and evaluation has evolved into a science and art during the last decade and is a multidisciplinary field. Most scientists and engineers, who specialize in environmental modeling and analysis, should be familiar with the basic concepts in this approach and also the terminology used in this field. Following the format given earlier, the definitions of the terminology commonly used in this field can be found in Appendix 2 (USEPA 1987, 1988, 1991, 1995; 2005; Louvar and Louvar 1998; Hogan 2000). In addition to the definitions given above, the acronyms and abbreviations given in Appendix 1 are commonly used in risk analysis literature. It would be important for the reader to become familiar with their definitions.

Environmental risk can be defined as the likelihood of the occurrence of adverse health effects to an individual, a population, or an ecosystem, based on his or her exposure to hazardous chemicals. Environmental risk assessment, which is the scientific methodology to evaluate environmental health risk or ecological risk, has become an important tool to support the planning and decision-making processes involved in environmental health risk management of populations or ecosystems. It is important to distinguish the risk assessment and risk management issues from one another as shown in Fig. 1.9. The environmental health risk assessment process is composed of four main stages:

  i. Hazard identification;
 ii. Dose–response assessment;
iii. Exposure–dose assessment; and,
 iv. Risk characterization stages.

The hazard identification stage involves the description of a particular chemical's capacity to adversely effect, at some dose, the health of living organisms (USEPA 1987, 1991, 1995, 2005). This stage of assessment is sometimes referred to as the weight-of-evidence classification, which establishes the existence of

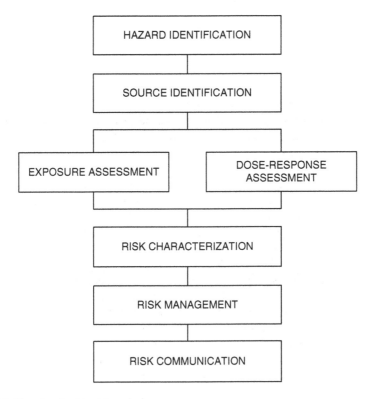

Fig. 1.9 Steps involved in risk analysis

hazards at a specific site. Dose is the amount of a substance available for interaction with metabolic processes or biologically significant receptors after crossing the exposure surface of an organism. Thus exposure is external and dose is internal. In the exposure–dose assessment stage, one utilizes models and analytical or numerical methods to evaluate the transformation and transport processes in environmental media for the purpose of estimating the exposure–dose levels that may reach humans and produce toxic hazards. This evaluation needs to be completed for multiple environmental pathways. Thus, exposure–dose assessment is the stage wherein the intensity, frequency, and duration of human exposure to chemicals are estimated through the use of models. In this text we will focus on multimedia environmental pathway transformation and transport models, which may be used in the exposure analysis stage. Thus, dose calculations are based on the chemical concentration in the ambient environment. Once the concentration distribution of the chemical in the ambient environment is determined using environmental models, an internal dose can be calculated. For example, potential dose for an intake process (inhalation or ingestion) can be defined as the integration of the chemical intake rate over time.

$$D_{pot} = \int\limits_{t_2}^{t_1} C(t)IR(t)dt \qquad (1.4)$$

where the concentration distribution in the ambient environment is $C(t)$, $IR(t)$ is the intake rate and $(t_1 : t_2)$ is the exposure period.

The dose–response assessment stage involves the characterization of the relation between the dose of a chemical and the incidence of adverse health effects in the exposed population. Finally, the risk characterization step is based on the previous three steps and involves the quantitative estimate of risk. In this phase, principal uncertainties and their consequences, are evaluated and environmental exposure factors are tied in with laboratory studies to understand health risk effects. Because of the uncertainties involved in all of these stages, effective application of simulation tools in environmental health or ecological risk assessment is both a quantitative and qualitative process, and can be considered to be a combination of both science and art. A successful health or ecological risk assessment study requires knowledge of scientific principles and mathematical methods that includes uncertainty, combined with expert insight in the assessment process, often to be provided within the framework of a multidisciplinary team effort. In summary, the purpose of exposure–dose analysis and risk assessment is to provide complete information to risk managers, policy makers and regulators, who in turn complete the risk evaluation by considering the various courses of action which can be taken to control the risks. A good example of this analysis is provided in Chapter 9 for an ecological risk assessment case study.

Primarily based on risk assessment results, a risk management process outlines the sequence of steps that may be taken to resolve and diminish the adverse effects of environmental chemicals on humans or other living organisms. It involves evaluation of complex factors such as availability and feasibility of the technology to prevent or control risk, cost of risk and cost of risk prevention or control, exposed population, level of exposure and public reaction. Risk communication is also an important aspect of environmental health risk assessment studies. Nowadays, in a modern society, individuals are more aware and also are more concerned with their ambient environments and with the threats that environment may bring to their health and safety. Their concerns are extremely important and thus should be handled by professionals who are trained in the health risk communication field. A discussion and review of these topics are out of the scope of the topics that will be covered in this book.

USEPA recommends that the risk assessment evaluation for carcinogenic and noncarcinogenic health effects be evaluated using different methods (USEPA 1987). USEPA recommends the use of a two-step evaluation to determine carcinogenic effects: a weight of evidence classification followed by slope factor calculations. In the weight of evidence step one evaluates the likelihood that a chemical is a human carcinogen. The tables of chemical compounds and their carcinogenic level classification can be found in USEPA (1991). When a chemical compound passes the first classification as a carcinogen, than the slope factor method is used to determine the life time probability of an exposed individual's risk of developing cancer due to that exposure (Cohrssen and Covello 1989; Louvar and Louvar 1998). For carcinogenic health effects risk assessment, dose–response models are used to extrapolate risks measured in high-dose animal experiments to the much lower doses typical of human environmental exposures. The dose–response models used for this purpose may fit the experimental data well. However, extrapolated values for low dose-risk estimates tend to differ by several orders of magnitude. Further, these estimates are based on the assumption of linear variation of increased risk with dose at low doses. This implies that any low dose would result in increased risk, no matter how small the dose is. This is a non-threshold approach in which there is no small dose beyond which one does not expect to have risk (NRC 1977). These are some of the weaknesses of this approach as discussed in the literature extensively (Crump et al. 1976; Crump 1984, 1985; Crump and Howe 1985). The carcinogenic risk levels are usually expressed in terms of the chance that an individual will develop cancer due to a 30-year exposure within 70-year lifetime (RMI).

As opposed to carcinogenic health effects, the noncarcinogenic health effects risk assessment represents more challenges and relatively less effort has been devoted to this task. One reason for this is noncarcinogenic effects covers a wide range of responses that one has to identify and each response has a wide range of severity level responses that has to be considered. For example the range of responses should cover adverse effects on specific organs, organ systems, reproductive capacity etc., and severity can be described as mild and reversible to high and irreversible or life threatening. The approach proposed here is the use of benchmark dose (BMD) approach in which the no observed adverse effect level (NOAEL) is replaced with a BMD. The BMD is determined by a dose–response model (USEPA 1995).

The USEPA uses the risk assessment steps outlined above to determine the required clean-up levels in groundwater, soil or surface water at contaminated sites that falls under the supervision of the federal government. These sites are commonly identified as Superfund sites under the Comprehensive Environmental Response, Compensation and Liability Act of 1980 (CERCLA). Based on the U.S. Safe Drinking Water Act and its amendments, the USEPA is directed to establish maximum contaminant level goals (MCLG) for drinking water supplied by public water agencies. A MCLG is a non enforceable goal set at a level to prevent known or anticipated adverse health effects for populations exposed to these chemicals, which include a considerable margin of safety. Maximum contaminant levels (MCL), on the other hand represent the enforceable standards. These may be considered to be primary standards, which are based on health risks. The secondary maximum concentration levels (SMCL) are those levels which are considered for some common chemicals for added safety. Thus, in support of risk assessment studies, the cancer-risk levels associated with exposures to various chemicals have to be established. This is the task of toxicologists, in which they utilize extremely conservative methods to identify cancer risks to humans. Usually in these studies, rodents are fed a diet containing large amounts of synthetic chemicals at what is called the maximum tolerated dose (MTD). If such a diet increases the cancer rate in rodents, the results are extrapolated to low doses to which humans may be exposed in natural or manmade environments. This approach has met with some criticism in the literature (Sagan 1994; Ames 1995). In this approach, it is also assumed that if a chemical is carcinogenic at high dose; it is also carcinogenic to some degree at any level of exposure. This assumption has also been challenged as being unsound (Goldman 1996). As a result of these assumptions, the MCLG for a compound that has been shown to be a rodent carcinogen is set by the USEPA as zero. Most recent drinking water standards that are promulgated by the USEPA are given in Appendix 4.

1.6   Environmental Epidemiology

Epidemiology is a field of science in which scientists seek to understand the links between infectious diseases and the manner in which they may spread in a community or population. Given the growing attention being paid to environmental concerns and health implications of environmental pollution, the methods and models used in the field of epidemiology have also been extended to understand the links between environmental pollution and health effects. The outcome is the science of "environmental epidemiology," which is defined by the National Research Council (NRC 1991) as:

> The study of the effect on human health of physical, biologic, and chemical factors in the environment, broadly conceived. By examining specific populations or communities exposed to different ambient environments, environmental epidemiology seeks to clarify the relationship between physical, biological or chemical factors and human health.

The environmental epidemiology field brings about methodologies to demonstrate a relationship between components of environmental pollution and one or more specific health effects. It is important for the professionals working in the environmental health field to familiarize themselves with various concepts and methods employed in this field to be able to understand the outcomes of epidemiologic studies and use them in environmental health analysis. For this purpose the following references are recommended, (Monson 1980; WHO 1983; Goldsmith 1986; NRC 1991; English 1992; Terracini 1992; Misch 1994). The list of terms and their definitions given in Appendix 2 are provided in this book to familiarize the reader with the terminology used in the environmental health field as a starting point.

1.7   The ACTS and RISK Software

The Analytical Contaminant Transport Analysis System (ACTS) and the Exposure Risk Analysis System (RISK) software are developed to provide a user friendly platform to implement the models discussed in this book. These two software systems provide a computational platform to health risk assessors, environmental engineers, risk managers and other decision makers to access and use the building block analytical models that are available in the public domain literature in environmental modeling and health risk assessment fields. Through the use of the ACTS software platform, users will be able to evaluate the steady state and time dependent behavior of contaminants in the environment and associate these contaminant levels with contaminant concentrations at human contact level for air, surface water and groundwater pathways (Maslia and Aral 2004; Anderson et al. 2007). The numerical results obtained from these simulations can then be linked to human exposure models through inhalation, dermal contact, ingestion and other pathways using the RISK computational platform. These two software tools, which use analytical methods to evaluate transformation and transport of chemicals and exposure analysis, also provide tools for uncertainty analysis for each pathway utilizing Monte Carlo methods (Maslia and Aral 2004). In this section a general review of these two software tools are given. Technical aspects of the models used and the mathematical methods utilized in the solution of these models are covered in more detail in the following chapters. The text of this book is designed to serve as a technical document for the ACTS and RISK software packages, as well as a standalone document in which the modeling principles, limitations and applications of commonly used models in the environmental health field are described in detail.

ACTS software utilizes analytical solution techniques to solve the mathematical models which describe transport and transformation of contaminants in multimedia environments. The simplifying assumptions required to obtain the analytical solutions may limit the complexity of the systems which can be represented by these models. The environmental transformation and transport models described in ACTS may be used for site-specific applications using site-specific parameters, boundary conditions and specific properties of contaminants. For more complex applications, it may be beneficial to use ACTS as a "screening level" tool which may allow the user to obtain a general understanding of the system behavior. For those cases, a more detailed model should then be used for more elaborate analysis of the system under study.

The ACTS software is based on the most commonly used models that are available in the literature for the air, groundwater and surface water pathway analysis. All pathway calculations are also linked to a Monte Carlo analysis module through which uncertainty analysis can be performed for all models considered in ACTS (Fig. 1.10).

To analyze cases involving uncertainty and variability of input parameters, Monte Carlo simulation codes are dynamically linked with all pathway models covered in the ACTS software. In a deterministic approach mode, "single point" values are specified for model parameters such as velocity, diffusion coefficient, retardation coefficient and width parameters V, D, R, and W (Fig. 1.11a). In this approach results are obtained in terms of single-valued output for concentration at selected spatial coordinates and time. In the Monte Carlo mode, input parameters (all or a selected subset) of a particular model may be characterized in terms of

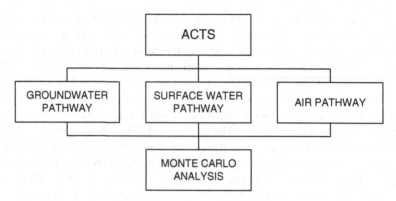

Fig. 1.10 ACTS software framework

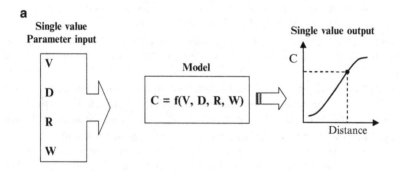

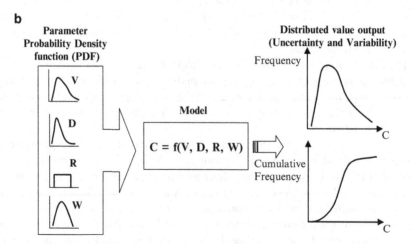

Fig. 1.11 Conceptual framework for: (**a**) deterministic analysis, (**b**) probabilistic analysis

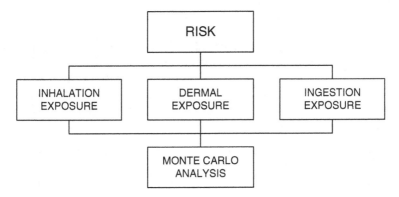

Fig. 1.12 RISK software framework

probability density functions (PDF) provided in the ACTS software. In this approach, the results are obtained in terms of distributed value outputs that can be used to characterize uncertainty and variability (Fig. 1.11b). This unique feature of the ACTS software enables the user to conduct probabilistic analysis without having to export input parameters to, and rely on, external or third party software to conduct this analysis. Once the PDFs for the selected input parameters are generated, users can conduct the transformation and transport simulations using any of the analytical models incorporated into ACTS. In Monte Carlo mode, statistical distributions of the exposure concentrations are obtained at a particular location in the modeled domain and at a specified time.

The RISK software is structured similarly to the ACTS software (Fig. 1.12). The most important aspect of the RISK software is that, although it is prepared as independent software, it can also be dynamically linked to the ACTS software. That is, the results generated from transformation and transport analysis models of the ACTS software can be directly accessed by the RISK software and the mean, maximum or average contaminant concentrations obtained at a point in space and over a time period can be used in the exposure models of the RISK software. In this context, the concentration data, which is based on the applications developed in the ACTS software, is considered to be the human contact level concentration in the RISK software. In this approach the RISK software calculations will be based on the results obtained from ACTS. However, the user may also choose to use a concentration level which is obtained from another source. In the latter case, the RISK software becomes independent of the ACTS software, and for those applications the user will have to enter the exposure concentrations into the model as external input data. Exposure models considered in the RISK software are grouped under three categories; inhalation exposure pathway, dermal exposure pathway and ingestion exposure pathway. Similar to the ACTS software, a Monte Carlo module is dynamically linked to all modules of the RISK software to provide uncertainty analysis. The three exposure pathways considered in the RISK software include several subcategory exposure models such as: ingestion of drinking water, ingestion while swimming, direct dermal contact or dermal absorption via soil sediment or dust, intake of chemicals via soil, sediment and dust, air intake and food intakes such as through fish, shellfish, vegetables and other produce, meat, eggs and dairy products. A flow chart of the modeling system of the RISK software is shown in Fig. 1.12.

In the development of the ACTS and RISK software, emphasis has been placed on the creation of a unified, user-friendly, WINDOWS™ based software framework, with the capability to perform uncertainty analysis for all models under this system. Important features and functions that are currently performed by the ACTS and RISK software systems include:

i. Use of WINDOWS™ based application environment in all aspects of the software;
ii. Capability to read input data files from earlier runs and interchangeably use the output files in each software;
iii. Use of WINDOWS™ based utility programs such as NOTEPAD.EXE or WRITE.EXE or any other compatible program to view, edit or print database generated;
iv. Allocation of default values to some input parameters/variables through chemical data base file;
v. Capability to edit chemical database to develop customized databases;
vi. Capability to view the results in customized graphic formats;
vii. Dependent on user-selected options;
   (a) Simulation of emission rates from a source.
   (b) Simulation of contaminant dispersion in the air pathway.
   (c) Simulation of one-dimensional unsaturated zone transport.
   (d) Simulation of one-, two- and three-dimensional saturated zone transport with constant and variable dispersivity models.
   (e) Simulation of parent daughter byproducts in subsurface analysis.
   (f) Simulation of in-stream or estuary concentrations due to contaminant loading using several surface water mixing zones.
viii. Capability to analyze the effects of uncertainty in input parameters for all pathways;
ix. Internal generation of random distributions for Monte Carlo simulations;
x. Performance of statistical analyses of Monte Carlo simulations;
xi. Graphical presentation of Monte Carlo analysis results;
xii. Dynamic linking of ACTS software results with RISK software for exposure analysis;
xiii. Exposure analysis through three primary pathways, which are inhalation exposure, dermal exposure and ingestion exposure. These exposure pathway models include other subcategory models based on the specific routes of exposure considered; and,
xiv. Dynamic linking of Monte Carlo module in the RISK software for uncertainty analysis.

Using these capabilities of the ACTS and RISK software, environmental characterization and health risk assessment studies can be developed for the environmental and exposure pathway models discussed in this text.

1.8 Outline

The purpose of this book is to provide the reader with fundamental principles as well as practical tools to implement the environmental transformation and transport and exposure models which are used in the exposure analysis of environmental health risk assessment studies. In this context, the theoretical background and physical interpretation necessary to understand, select and implement these models are reviewed in sufficient detail so that the text may be used as an independent reference in environmental health risk assessment studies, as well as a text book in undergraduate or graduate courses. In Chapter 2 fundamental principles of environmental modeling are discussed in sufficient detail to provide the reader with the necessary background on which the following chapters are based. Chapter 2 starts with definitions and classification of models. This is followed by a review of calibration, validation and verification issues associated with mechanistic models. In this chapter the topics on model building and model selection are also discussed in sufficient detail to provide the reader with an understanding of the pitfalls of modeling as well as importance of modeling in environmental studies. This is an extremely important component of the art and science of modeling, which needs to be addressed. In Chapter 3 laws of conservation and mass balance are discussed. This leads to primary transformation and transport processes and parameters that are essential components of these mechanisms. In Chapter 4 air pathway models are discussed. In this chapter, first emission models are covered and then these models are linked to Gaussian and Box dispersion models which are the most commonly used modeling techniques in air pathway analysis. This chapter and the following other environmental pathway modeling chapters are organized in a uniform format so that it will be very easy for reader to follow and understand the common features and differences of these environmental pathway models. Within each chapter, in which a different environmental pathway modeling group is described, the basic principles governing that environmental pathway models are reviewed as an introduction. Based on this review the chapter continues with the description of the specific models themselves. This is the common format of each chapter in this book. Chapter 5 contains a description of the saturated and the unsaturated groundwater transformation and transport models. The surface water pathway models are discussed in Chapter 6. In this chapter, near field, far field, river and estuarine modeling and sediment transport processes are discussed. In Chapter 7 background information on statistical methods and Monte Carlo methods are discussed with the goal of providing a unified risk assessment platform for all models discussed in the previous chapters. This chapter also includes a basic introduction to the concepts of statistical analysis for completeness. In Chapter 8 fundamental principles of health risk assessment are reviewed. In this chapter the methods used in exposure through inhalation, dermal contact and ingestion are presented. In Chapter 9 a site specific analysis of ecological risk assessment study is given. The overall text is supplemented with several appendices to provide the reader with a unified source in environmental and exposure modeling and risk analysis. In Appendix 1 definitions of the acronyms that are commonly used in the environmental health literature are given. In Appendix 2 more detailed definitions of the numerous terms used in the environmental health field are provided. In Appendix 3 the detailed description of the interface developed for the ACTS and RISK software is presented. In Appendix 4 current MCLG, MCL, SMCL levels of several chemical are provided as a reference. Appendix 5 provides information on common properties for water and conversion factors that may supplement the studies.

References

Ames BN (1995) Cancer prevention – what assays are needed. J Cell Biochem 187–197

Anderson BA, Maslia ML et al (2007) Probabilistic analysis of pesticide transport in shallow groundwater at the Otland Island Education Center, Otland Island, Georgia. Agency for Toxic Substances and Disease Registry (ATSDR), Atlanta, GA, 48 p

Aral MM (2005) Perspectives on environmental health management paradigms. In: 1st International conference on environmental exposure and health, WIT Press, Atlanta, GA

Aral MM (2009) Water quality, exposure and health: purpose and goals. Water Qual Expo Health 1(1):1–4

Ballantine B, Sullivan JB (1992) Basic principles of toxicology. In: Sullivan JB, Krieger GR (eds) Hazardous materials toxicology: clinical principles of environmental health. Williams & Wilkins, Baltimore, MD, pp 9–23

Ballantine B, Marrs TC et al (1993) Fundamentals of toxicology. In: Ballantine B, Marrs T, Turner P (eds) General and applied toxicology. M. Stockton Press, New York, pp 3–38

Bear J (1972) Dynamics of fluids in porous media. American Elsevier Pub. Co., New York

Bloschl G, Sivapalan M (1995) Scale issues in hydrological modeling: a review. In: Kalma JD, Sivapalan MC (eds) Scale issues in hydrological modeling. Wiley, Chichester, UK, pp 9–48

Clark MM (1996) Transport modeling for environmental engineers and scientists. Wiley, New York

Cohrssen JJ, Covello VT (1989) Risk analysis: a guide to principles and methods for analyzing health and environmental risks. Executive Office of the President of the U.S., Washington, D.C

Colby M (1990) Environmental management in development. The Evolution of Paradigms World Bank Discussion Papers D.C, Washington, D.C

Crump K (1984) A new method for determining allowable daily intakes. J Fundam Appl Toxicol 4:854–871

Crump K (1985) Mechanism leading to dose-response models. In: Ricci P (ed) Principles of health risk assessment. Prentice Hall, Englewood Cliffs, NJ, pp 321–372

Crump K, Hoel D et al (1976) Fundamental carcinogenic processes and their implications to low dose risk assessment. Cancer Res 36:2973–2979

Crump K, Howe R (1985) A review of methods for calculating confidence limits in low dose extrapolation. In: Krewski D (ed) Toxicological risk assessment. CRC Press, Boca Raton, FL

Dagan G (1986) Statistical theory of groundwater flow and transport: pore to laboratory, laboratory to formation, and formation to regional scale. Water Resour Res 22(9):120–134

Eaton DL, Klassen CD (1996) Principles of toxicology. In: Klassen CD (ed) Casarett and Doull's toxicology: the basic science of poisons. McGraw Hill, New York, pp 13–34

English D (1992) Geographical epidemiology and ecological studies. In: Elliot P, Guzick J, English D, Stern R (eds) Geographical and environmental epidemiology: methods for small-area studies. Oxford University Press, New York, pp 3–13

Fishman GS (1996) Monte Carlo: concepts, algorithms, and applications. Springer, New York

Goldman M (1996) Cancer risk of low-level exposure. Science 271(5257):1821–1822

Goldsmith JR (1986) Environmental epidemiology: epidemiological investigation of community environmental health problems. CRC Press, Boca Raton, FL

Hogan KA (2000) Characterization of data variability and uncertainty: health effects assessments in the integrated risk information system (IRIS). U.S. Environmental Protection Agency, Office of Research and Development, National Center for Environmental Assessment, Washington Office, Washington, D.C

Kentel E, Aral MM (2004) Probabilistic-fuzzy health risk modeling. Stoch Environ Res Risk Assess 18(5):324–338

Kentel E, Aral MM (2005) 2D Monte Carlo versus 2D fuzzy Monte Carlo health risk assessment. Int J Stoch Environ Res Risk Assess 19(1):86–96

Kosko B (1997) Fuzzy engineering. Prentice Hall, Upper Saddle River, NJ

Louvar JF, Louvar BD (1998) Health and environmental risk analysis: fundamentals with applications. Prentice Hall, Upper Saddle River, NJ

Maslia ML, Aral MM (2004) ACTS – a multimedia environmental fate and transport analysis system. Pract Periodical Hazard Toxic Radioact Waste Manage-ASCE 8(3):1–15

Meharg AA (2005) Venemous Earth: how arsenic caused the world's worst mass poisoning. Macmillan, London

Misch A (1994) Assessing environmental health risks. In: Starke L (ed) State of the world, 1994: a worldwatch institute report on progress towards sustainable society. W. W. Norton, New York, pp 117–136

Moeller DW (1997) Environmental health. Harvard University Press, Cambridge, MA

Monson RR (1980) Occupational epidemiology. CRC Press, Boca Raton, FL

NRC (1977) Drinking water and health. Safe Drinking Water Committee. N. R. Council. National Academy of Science, Washington, D.C

NRC (1991) Environmental epidemiology – public health and hazardous wastes. N. R. Council. National Academy Press, Washington, D.C

Ottobani MA (1991) Toxicology – a brief history. In: Ottobani MA (ed) The dose makes the poison. Van Nostrand-Reinhold, New York, pp 29–38

Rosenbluth A, Wiener N (1945) The role of models in science. Phil Sci XII(4):316–321

Rubinstein RY (1981) Simulation and the Monte Carlo method. Wiley, New York

Sagan L (1994) A brief history and critique of the low-dose paradigm. In: Calabrese EJ (ed) Biological effects of low level exposures. CRC Press, Boca Raton, FL

Schnoor JL (1996) Environmental modeling: fate and transport of pollutants in water, air, and soil. Wiley, New York

Sullivan JB, Krieger GR (1987) Introduction to hazardous material toxicity. In: Sullivan JB, Krieger GR (eds) Hazardous materials toxicology: clynical principles of environmental health. Wiliams & Wilkins, Baltimore, MD, pp 229–238

Terracini B (1992) Environmental epidemiology: a historical perspective. In: Elliot P, Cuzick J, English D, Stern R (eds) Geographical and environmental epidemiology: methods for small-area studies. Oxford University press, New York, pp 253–263

Till JE, Meyer HR (1983) Radiological assessment: a textbook on environmental dose analysis. U.S. Nuclear Regulatory Commission (USNRC), Washington, D.C

USEPA (1987) The risk assessment guidelines of 1986. Office of Health and Environmental Assessment, Washington, D.C

USEPA (1988) Superfund exposure assessment manual. Office of Health and Environmental Assessment, Washington, D.C

USEPA (1991) Health effects assessment summary tables. Office of Health and Environmental Assessment, Washington, D.C

USEPA (1995) The use of benchmark dose approach in health risk assessment. Office of Health and Environmental Assessment, Washington, D.C

USEPA (2005) IRIS database for risk assessment. http://www.epa.gov/iris/index.html, February 1, 2010

USNRC (1978) Liquid pathway generic study: impacts of accidental radioactivity releases to the hydrosphere from floating and land-based nuclear power plant. Nuclear Regulatory Commission, Washington, D.C

WHO (1983) Guidelines on studies in environmental epidemiology. World Health Organisation, Washington, D.C

Williams PL, James RC et al (2000) Principles of toxicology: environmental and industrial applications. Willey Interscience, New York

Chapter 2
Principles of Environmental Modeling

Everything should be made as simple as possible, but not simpler.
Albert Einstein

We have three primary scientific tools at our disposal to evaluate transformation and transport processes in the environment or to find solutions to environmental pollution problems and make decisions based on these solutions. These are, in no particular order: (i) direct field observations; (ii) laboratory scale tests and physical modeling studies; and, (iii) mathematical modeling. We recognize that transformation and transport processes that may occur in the environment and the accurate characterization of these processes both in the physical and also the mathematical domain are extremely complex. Thus, each of these tools has its appropriate place and mutually supporting role, as well as advantages and disadvantages of its use in understanding and solving environmental pollution problems.

It is well established in the literature that field observations tend to be costly but necessary. They are commonly used after the primary symptoms of the problems emerge at a contamination site. In this sense, they are extremely useful in characterizing the extent of the environmental problem, identifying its bounds or in evaluating whether the proposed remedial strategies are contributing to the solution of the environmental problem at a specific site. Laboratory studies, on the other hand, may be only useful in understanding the basic principles governing the problem at a micro or molecular scale. Findings and knowledge gained at this scale may experience significant problems in up-scaling the results to the field-scale analysis. Nevertheless, laboratory studies are extremely useful for both solving problems and for understanding micro scale issues at various stages of environmental pollution investigations and remediation.

In this book, among other topics, we will focus our attention on the use of mathematical modeling techniques in evaluating environmental transformation and transport processes. Thus it is important that we discuss problems we may encounter during model building and application, and the expectations we may have from a modeling study in an environmental application. First we should agree that

M.M. Aral, *Environmental Modeling and Health Risk Analysis (ACTS/RISK)*,
DOI 10.1007/978-90-481-8608-2_2, © Springer Science+Business Media B.V. 2010

mathematical models cannot help us in the problem recognition stage of an environmental pollution problem. However, they are very useful tools in the "gaining control" and "finding solutions" stages of our problem solution spectrum. They are cost effective and can be easily set up to test "what if" scenarios associated with a remedial application or a contamination problem. This cannot be easily studied with the other two scientific tools. The downside is the approximate nature of these tools which should always be kept in mind when their outcome is utilized. The level of contribution of each of these three tools to an analysis throughout the environmental problem solving spectrum is shown in Fig. 2.1.

Mathematical models are an abstraction of the environmental system and they are based on our understanding of the physical principles that govern the system. Since models are always going to be an abstraction of a system or a physical process, their outcome should always go through a careful and detailed interpretation stage before the results obtained from a model are determined to be representative of the behavior of the process or the system modeled (Fig. 2.2).

The purpose of mathematical model building and modeling is to simulate the behavior of the environmental system being modeled. Models are built to represent the system behavior in a controlled and cost effective computational environment. In this sense, modeling has become a common building block of most scientific applications. Using this tool we may observe, analyze, synthesize and rationalize the behavior of these systems under controlled conditions, and also we may evaluate the performance of the proposed solutions to an environmental problem. A common feature of all models is that they are all based on the "concept" of

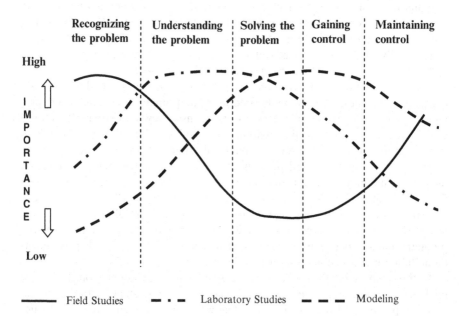

Fig. 2.1 Importance of scientific tools in assessing, evaluating and solving environmental problems

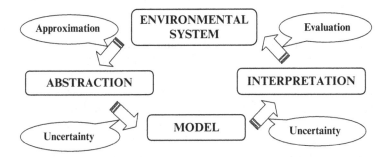

Fig. 2.2 Principles of modeling philosophy

simplification of the environmental system they are built to represent. This simplification may be achieved either through reducing the dimensionality of the system, elimination of less important processes that govern or affect the system, or through the introduction of simplified definitions for the parameters and variables that are used to describe the system. All of these or a selected subset of these simplifications are always observed in models built to represent an environmental process or an environmental system. Before we describe and make use of the models that are included in this text and also in the ACTS and RISK computational platforms, it is important that we review the modeling terminology from this perspective since it is necessary and extremely important for the reader to understand the limitations of models and modeling procedures in general. Otherwise, models or modeling may end up becoming a dangerous tool if their output is interpreted as the absolute truth without regard to the inherent simplifications and limitations they may have, or used as if they represent the environmental system under all circumstances. As a rule of thumb, modeling should always be considered to be a cost effective, efficient but approximate substitute for observing the modeled system behavior in its natural environment. Since observation of a process cannot always be achieved in a timely and cost effective manner, the models are here to stay among our scientific arsenal of tools as an important and alternative method.

The three evaluation tools identified above also differ from one another in the instruments that they may use to perform the analysis. In this sense, field study tools and laboratory tools are more closely related. Both of these methods may use electronic instrumentation to record and measure macro scale or micro scale processes. To provide a systematic procedure, these instruments may be linked to a computer or the observations can be done manually. On the other hand, computers are an essential component of all mathematical modeling studies. The language used in this analysis is primarily the language of mathematics. The interpretation of mathematics in the computer is done through coded systems, which nowadays can take the form of object or class oriented computer programming languages. As a simple definition one can say that a computer program written in any language to solve a mathematical problem is an orderly collection of coded instructions to the computer which perform certain mathematical tasks described in the mathematical model. A collection of coded programs for an application are commonly identified as software. The ACTS and RISK computational platforms, in this sense, can be identified as software that can be used in the modeling of multimedia environmental transformation and transport problems and health risk analysis.

Finally, the analysis tools described above should always be used in coordination with one another. Field studies should support the laboratory studies or vice versa, and mathematical modeling should support both of these efforts and vice versa. The advantages of any one tool should be exploited to the utmost for the benefit of finding a satisfactory solution to the problem. The outcome of each tool should be checked and verified with the outcome of the other tool. In this sense, these tools should be viewed as complementing, rather than competing scientific methods.

2.1 Modeling Principles

Principal steps involved in modeling and the uncertainty and approximations introduced at each step are summarized in Fig. 2.2 in their simplest form. As a preliminary definition, one can say that to model is to abstract from the natural system a description which addresses a question we have posed for the system. All models are developed to answer a specific question about the system outcome. The use of models in a specific application cannot and should not go beyond the question posed during the model development stage. This is an inherent approximation and limitation that is involved in all models. After this stage several other uncertainties are introduced in model coding and analysis. Some of these uncertainties are associated with mathematical representations used in modeling and others are related to the choice of model parameter characterization during implementation. When the model is used in the simulation phase it may produce a significant amount of output. The evaluation of this output is identified as the interpretation stage. Thus the overarching goal of mathematical modeling is first to come up with an abstract representation of an environmental system and to characterize this abstraction in a mathematically consistent manner such that it yields easy to use and understandable representations of the outcome, and second to use the outcome to interpret the behavior of the modeled system within the bounds of the model. Within this sequence, approximations and uncertainties are introduced to the analysis at each stage as shown in Fig. 2.2.

A common aspect of all mathematical models is that there is an input and an output component. Outputs are tied to inputs in some mathematical sense which describes the behavior of the abstracted physical problem. Since all models are approximate representations of a natural system, they are commonly designed to accept only a subset of all possible inputs an environmental system may have. Consequently models can only generate a subset of outputs that is expected from an environmental system. In other words we can never see the complete output or picture of the modeled system. To the extend that the inputs are limited the outputs will be limited as well.

When completed, models are used in simulation. Simulations are done to provide the data necessary in decision making or in evaluating the behavior of the system that is modeled. Decision making is based on simulation results and simulations themselves should not be interpreted as decision making. Human interaction or other heuristic mathematical models are always necessary in decision making which will be based on the outputs obtained from a model. Simulation results generated by models only provide us with the pieces of the puzzle that will help us make the appropriate decision. Evaluating the behavior of the modeled system should also be interpreted the same way. Simulation output only gives us the pieces of the puzzle needed to evaluate the system behavior.

Developing abstracted conceptual systems and a computational code for the conceptual systems is the scientific part of the modeling effort which may introduce scientific uncertainties (Lemons 1996). Simulation can be identified as the labor intensive part. Interpretation of the outcome and decision making can be considered to be the artistic part of the overall modeling effort (Fig. 2.2).

Fallout in modeling is the tendency to model in too much detail rather than modeling a finite manageable abstraction. The key to avoid this pitfall is to model around a question that needs to be answered rather than shooting for a universal representation. A simple model can always be fine tuned (calibrated) to overcome the approximations introduced through simplification. As a rule of thumb the following are key elements of a successful modeling effort:

  i. Understand the problem and clearly state the question that needs to be addressed.
 ii. Evaluate existing models first, do not re-invent the wheel.
iii. Create a conceptual model that is logical and represents the conceptual model in consistent mathematical terms.
 iv. In developing the model involve the user or think like a user.
  v. Simplify the conceptual model, its mathematical interpretation and its user interface. This may lead to a trial and error process. Don't be shy of remodeling.
 vi. When complete make sure that the model satisfies the objective and mission of the effort (see item 1).
vii. Design the simulations such that they provide answers to the question posed. Do not expect answers beyond the questions posed.
viii. Always remember that the purpose of modeling is the knowledge gained from a model and not the models themselves.

2.2   Model Building and Model Types

In model building the starting point should always be the identification of the goals of the modeling study. In this context, the following alternative goals can be cited:

  i. The modeling study is going to be a scientific study in which different hypotheses regarding the governing principles of the study will be tested, dominant processes of the problem will be identified, bounds of the parameter ranges that define these processes will be quantified.

ii. The modeling study will be used to characterize a study area, i.e. to determine the site specific parameters that are associated with the processes included in the model.

iii. The model will be based on well established basic principles and will be used as a predictor either to reconstruct a past event or simulate the future behavior of an environmental process at a site.

iv. The model will be used as an imbedded predictor (slave application) within a master application and will be used repeatedly to supply data to the master application. Simulators used in optimization models or statistical applications (Monte Carlo analysis) fall into this category and may include the goals identified in item 3.

v. The model will be used to support engineered decisions that will be made at a site and the purpose of modeling is the evaluation of the performance of these decisions.

Given the list of goals stated above, we should expect the following characteristics to be the dominant features of the model built. In case 1 the model should be considered to be modular. The construction and solution method of the model should allow for inclusion or exclusion of certain sub-processes to the model with relative ease. Complexity of the model is of no concern in these applications. The purpose is to include all possible and important sub-processes into the model. In case 2 the model will be used in the inverse modeling sense. In these applications, independent parameters of the model are treated as unknowns and dependent variables are treated as known variables and the solution process is based on the intrinsic relation between the independent and dependent variables. These models are not expected to include many independent parameters; otherwise, the solution becomes impossible. These models rely heavily on accurate field data on dependent variables. In case 3 the model will be used as a predictor. In this case the model should include all the dominant sub-processes of the problem studied, independent of the availability of accurate definitions of the parameters that are necessary to define these sub-processes. During simulation these parameters will be varied anyway, and the model output sensitivity with respect to these parameters will be documented. In case 4 the model should yield results efficiently with minimal computation time. For this to happen one may either resort to closed form solutions (analytical) or simplified models that may not include complex sub-processes which may exist in the overall system. In this case, as another simplification alternative, one may choose to represent complex processes in their simplest approximate forms. For example, in contaminant transformation and transport analysis one may either choose not to include chemical reactions, that is only simulate transport of a conservative chemical behavior, or represent this chemical reaction as a first order reaction for a single species application. These are all acceptable simplifications for a class of applications. For case 5 the model will be used to test the "what if" scenarios with respect to an environmental decision that will be made at a site. In this sense, the model should definitely include the best and most accurate definition of the sub-process that is being evaluated at the site. Secondary sub-processes that may not influence the main process may be given lesser importance in the construction of the model. In all of these cases the dimensionality of the model is determined based on the available data and the complexity desired by the model builder. Whatever the goal of the modeling study is, one always has to recognize that the tool at hand is an approximate representation of the process that is being modeled.

From the perspective of inclusion of some mathematical reasoning into the analysis of system behavior, as a general rule, the three procedures discussed above are available: (i) physical modeling (laboratory); (ii) empirical modeling (laboratory and field scale); and, (iii) computational modeling. In physical modeling the natural system being modeled is duplicated by a scaled model which is geometrically and dynamically similar to the large scale system. In this case the mathematical processes are used to arrive at similarity laws that are based on the similarity of the force ratios which govern the behavior of the natural system. Observations are conducted on the scaled model and the results are projected to the large scale system, again using the same similarity laws. Mathematical reasoning behind empirical models is based on induction supported by the data collected in field or laboratory studies. In a sense, the empirical approach represents our declaration that the system modeled is very complex, or not fully understood, and that the only alternative left for us is to represent the system by the use of a black box approach. In some cases, the empirical equations that are developed may even end up being dimensionally non-homogeneous, such as the case of the well known Manning's equation in open channel flow analysis. This is a further an indication that the natural process modeled is not well understood. Sometimes modelers get around the issue of dimensional non-homogeneity by attributing dimensions to the proportionality constants that are used in the empirical model. This of course may lead to a dimensionally homogeneous equation but does not resolve the issue of how well we understand the process that is modeled. Some of these models are so well established in the technical literature that we do not question their validity, such as the Manning's equation used in open channel flow analysis, which is sometimes inhibiting. In other cases statistical methods are used to verify the predictions made from these models.

Finally, computational models (mechanistic modeling) are based on deductive reasoning. Derivation of these models is tied to fundamental principles that govern the system. In these models, more often than not, it is impossible to include all sub-processes affecting the behavior of a complex system. Thus, as stated earlier, these models commonly include simplifying assumptions which should be accounted for when they are put to use. In this sense, although these models are generic models, i.e. can be used in any large or small scale modeling study, we use calibration methods to overcome this deficiency and adjust the model response to a site or an application to represent a specific behavior. A classification of mathematical models is given in Fig. 2.3.

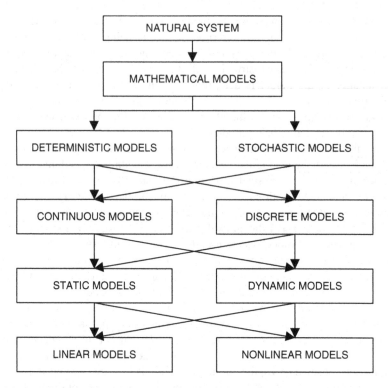

Fig. 2.3 Classification of mathematical models

The distinction in this classification is that deterministic models always produce the same output for a given input. On the other hand stochastic, a word of Greek origin which is synonymous with "randomness" and means "pertaining to chance," describes models in which a random set of inputs producing set of outputs that are interpreted statistically. Thus, stochastic is often used as the counterpart of the modeling exercise which is "deterministic," which means that random phenomena are not involved. Continuous models are based on the general mathematical property obeyed by mathematical objects and imply expressions in which all elements of the objects are within a neighborhood of nearby points. The continuity principle applies to dependent as well as independent variables of a mathematical model and implies smoothly varying properties, i.e. at least continuous first derivatives. Their counterpart is discrete models in which mathematical objects are not continuous and abrupt variation of parameters is expected. Static and dynamic refer to the dependence of the model on the independent variable "time". Static models are time independent and dynamic models are time dependent. Mathematical models that satisfy both the principles of additivity and homogeneity are considered to be linear models. These two rules, the additivity and homogeneity – taken together, lead to the possibility of the use of the principle of superposition. Nonlinear models are mathematical systems in which the behavior of the system is not expressible as a linear operation of its descriptors. Nonlinear models may exhibit behavior and results which are extremely hard (or impossible) to predict under current knowledge or technology.

Mathematical model building is a complex process. However, a systematic path to successful model building can be defined and this path should be followed to avoid common mistakes that may render the overall effort fruitless. Following the commonly accepted principles, a model building path is given in Fig. 2.4. The modeling framework, as identified in Fig. 2.4, includes standard checks and balances that should be used in model building, no matter what the purpose of the model may be. Remodeling is always an integral path of this process to improve on what is being built.

2.3 Model Calibration, Validation, Verification and Sensitivity Analysis

Since all models are simplifications of a complex system they need to be calibrated and verified before they are used in simulation. Validation and sensitivity analysis of models is also another concept that needs to be addressed and clarified. The literature on the definition and use of these concepts is abundant and sometimes confusing. Most of the confusion is associated with the concept of validation of models (Gentil and Blake 1981; Tsang 1991; Mayer and Butler 1993; Power 1993; Oreskes et al. 1994a, b; Rykiel 1996). For example validation is sometimes considered essential (Power 1993) and sometimes validation of models is considered impossible (Starfield and Bleloch 1986; Oreskes et al. 1994a, b), and some technicians of this field indicate that models can only be invalidated (Holling 1978; McCarl 1984). Due to this confusion and conflicting definitions it is appropriate to review the meaning of these terms as well as the interpretation of the very important terms "calibration" and "sensitivity analysis" from a mathematical modeling perspective.

Model Calibration: Models include parameters and constants that need to be associated with values. These parameters are used as input to the mathematical models to produce numerical output. Ideally, these parameters should have a good definition and a physical basis for the environmental system studied. Usually these parameters either are calculated using the mathematical representation of this physical basis, or they are measured in field or laboratory studies. More often than not, however, the values of these parameters are unknown or only known approximately. Thus a range of these parameters can be input to a model to yield the best outcome when compared to an observation made in a field or laboratory study. Thus, appropriate values of the parameters are needed in the model to achieve the appropriate output that is observed at a site. Calibration of a model can then be identified as the stage where we adjust the parameters of the mathematical model such that the model agreement is maximized with respect to the observation data we have on the modeled system. In this sense, model calibration is fine tuning the

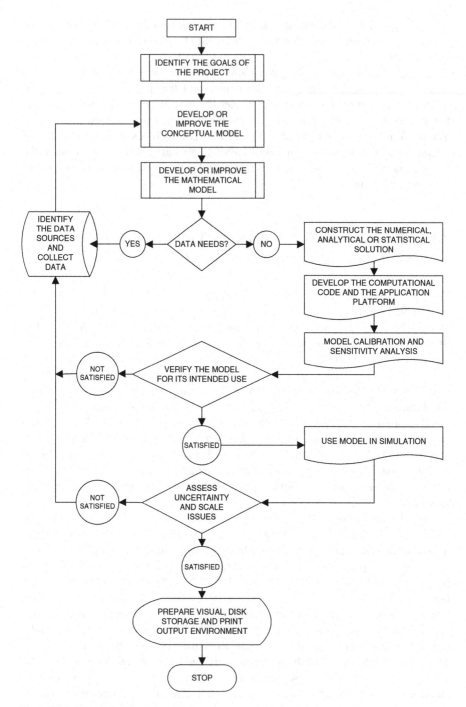

Fig. 2.4 Model building framework

model to a set of data on the natural system. Calibration of a model can be done manually, i.e. by trial and error adjustment of model parameters or it can be automated using stochastic procedures. Success in calibration, or lack of it, may yield information on how reasonable the modeler was in conceptualizing the natural system and mathematical representation of the conceptualized system. If a model fails to calibrate, it may mean that the conceptualization and mathematical representation stages need to be revisited. This also emphasizes the importance of remodeling in model development (Fig. 2.4). Calibration should not be interpreted as an inverse modeling technique which is used in parameter identification problems. Calibration procedure basically readies a model for its further use in simulation.

Model Verification: The confusion pointed out earlier may originate from the way we use the words 'verify' and 'validate'. In ordinary language, they are synonymous. From the perspective of modeling terminology these two words are used to describe two distinct concepts. Verification is a demonstration that the modeling formalism is correct. There are two types of verification avenues in modeling: (i) mechanical; and, (ii) logical. The former is associated with the debugging process of a computer program and in mathematical models, which shows that the mathematics and their numeric calculations are mechanically correct. A more important and difficult verification issue is the latter: showing that the program logic is correct. Some logical errors in a model may only appear under special circumstances that may not routinely occur in an application. Thus, these errors may not be recognized in routine applications of the model. Verification is thus a technical matter that identifies how faithfully and accurately ideas are translated into a computer code or mathematical formalisms (Law and Kelton 1991). In the case of large (complex) models, it is extremely difficult to verify that the model is entirely accurate and error free under all circumstances. Models are thus generally verified for the normal circumstances in which they are expected to be applied, and such verification is presumed inapplicable if the model is run outside this range. It is important to distinguish verification logic which relates to program operation from conceptual model logic which refers to the ecological logic used in structuring the model. Verification of models is needed in both aspects.

In summary, verification of a model is the stage at which we quantify the predictive capability of a mathematical model. This may be accomplished through a comparison of the output obtained from a model, which is based on input data, or with a set of observation data we have on a natural system which is based on the same input data. It is important to note that the observation data used in the calibration stage should be distinctly different from the data set used in the verification stage. That is, the data used for verification should be such that the calibration parameters should be fully independent of the verification data. The verified model can then be used for forecasting.

Model Validation: The absolute validity of a model can never be determined (NRC 1990). This statement is a strong reference to the impossibility of validation of a model. This reference to the impossibility of validation of models is somewhat relaxed in a statement in which Hoover and Perry state that: "The computer model is verified by showing that the computer program is a correct implementation of the logic of the model. Verifying the computer model is quite different from showing that the computer model is a valid representation of the real system and that verified model does not guarantee a valid model" (Hoover and Perry 1989), which implies that "validity" of a model is a possibility. To clear this confusion we need to expand on these definitions.

The term model uncertainty which is linked to model validation is used to represent lack of confidence that the mathematical model is a "correct" formulation of the problem solved. Model uncertainty exists if the model produces an incorrect result even if we input the exact values for all of the model parameters. The best method for assessing model uncertainties is through model validation (Hoffman and Hammonds 1994), a process in which the model predictions are compared to numerous independent data sets obtained. Thus, as is the case with verification, validation is better understood as a process that results in an explicit statement about the behavior of a model. A common definition of validation can be the demonstration that a model, within its domain of applicability, possesses satisfactory accuracy consistent with the intended application of the model (Sargent 1984; Curry et al. 1989). This demonstration indicates that the model is acceptable for use. But that does not imply that it represents the absolute truth for the system modeled, nor even that it is the best model available. For operational validation, this demonstration involves a comparison of simulated data with data obtained by observation and measurement of the real system. Such a test cannot demonstrate the logical validity of the model's scientific content (Oreskes et al. 1994b). Validation only demonstrates that a model meets some specified performance standard under specified conditions. It is often overlooked that the "specified conditions" include all implicit and explicit assumptions about the real system the model represents as well as the environmental context it covers. That is, that a model is declared validated only within a specific context, is an integral part of the certification. If the context changes, the model must be re-validated; however, that does not invalidate the model for the context in which it was originally validated (Rykiel 1996). Validation is a "yes" or "no" proposition in the sense that a model does or does not meet the specified validation criteria. These criteria may include requirements for statistical properties (goodness-of-fit) of the data generated by the model, and thus are not necessarily deterministic. Ambiguous situations may develop when the model meets some but not all of the criteria. The criteria may need to be prioritized, and the model may be validated with respect to these priorities. Because modeling is an iterative process, validation criteria may evolve along with the model. This is more typically the case with scientific research models than with engineering models. From a technical perspective, a valid model is the one whose scientific or conceptual content is acceptable for its purpose.

Sensitivity Analysis: Sensitivity analysis, on the other hand, can be considered to be a component of simulation through which the modeler evaluates the response of the model to changes in input parameters or boundary conditions of the model.

Sensitivity of model response to the input data and parameters of the model and the model output obtained is critical and must be quantified both during calibration and verification stages. Through this process, discrepancies between the model output and observation must be minimized to the extent that is possible by identifying and minimizing sources of error. These error sources include measurement errors, conceptual error in model development and approximation errors that may exist in mathematical representations. The goal of sensitivity analysis is to estimate the rate of change in the output of a model with respect to changes in model inputs or parameters. This knowledge is important for:

i. Evaluating the applicability range of the model developed;
ii. Determining parameters for which it is important to have more accurate values; and,
iii. Understanding the behavior of the system being modeled at critical points of solution – possibly at singular points.

The choice of the method of sensitivity analysis depends on:

i. The sensitivity measure employed;
ii. The desired accuracy in the estimates of the sensitivity measure; and
iii. The computational cost involved in calculating the error.

Consider a contaminant transport model in which several parameters P_i characterize the contaminant concentration C as a continuous function in a linear mathematical function, $C = f(P_1, P_2, P_3, ..., P_n)$ from which some reference value of C can be calculated, $C_o = f(P_1^o, P_2^o, P_3^o, ..., P_n^o)$. For this case some of the more common sensitivity measures S_{ij}, which can be used, are:

Local gradient measure: $$S_{ij} = \frac{\partial C_i}{\partial P_j^i}$$

Normalized gradient measure: $$S_{ij} = \frac{\partial C_i}{\partial P_j^i} \frac{P_j^i}{C_i}$$

Normalized variance measure: $$S_{ij} = \frac{\partial C_i}{\partial P_j^i} \frac{std\{P_j\}}{std\{C_i\}}$$

Expected value measure: $$S_{ij} = C_i[\mathrm{E}(P_i)]$$

Extreme value measure: $$S_{ij} = \left\{ \max C_i\left(P_j^i\right), \min C_i\left(P_j^i\right) \right\}$$

Normalized response measure: $$S_{ij} = \left(C_o - C_i\left(P_j^i\right)\right) \Big/ C_i(P)$$

Average response measure: $$S_{ij} = \sum_j C_i\left(P_j^i\right) \Big/ \sum_j P_j^i$$

(2.1)

where E is the expected value measure and the expected value of P_i is the mean value of parameters P_i.

Based on the choice of the sensitivity measure and the variation in the model parameters, methods of sensitivity analysis can be broadly classified into one of the following categories:

i. Variation in parameters or model formulation: In this approach, the model is run for a set of sample points (different combinations of parameters of concern) or with straightforward changes in model structure (e.g., in model resolution). Sensitivity measures that are appropriate for this type of analysis include the response from arbitrary parameter variation, normalized response and extreme value measure. Of these measures, the extreme values are often of critical importance in environmental applications.
ii. Sensitivity analysis over the solution domain: In this case the sensitivity involves the study of the system behavior over the entire range of parameter variation, often taking the uncertainty in the parameter estimates into account.
iii. Local sensitivity analysis: In this case the model sensitivity to input and parameter variation in the vicinity of a sample point(s) is evaluated. This sensitivity is often characterized through gradient measures.

The discussion of the terms calibration, verification, validation and sensitivity analysis given above outlines the basic principles involved in any modeling and model development effort. There are numerous models that are available in the scientific literature which may be used to analyze a multitude of physical processes. These models are sometimes identified as off-the-shelf models from which the users may download a code and implement it in a specific application that is of interest to the user. Here, it is important to note that the user must be fully aware of the limitations and the application range of the model used for the intended purpose. In certain cases some of these models have become so common in the literature that we no longer truly check the application rage of the model downloaded and we do not verify if the model truly fits the physical problem being modeled. In certain cases there are model applications in which the physical system modeled is restricted just to fit the system into a readily available off-the-shelf model. This practice can be characterized as fitting a physical system to a model rather than fitting a model to a physical system. This approach in modeling should be avoided at all times, at all cost. One should never try to define a physical system based on the limitations of the model that may be readily available. One should always remember the hierarchical steps involved in modeling. The description of the physical system always comes first, while the development of the model to describe the system follows behind.

2.4 Model Scales, Error and Uncertainty

The term "scale" refers to the characteristic spatial or temporal dimensions at which entities, patterns, and processes can be observed and characterized to capture the important features of an environmental process. Borrowing from cartography concepts, as environmental modelers we define scale as having two components: grain and extent. The former corresponds to the smallest spatial and temporal sampling units used to gather a series of observations or perform a computation. Extent is the total area or time frame over which observations or computations related to a particular grain are made (O'Neill and King 1998). For example, this may be defined for an observation of a hydrologic process, or it may be defined for a modeled environment (Klemes 1983; Bloschl and Sivapalan 1995; Singh 1995). All environmental processes, large-scale or small-scale, have their own characteristic scales of reference, which are necessary to capture details of the processes modeled or observed. Independent of the size of the model used, all environmental models, as covered in this book, are based on some mathematical representation of a physical process which is scale dependent (Gupta et al. 1986). When analysts use large-scale models to predict small-scale events, or when small-scale models are used to predict large-scale events, problems may arise (Fig. 2.1).

From groundwater flow and contaminant transport models to flow and transport in river channel networks to overland flow in a watershed or air shed models, the environmental processes occur over a wide range of scales and may span about ten orders of magnitude in space and time. When we attempt to model an integrated system the first question one should ask is: "if it is necessary to link all components of the environmental cycle into one system model?" The answer to that question should not be based on whether these components are separable or not. In a global sense they are not. However, the answer to that question should be based on whether one wants to separate them or not depending on the goals of the project and the importance of the contribution of the sub-processes to the understanding and evaluation of that goal. For example, if one is not interested in observing or reflecting the effect of one subcomponent on the other, then one can easily isolate an environmental process and analyze that subcomponent alone. For example, there are numerous groundwater flow and contaminant transport models which are extensively used in the literature just to study groundwater systems (McDonald and Harbaugh 1988; Aral 1990a, b). In their analysis, groundwater would receive input from surface water, but the reverse influence cannot be considered. On the other hand, if the simulation of multipathway interaction of an environmental process is the goal, than an integrated systems modeling approach is a must, and therein one encounters the difficulties of integration over scales (Gunduz and Aral 2005).

The transfer of data or information across scales, or linking sub-process models through a unified scale, is referred to in the literature as "scaling." Up-scaling consists of taking information from smaller scales to derive processes at larger scales, while downscaling consists of decomposing information at one scale into its constituents at smaller scales (Jarvis 1995). In the context of absolute space and time, scaling primarily involves a change in the geometric and temporal structure of the data and their corresponding attributes. In using the term "absolute scale" here we are referring to the definitions used in an Eulerian coordinate system in which distances between points in time and space are well defined geometric and differential entities. Thus, linking sub-process parameters within the well defined rules can be considered to be objective and to be independent of one's viewpoint or frame of reference in solving a problem. From a relative perspective, scaling becomes a more complex task than it would be in an absolute framework. In a relative scale framework one focuses on the sub-environmental processes and defines space and time as a measure of the relationship between these sub-processes. In a way one can interpret this definition as a Lagrangian frame of reference.

The relative scales concept represents the transcending concepts that link processes at different levels of space and time. It entails a change in scale that identifies major factors operational on a given scale of observation, their congruency with those on lower and higher scales, and the constraints and feedbacks on those factors (Caldwell et al. 1993). With this definition, one can observe that two processes that occur in close proximity by the definition of an absolute scale may be very distant from one another in terms of a relative scale sense. An example could be the case of the two hydrologic processes, overland flow and saturated groundwater flow, that normally are separated by an unsaturated zone. These two hydrologic processes could be close to each other in an absolute sense, but in terms of their interaction with one another, they could be very distant in a relative space and time frame of reference, due to limiting transfer rates that may exist in the unsaturated zone. In such cases, when scaling is considered the relative frame of reference should take precedence.

As expressed by Jarvis (1995), what makes scaling a real challenge is the nonlinearity between processes and variables scaled, and the heterogeneity in the properties that determine the rates of processes in a relative frame of reference. Therefore, it is important to realize that scaling requires an understanding of the complex hierarchical organization of the geographic and temporal worlds in which different patterns and processes are linked to specific scales of observation, and in which transitions across scales are based on geographically and temporarily meaningful rules (Marceau 1999).

Scaling and its effects on environmental modeling are commonly linked to the heterogeneity of the system modeled. However, this link should also include the refinement necessary to resolve the mathematical nonlinearities incorporated into an environmental process. Scale differences necessary to resolve nonlinearities, such as the nonlinearities introduced by the dependence of the higher order chemical reaction terms on rate constants as opposed to the easily solved differential equation that accompanies the first order reaction rates can be given as an example. Thus nonlinearity and heterogeneity are the two important factors that need to be considered in scaling. The greater the degree of heterogeneity and nonlinearity, the smaller the scale one would have to use to represent such variability or resolve such nonlinearity.

The other component of scaling effect arises in the interpretation of field data. Integrated environmental models use a variety of parameters to represent the characteristics of an application domain. However, data on large scale domain parameters are often limited. The task is then to transform this spatially limited data to a scale which can be used as an input in large scale applications. The question to answer here is what scale one should use to represent this data without losing accuracy during the extrapolation process. As the spatial scale of the model increases from a small area to a large area, the extrapolation of limited spatial data to a large scale system would introduce errors in the analysis from the start, which should be avoided.

An optimum scale of an integrated model should then reflect the "functional scale" (Aral and Gunduz 2003), that provides a compromise between the resolution of nonlinearities of the mathematical model, availability and extrapolation of data and the heterogeneity of the system. Thus, in environmental modeling, in order to resolve scale and scaling problems, one should first attempt to answer the following fundamental questions:

i. What is the appropriate scale of study for a particular hydrologic sub-process in the study?
ii. How close these sub-processes are in a relative frame of reference?
iii. How can one accurately transfer the necessary information from one process scale to another for closure?

When answering these questions we end up with a so called compromised scale which we identify as the functional scale (Aral and Gunduz 2003).

Scales of Sub-processes: Different scales of space and time govern the flow and transport phenomena in the environmental cycle. For an integrated environmental model these scales vary by several orders of magnitude in terms of the idealization of the solution domain, the computational step size and the simulation extent that is necessary to capture the important aspects of the process modeled as well as the proper scales that are necessary to interpret the input data.

One important aspect of integrating various sub-processes is the selection of the method applied to solve the equations that define the system. In this regard, coupling via iterative solution and coupling via simultaneous solution are the most advanced levels of solving the sub-processes in an integrated fashion. In iterative solutions, each sub-process model is solved separately and integrated sequentially by using the contributions from the other sub-processes. When each sub-model is solved, the common parameters linking these systems are checked for convergence (i.e., deviation from the previous solution). If the solutions of these common parameters are not sufficiently close, the solution procedure is repeated until the differences between subsequent solutions are below a pre-determined convergence criteria. This iterative coupling approach is slow, especially when more than two sub-processes are linked together. On the other hand this approach would be less restrictive from the perspective of scaling concerns since each sub-process can be analyzed within its own scale.

In the simultaneous solution approach, all sub-process models are solved together using a common idealization scale and a common time step. In this approach all sub-model solution matrices are grouped in a single matrix structure and solved at once. Hence, this method requires the use of the smallest idealizations and smallest time step of all sub-models, which may be impractical for the coupling processes requiring idealization and time steps from the two extremes. For example linking the two processes such of saturated groundwater flow and transport and the unsaturated groundwater flow and transport falls into category. Attempting to solve such a system simultaneously results in small idealization scales and time steps and creates incompatibility between systems. For example, unsaturated flow requires small time steps in the order of seconds to describe the vertical movement of moisture in the unsaturated domain whereas the groundwater flow can be run with time steps in the order of days. If a simultaneous solution technique is used to couple these two systems, then the entire system would need to be run with the time step of the unsaturated zone. This condition is computationally costly and inefficient for the groundwater flow and contaminant transport simulations. On the other hand, this approach is more accurate than the iterative method since it does not involve improvement of the solution by iterating on the common parameters of the two sub-models (Gunduz and Aral 2003a, b, c, d). Thus the wide array of time scales required to simulate efficiently the flow and transport processes in the environment is the most important problem of environmental modeling. The incompatibility of the sub-process time scales makes the overall coupling of the system difficult and sometimes impractical.

Suggested Solutions to Scaling Problems in Integrated Environmental Modeling: In large scale environmental modeling, the scale issues and up-scaling or down-scaling difficulties outlined above must be resolved if we are to develop an integrated representation of these processes. Technicians in the field of modeling believe that these problems can be resolved through some compromises. In order to develop an order of importance list of compromises that can be considered, the modeler has to introduce concepts such as:

    i. Order of importance;
   ii. Domain of importance;
  iii. Functional scales; and,
  iv. Hybrid modeling concepts.

In an integrated modeling effort, the order of importance ranking of different sub-processes can be achieved by the analysis of the data associated with the environment under study. For example in an environment where the groundwater table is high and the unsaturated flow zone thickness is very small, it may not be a significant loss of accuracy if the unsaturated zone is not modeled as a distributed model but instead is represented in terms of lump parameter models. Similar order of importance analysis evaluation can be made for overland flow as well as for the contaminant transport modeling. In arid regions or for rainfall events which are not significant, the contribution of this component may also be represented in terms of lumped parameter models rather than distributed parameter models. However, in all cases the groundwater flow zone and the river channel flow zone will play an important role in the overall watershed hydrology and should be included in the analysis in terms of distributed models for improved accuracy of representation of these sub-processes in the integrated environmental model.

The domain of importance concept arises from the analysis of the type of the problem solved. For example, if the concern is the transport of a certain contaminant source in the watershed, and if this source is not located in the unsaturated zone, then modeling the hydrologic processes in the unsaturated zone in detail with the use of distributed models may not be necessary. Similarly, if it is known that the flux of water between the unsaturated and the saturated zones is negligible, there is no need to complicate the analysis by including the unsaturated zone. On the contrary, there may not be any need to model the saturated groundwater flow when the top few meters of the soil column are of concern to the modeler and the groundwater table is at a much deeper elevation. Such simplifying judgments are a direct consequence of the available data for the domain modeled and are essential components of engineering evaluations to be made in a modeling study.

The functional scales concept is associated with the limitations of the integrated domain scales. If all sub-processes are important in an integrated environmental modeling effort and the use of distributed models is the goal, then one has to analyze the final time and space scales that are necessary to combine these models in an integrated system. At that point one may clearly see that this is not possible given the computational difficulties or long computation times required to solve the system. In such cases a compromise, as described earlier, is again the only solution.

Data availability is another limiting aspect of the integrated large scale environmental modeling studies. More often than not, field data is not available to justify the use of a distributed model at a large scale. This may be observed at a sub-process scale, in which case there is no reason to force a distributed model application for that sub-process as well. Otherwise, unforeseen errors will be introduced to the modeling effort. The availability of the alternative models, which range from simplified to more detailed system representations, or from small scale to large scale models, aids in evaluating the applicability of the low resolution models. If the results of the low resolution models (either in detail or in scale complexity) agree closely with those of the high resolution models, then the low resolution models are preferable, since they typically require lower computational resources and lesser input data.

Given the limitations on computational resources, computational methods and data limitations, the outcome of the integrated modeling compromises, as discussed above, is clearly to direct the modeler towards the use of hybrid models in integrated environmental modeling. In these models, lumped parameter models are used along with distributed parameter models to develop an integrated system.

Uncertainty and Error: The discussion above leads to uncertainty and error associated with environmental models and modeling (Figs. 2.2 and 2.5). Uncertainty in transformation and transport models arises in the following two stages of modeling: (i) model conceptualization or model building; and, (ii) model application. As mentioned above, model building uncertainty arises under several conditions, including the following:

i. When alternative sets of scientific or technical assumptions for developing a model exist (model structure);
ii. When models are simplified for purposes of tractability (model detail – inclusion or exclusion of sub-processes); and,

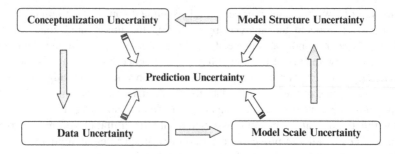

Fig. 2.5 Uncertainty sources in modeling

iii. When a coarse discretization and of data is used to reduce the computation demands of the model (model resolution – scale issues and statistical uncertainty).

The uncertainties and errors in simulation may arise from uncertainty in model inputs or parameters (i.e., parametric or data uncertainty). When a model application involves both model and data uncertainties, it is important to identify the relative magnitudes of the uncertainties associated with data and model formulation. Such a comparison is useful for focusing resources where they are most appropriate (e.g., data gaps versus model refinement).

Uncertainties in model parameter estimates may stem from a variety of sources. Even though many parameters could be measured or calculated up to some degree of precision, there are often significant uncertainties associated with their estimates. Some uncertainties and errors can be identified as:

i. Random errors in analytic devices used in field and laboratory measurements;
ii. Systematic biases that occur due to imprecise calibration;
iii. Extrapolation of data from one scale to another; and,
iv. Inaccuracy in the assumptions used to infer the actual quantity of interest from observations of a "surrogate" parameter or estimation of parameters based on mildly representative samples.

Uncertainty analysis should not be confused with sensitivity analysis. In uncertainty analysis one attempts to describe the entire set of possible outcomes of a model together with their associated probabilities of occurrence. In sensitivity analysis one determines the relative change in model output given changes in model input values.

Model errors can be evaluated by analyzing the variation in dependent variables in the model based on the variation of the independent variables of the model, i.e. the parameters of the model. Taylor series analysis is commonly used in this analysis. Since Taylor series will be used in several different contexts in this book it is appropriate to introduce a review of this topic.

A Taylor series is the sum of functions composed of continually increasing derivatives. For a dependent variable such as contaminant concentration $C(P)$, which depends on only one independent parameter P, the value of the function $C(P)$ at points near P_o can be approximated by the following Taylor series,

$$C(P_o + \Delta P) = C(P_o) + \frac{\Delta P}{1!}\left(\frac{dC}{dP}\right)\Bigg|_{P_o} + \frac{(\Delta P)^2}{2!}\left(\frac{d^2C}{dP^2}\right)\Bigg|_{P_o} + \frac{(\Delta P)^3}{3!}\left(\frac{d^3C}{dP^3}\right)\Bigg|_{P_o}$$
$$+ \cdots + \frac{(\Delta P)^n}{n!}\left(\frac{d^nC}{dP^n}\right)\Bigg|_{P_o} + R_{n+1}$$

(2.2)

in which P_o is some reference value of the parameter P, ΔP is the increment in the parameter P and $(P_o + \Delta P)$ identifies the point where the concentration C is to be evaluated $C(P_o + \Delta P)$ and R_{n+1} represents the remainder terms of a Taylor series expansion. In Eq. (2.2) the derivatives of $C(P)$ are evaluated at P_o. Using the definition above a first order approximation can be defined by keeping the terms of the Taylor series up to and including the first derivative as follows,

$$C(P_o + \Delta P) \approx C(P_o) + \Delta P\left(\frac{dC}{dP}\right)\Bigg|_{P_o}$$

(2.3)

Similarly, the second and third order approximations to Taylor series are given by

$$C(P_o + \Delta P) \approx C(P_o) + \Delta P\left(\frac{dC}{dP}\right)\Bigg|_{P_o} + \frac{(\Delta P)^2}{2!}\left(\frac{d^2C}{dP^2}\right)\Bigg|_{P_o}$$

(2.4)

and

$$C(P_o + \Delta P) \approx C(P_o) + \Delta P\left(\frac{dC}{dP}\right)\Bigg|_{P_o} + \frac{(\Delta P)^2}{2!}\left(\frac{d^2C}{dP^2}\right)\Bigg|_{P_o}$$
$$+ \frac{(\Delta P)^3}{3!}\left(\frac{d^3C}{dP^3}\right)\Bigg|_{P_o}$$

(2.5)

respectively. The accuracy of a Taylor series approximation improves as the order of the Taylor series increasesas shown in Eqs. (2.3) through (2.5). In these equations an approximate relationship is implied since the remainder terms of the Taylor series are omitted. Referring back to Eq. (2.3), we can associate the point P_o with the mean value of the parameter distribution P. Accordingly, the Eq. (2.3) will represent the value of C(a space is needed here such as) C around the mean value of P. We can now write an equation for the variance of the concentration C, using the definition of variance of $C(P)$ about the mean P_o, $S^2(C(P_o))$,

$$S^2(C(P)) = S^2(P)\left(\frac{dC}{dP}\right)^2\Bigg|_{P_o}$$

(2.6)

where $S(P)$ is the sample standard deviation, and $S^2(P)$ is the sample variance around the mean P_o. Eq. (2.6) implies that the variance in the dependent variable (uncertainty) is a function of the variance (uncertainty) in the parameter P, the sensitivity of the dependent variable to the changes in the parameter P around its mean, $\left(\frac{dC}{dP}\right)^2\big|_{P_o}$ and the variance in the parameters $S^2(P)$.

For a multivariate relationship, $C(P^i)$, $i = 1, 2, 3, ..., n$ the first order Taylor series expansion, Eq. (2.3), can be written as,

$$C\left(P_o^1 + \Delta P^1, P_o^2 + \Delta P^2, P_o^3 + \Delta P^3, ..., P_o^n + \Delta P^n\right)$$

$$\approx C\left(P_o^1, P_o^2, P_o^3, ..., P_o^n\right) + \sum_{i=1}^{n} \Delta P^i \left(\frac{\partial C}{\partial P^i}\right)\bigg|_{P_o^i} \tag{2.7}$$

which yields the variance relation,

$$S^2\left(C\left(P_o^1, ..., P_o^n\right)\right) \approx \sum_{i=1}^{n} S^2\left(P_o^i\right)\left(\frac{\partial C}{\partial P_o^i}\right)^2$$

$$+ 2\sum_{j=1}^{n-1}\sum_{i=j+1}^{n} \left(\frac{\partial C}{\partial P_o^i}\right)\left(\frac{\partial C}{\partial P_o^j}\right) S(P_o^i)S(P_o^j)\Phi(P_o^i, P_o^j) \tag{2.8}$$

where P_o^i is the mean of the ith parameter, $S\left(P_o^i\right)$ and $S^2\left(P_o^i\right)$ are the standard deviation and the variance of the ith parameter around its mean respectively, $S^2\left(C\left(P_o^i\right)\right)$ is the variance of $C(P^i)$ around the means P_o^i, $\Phi\left(P_o^i, P_o^j\right)$ is the correlation coefficient in a linear least squares regression between the parameters P^i and P^j (Crow et al. 1960; Reckhow and Chapra 1983; Bogen and Spear 1987; Ayyub and McCuen 1997; Conover 1999).

Monte Carlo analysis is another method used to evaluate parameter sensitivity to solution. Since this approach is used extensively in the ACTS and RISK software we will review this topic in more detail in Chapter 7.

2.5 Methods of Solution

Some mathematical models are relatively simple and their solution can be achieved using analytical methods, sometimes referred to as a closed form solution. Numerical calculation based on an analytical solution can be exact or approximate. Its accuracy depends on the complexity of the analytical solution. More complex models may require numerical solution which are all inherently approximate solutions to the problem. Both solutions will require computer based calculations to relate the model inputs to model outputs.

As indicated above statistical models and statistical calculations are also a necessary component of a modeling exercise. If not explicitly used in the modeling itself, statistical methods will become an important component in the sensitivity, calibration and verification phases of the modeling exercise.

In the case of the ACTS and RISK software analytical solutions will commonly be employed, since the models included in these software platforms are considered as screening models and in that sense are simpler representations of the modeled system. To perform sensitivity analysis the ACTS and RISK software also includes a Monte Carlo module in all models where the models can be run in a stochastic mode.

2.6 Modeling Terminology

The modeling field is quite a diverse field of science. It is important for the professionals working in the environmental health field to familiarize themselves with various concepts and methods employed in this field to be able to understand the outcomes and limitations of environmental modeling and use them in environmental health analysis appropriately. For this purpose a review of the following references are recommended, (Gentil and Blake 1981; USEPA 1984; Starfield and Bleloch 1986; Hoover and Perry 1989; Law and Kelton 1991; Tsang 1991; Mayer and Butler 1993; Oreskes et al. 1994b; Lemons 1996; Schnoor 1996; Abdel-Magid et al. 1997; Saltelli et al. 2000; Anderson and Bates 2001; Nirmalakhandan 2002; Aral and Gunduz 2003). The acronyms used in this field are given in Appendix A of this book. The list of terms and their definitions given in Appendix B are also included in this book to familiarize the reader with the terminology used in the environmental modeling field as a starting point.

References

Abdel-Magid IS, Mohammed A-WH et al (1997) Modeling methods for environmental engineers. CRC Lewis, Boca Raton, FL

Anderson MG, Bates PD (2001) Model validation: perspectives in hydrological science. Wiley, West Sussex, England

Aral MM (1990a) Groundwater modeling in multilayer aquifers: steady flow. Lewis, Chelsea, MI

Aral MM (1990b) Groundwater modeling in multilayer aquifers: unsteady flow. Lewis, Chelsea, MI

Aral MM, Gunduz O (2003) Scale effects in large scale watershed modeling. In: Singh V, Yadava RN (eds) Advances in hydrology. Proceedings of the international conference on water and environment (WE-2003), Bhopal, India, pp 37–51

Ayyub BM, McCuen RH (1997) Probability, statistics, and reliability for engineers. CRC Press, Boca Raton, FL

Bloschl G, Sivapalan M (1995) Scale issues in hydrological modeling: a review. In: Kalma JD, Sivapalan MC (eds) Scale issues in hydrological modeling. Wiley, Chichester, UK, pp 9–48

Bogen KT, Spear RC (1987) Integrating uncertainty and interindividual variability in environmental risk assessments. Risk Anal 7:427–436

Caldwell MM, Matson PA et al (1993) Prospects for scaling. In: Ehleringer JR, Field CB (eds) Scaling physiological processes: leaf to globe. Academic, San Diego, CA, pp 223–230

Conover WJ (1999) Practical nonparametric statistics. Wiley, New York

Crow EL, Davis FA et al (1960) Statistics manual. Dover Publications Inc, New York

Curry GL, Deuermeyer BL et al (1989) Discrete simulation. Holden-Day, Oakland, CA

Gentil S, Blake G (1981) Validation of complex ecosystems models. Ecol Model 14(1–2):21–38

Gunduz O, Aral MM (2003a) Simultaneous solution of coupled surface water/groundwater flow systems. In: Brebbia CA (ed) River basin management II. WIT Press, Southampton, UK, pp 25–34

Gunduz O, Aral MM (2003b) An integrated model of surface and subsurface flows. In: Brebbia CA (ed) Water resources management II. WIT Press, Southampton, UK, pp 367–376

Gunduz O, Aral MM (2003c) A simultaneous solution approach for coupled surface and subsurface flow modeling, multimedia environmental simulations laboratory. School of Civil and Environmental Engineering, Georgia Institute of Technology, Atlanta, GA, 98 pp

Gunduz O, Aral MM (2003d) Hydrologic modeling of the lower Altamaha River Basin. In: 2003 Georgia water resources conference, University of Georgia, Athens, GA

Gunduz O, Aral MM (2005) River networks and groundwater flow: a simultaneous solution of a coupled system. J Hydrol 301(1–4):216–234

Gupta VK, Rodriguez-Iturbe I et al (1986) Scale problems in hydrology. D. Reidel, Dordrecht, The Netherlands

Hoffman FO, Hammonds JS (1994) Propagation of uncertainty in risk assessments: the need to distinguish between uncertainty due to lack of knowledge and uncertainty due to variability. Risk Anal 14(5):707–712

Holling CS (1978) Adaptive environmental assessment and management. Wiley, New York

Hoover SV, Perry RF (1989) Simulation. Addison–Wesley, Reading, MA

Jarvis PG (1995) Scaling processes and problems. Plant Cell Environ 18:1079–1089

Klemes V (1983) Conceptualization and scale in hydrology. J Hydrol 65:1–23

Law AW, Kelton WD (1991) Simulation modeling and analysis. McGraw-Hill, New York

Lemons J (ed) (1996) Scientific uncertainty and environmental problem solving. Blackwell Science, Cambridge, MA

Marceau D (1999) The scale issue in social and natural sciences. Can J Remote Sens 25(4):347–356

Mayer DG, Butler DG (1993) Statistical validation. Ecol Model 68(1–2):21–32

McCarl BA (1984) An overview: with some emphasis on risk models. Rev Market Agric Econ 52:153–173

McDonald MG, Harbaugh AW (1988) A modular three-dimensional finite difference groundwater flow model. U.S. Geological Survey Techniques of Water Resources Investigations, Book 6, Chapter A1, 586 pp

Nirmalakhandan N (2002) Modeling tools for environmental engineers and scientists. CRC, Boca Raton, FL

NRC (1990) Groundwater models; scientific and regulatory applications. National Research Council, Washington, DC, p 303

O'Neill RV, King AW (1998) Homage to St. Michael: or why are there so many books on scale? In: Peterson DL, Parker VT (eds) Ecological scale, theory and applications. Columbia University Press, New York, pp 3–15

Oreskes N, Belitz K et al (1994a) The meaning of models – response. Science 264(5157): 331–341

Oreskes N, Shraderfrechette K et al (1994b) Verification, validation, and confirmation of numerical-models in the earth-sciences. Science 263(5147):641–646

Power M (1993) The predictive validation of ecological and environmental-models. Ecol Model 68(1–2):33–50

Reckhow KH, Chapra SC (1983) Engineering approaches for lake management. Butterworth, Boston, MA

Rykiel EJ (1996) Testing ecological models: the meaning of validation. Ecol Model 90(3): 229–244

Saltelli A, Chan K et al (2000) Sensitivity analysis. Wiley, West Sussex, England

Sargent RG (1984) A tutorial on verification and validation of simulation models. In: Proceedings of the 1984 Winter Simulation Conference. IEEE 84CH2098-2, Dallas, TX

Schnoor JL (1996) Environmental modeling: fate and transport of pollutants in water, air, and soil. Wiley, New York

Singh VP (1995) Watershed modeling. In: Singh VP (ed) Computer models of watershed hydrology. Water Resources Publications, Highlands Ranch, CO, pp 1–22

Starfield AM, Bleloch AL (1986) Building models for conservation and wildlife management. Macmillan, New York

Tsang CF (1991) The modeling process and model validation. Ground Water 29(6):825–831

USEPA (1984) Evaluation and selection of models for estimating air emissions from hazardous waste treatment, storage and disposal facilities. Office of Air Quality Planning and Standards, Research Triangle Park, NC

Chapter 3
Conservation Principles, and Environmental Transformation and Transport

As far as the laws of mathematics refer to reality, they are not certain;
and as far as they are certain, they do not refer to reality.

Albert Einstein

Every moment, a wide range of contaminants, in significant quantities, is either deliberately or accidentally released into the environment worldwide. The intensity of these releases has grown along with the economic, industrial and social development of the countries of the world since the early 1950s. Thus, more often than not the sources of these contaminants and the resulting environmental pollution are attributed to industrial activities. However, other activities such as the use of pesticides in agriculture, the uncontrolled disposal of waste and waste discharge, or the handling and disposal of contaminants in landfills that are not properly constructed or monitored, and many other similar activities also contribute to environmental pollution. The effects of this pollution on the ecosystem and on the health of populations have been documented in numerous technical publications world wide. In certain cases the levels of environmental pollution and health effects outcomes are alarming. The degradation of the environment and its adverse health effects is more pronounced in developed nations but developing nations are not immune to this problem. They will see this same degradation if the mistakes of the past are repeated and lessons are not learned, especially for those regions of the world where environmental regulations are not strictly enforced. It is also clear that, in a modern society, it is not possible to eliminate totally the release of the contaminants into the environment altogether. Thus, it is important to understand the principles and mechanisms that may be used in understanding and describing the transformation and transport of contaminants in the environment. The goal of

M.M. Aral, *Environmental Modeling and Health Risk Analysis (ACTS/RISK)*,
DOI 10.1007/978-90-481-8608-2_3, © Springer Science+Business Media B.V. 2010

achieving this knowledge would be to use this information in appropriate management practices to minimize the adverse effects of environmental pollution.

Contaminants in the environment move around in several pathways such as air, soil, plants, animals, groundwater and surface water. This movement, sometimes referred to as contaminant migration, involves numerous transformation and transport processes. Thus, in environmental modeling the word "transport" is associated with the question of "How do contaminants migrate or move around in the environment?" The transport mechanism is associated with the motion of contaminant particles within and with the ambient flow field. The flow field that needs to be considered may be associated with the transport conditions in the atmosphere or with the surface and subsurface environmental pathways. The word "transformation" is associated with the questions of "What happens to the contaminant particles in their path of migration, and what type of chemical or biological transformations might they undergo as they migrate?" The main concern here is the analysis of degradation, accumulation, chemical or biologic transformations, and change of phase processes that influence the contaminant concentrations in the ambient environment. The ambient environment is commonly described with the use of the generic word "environment". Both transformation and transport processes are necessary to create healthy environments. An environment can only recover from these stresses by the action of these processes. Without transformation and transport processes environmental pollution would remain localized and ever increasing, which would be a complete disaster.

In this book the principal environmental pathways focussed on are the air, groundwater and surface water pathways. In these pathways contaminants usually appear as a source mainly due to human activities. They eventually reach population clusters, i.e. receptors, through transformation and transport processes in each pathway. In most cases these three pathways are connected. The source to population (receptor) pathway can be managed and the adverse effects of environmental pollution on populations can be controlled but not totally eliminated. A typical source-to-receptor model is outlined in Fig. 3.1. In this figure the possible control points and measures that may help manage these control points are highlighted. Thus, environmental management or intervention may occur in four control points: (i) the source control; (ii) the transformation control; (iii) the transport control; and, (iv) population exposure control and management. In an environmental management process, all or a subset of these control points can be used and prevention measures can be implemented. These control points and measures will depend on the type and pathway of the contamination involved. The subject of this book is to study environmental pathway analysis and enumerate exposure risk assessment methods along these pathways. Remediation and source control measures are not the subject matter of this book.

Fundamental principles that describe the movement of contaminants in various environmental pathways are based on similar hydrodynamic and chemical fate and transport processes. A difference in the description of the transformation and transport processes in each pathway may arise if and when the magnitude and the importance of the process that characterizes the transformation and transport process

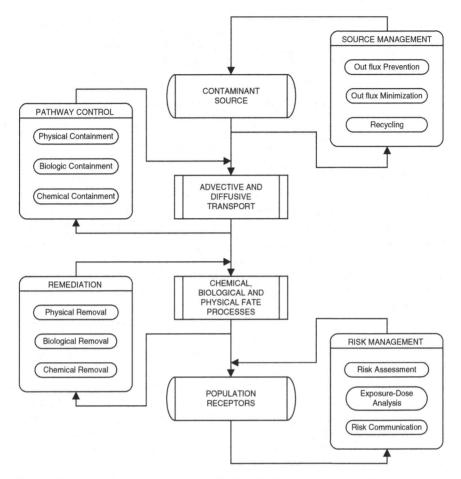

Fig. 3.1 Source-to-population environmental and health risk management paradigm

gets amplified or dampened due to the transport characteristics of the pathway in which the process is occurring. In this chapter we will review the basic principles that govern the environmental transformation and transport processes in general. Individual models of transformation and transport in air, groundwater and surface water pathways will be treated in more detail in the following chapters. Other environmental pathways such as plants or animals are not the subject matter of this book.

The analysis of transformation and transport of contaminants in various pathways such as air, groundwater and surface water pathways as independent pathways is somewhat artificial since the interactions between these pathways are naturally occurring and should be considered for closure and complete analysis. Common to all pathways are the conservation laws governing mass, momentum and energy. These are the three most important laws through which environmental transport can be described and studied. In this chapter we will work with several different forms of conservation principles. In each case the conservation principles will be

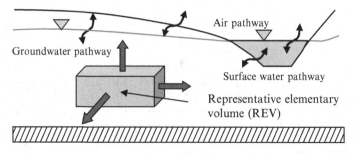

Fig. 3.2 Control volume approach (REV)

integrated over a control volume to describe the conservation law in terms of mathematical models. Differential equations are an integral part of the representation of these principles within a representative elementary volume (REV) which is sometimes identified as the control volume (Fig. 3.2). The one-, two- or three-dimensional (1D, 2D, 3D) domain of these problems will have a boundary. On these boundaries the boundary conditions of the problem must be defined for closure of the mathematical definition of the problem. If the problem is a time dependent application (dynamic model or unsteady state model) the mathematical representation of the problem as a complete model will also include an initial condition which would identify the initial state of the contaminant distribution in the solution domain at the start of the solution.

The solution to these mathematical models can be obtained through the use of analytical, numerical or statistical methods. Complex models usually require numerical solutions which will yield discrete solutions of the problem. Solutions that involve uncertainty or sensitivity analysis must employ statistical or stochastic methods. For simpler models, which are described as screening models in the ACTS software, the analytical solutions are more commonly used. In this case the solution of the differential equation model of the problem is represented as an analytical solution that is obtained through some integration process. Since the focus of this book is on the use of simpler environmental models, the analytical solution strategy will be the only approach discussed and used in the accompanying ACTS software. In a very few cases the matrix methods are incorporated into the solution strategy to render the analysis more meaningful and general.

The topics covered in this chapter are also treated in other reference books either as a general discussion or as pathway specific applications. For completeness the following references are recommended as a supplementary reading material of the subject matter that is covered in this chapter (Cheng. 1976; Anderson and Woessner 1992; Zheng and Bennett 1995; Clark 1996; Schnoor 1996; Sun 1996; Abdel-Magid et al. 1997, Weber and DiGiano 1996; Charbeneau 2000; Hemond and Fechner-Levy 2000; Bird et al. 2002; Nirmalakhandan 2002). In addition to the description of the conservation principles covered in each section below, the acronyms and abbreviations that are used in this field are given in Appendix 1. Further, a comprehensive list of definitions of the important terms used in conservation laws and mathematical models is given in Appendix 2. These two appendices will provide the reader with a summary of the terminology used in the environmental modeling field. It would be important for the reader to familiarize themselves with these definitions and abbreviations.

3.1   Transport Principles

Conservation principles are based on material checks and balance of some property in a REV. For example, quantifiable properties of a fluid medium, in reference to conservation principles, can be defined in terms of the mass, momentum and energy of the medium. These three properties are identified as extensive properties, as their values are additive in a REV. For example, addition of two volume elements of mass will double the mass in a REV. The same is true for momentum and energy. By definition, as opposed to extensive properties, intensive properties are not additive. For example, the intensive property of a fluid element such as density (mass per unit volume) will not double if two volumes of the same fluid element at the same temperature are added together. Accordingly, properties such as tempera-ture, pressure and density are examples of intensive properties, while heat, mass, momentum and energy are examples of extensive properties. In either case the goal of a conservation principle is to account for all of a property that initially exists in the REV and to track the change of these properties over space and time. Since mass, momentum and energy refer to different concepts in extensive properties of a fluid element it is important to review these definitions first. In working with these properties and understanding their transformation and transport definitions within the context of conservation principles, the proper understanding of the following terms that are used in the definition of these processes is important. We define advection as the transport of an entity by the moving fluid – bulk movement. Convection is associated with the transport of an entity due to density differences. Diffusion can be defined as the random walk of an ensemble of particles from regions of high concentration to regions of lower concentration. Conduction is the transfer of energy by the jostling motion of atoms through direct contact between atoms. This is an analogous form of diffusion. In this case, "heat" is the "property" that is transported by molecular motion. As can be seen, although some of these terms may be used as synonyms in common language, they imply and refer to completely different transport processes in technical language.

Contaminant Concentration Versus Mass: The concentration of a substance in a REV is defined as the quantity of substance, usually measured in mass per unit volume.

$$C = \frac{\text{quantity of substance}}{\text{REV}} = \frac{m}{\cancel{V}}; \quad \left[ML^{-3} \right] \tag{3.1}$$

In all references made to a property in this book, the dimensions of the property or the equation used to define the property are given in square parenthesis where $\left[M, L, T, \widehat{T}\right]$ refer to mass, length, time and temperature dimensions.

In Eq. (3.1) concentration has the dimensions of mass per unit volume. In standard international units concentration can be expressed as kg/m^3 or mg/m^3 (Weber and DiGiano 1996). This definition may change somewhat if the ambient environment changes. For example water at local barometric pressures can be considered to be an incompressible medium, whereas, air is a compressible medium. The compressibility of a medium is associated with the possible change of its volume as a function of the pressure or stress (force per unit area) exerted on it. Since volume is included in the definition of concentration (Eq. (3.1)) a possible change in volume as a function of pressure should be accounted for in the definition of concentration if the ambient environment is a compressible medium, i.e. air (see Chapter 4).

In water, concentrations can be expressed in milligrams or moles of substance per liter $\left(1\,L = 10^{-3}m^3 = 10^3 cm^3\right)$. The purpose of using moles to represent mass stem from the fact that it simplifies the translation or up-scaling of chemical equations that are defined at molecular or micro scales to the mass balance analysis that may be conducted at a macro or mega scale (see Fig. 1.7). At micro scale one molecule is 1 mol $(6.023 \times 10^{23}$ atoms) and the expression of the mass of a molecule is associated with the molecular weight of the atoms it contains. The chemical equation of a molecule can be used to calculate the molecular weight of the compound. For example the mass of 1 mol of carbonic acid $[H_2CO_3]$ is $[2 \times 1 + 1 \times 12 + 3 \times 16 = 62\,g]$ since the atomic weights of hydrogen, carbon and oxygen are 1, 12 and 16 g/mol respectively. In air, however, the concentrations may change not because the mass of the chemical in a REV is changing but because the volume defining that concentration may change as a function of pressure. Thus, in the case of air it is more appropriate to represent the concentration of substances as moles of substance per mole of air or mass of substance per mass of air under partial pressures. The same argument is also valid for other compressible solids media such as soils. For soils, concentration would again be defined as mass of substance per mass of soil medium. Of course the wetness or dryness of the soil medium should also be taken into account when considering compressibility effects.

If there are several constituents or if the substance appears in several species in a medium, the concentration of each species can be defined according to the equation,

$$C_i = \frac{\text{quantity of species } i}{\text{REV}} = \frac{m_i}{V}; \quad i = 1, 2, 3, ..., n; \quad \left[ML^{-3}\right] \qquad (3.2)$$

In Eq. (3.2) the subscript i refers to the mass of different species in the REV.

Advective Versus Diffusive Flux: The flux of some mass or concentration in some direction in a REV is defined as the quantity of that mass or concentration passing through a cross-section area perpendicular to the direction of movement per unit area per unit time. Thus, flux of a property, in this case mass, may be represented as,

$$q(flux) = \frac{m}{A\Delta t}; \quad \left[ML^{-2}T^{-1}\right] \tag{3.3}$$

The definition of flux given above may be associated with two different types of transport processes. Using the definition given in Eq. (3.3), if the transport process is associated with the bulk movement of the substance due to the advective velocity of the carrying fluid, then this flux can be related to the concentration times the velocity of the fluid medium as follows,

$$q_a \left(advective\ flux\right) = \frac{m}{\not V} \frac{\not V}{A\Delta t} = C\vec{V}; \quad \left[ML^{-2}T^{-1}\right] \tag{3.4}$$

where $\vec{V}$ is the velocity vector of the carrying fluid in some general direction (a vector quantity), A is the cross-section area perpendicular to the velocity direction and $\not V$ is the volume of teh REV. This transport process is identified as the advective transport.

In the environment substances can be transported in the absence of an ambient fluid flow field or bulk fluid velocity. Let's consider the simple example of a container separated by some permeable membrane with distinct contaminant concentrations C_1 and C_2 placed on either side of the membrane (Fig. 3.3). Let's further assume that $C_1 < C_2$ and that there is no net advective transport or ambient velocity between the two container chambers. This implies that, for no net advective transport, the fluctuating flow velocity $\mathbf{V}'$ in one direction is compensated by the velocity of the same magnitude in the opposite direction yielding zero velocity between the containers. The accounting of mass balance between these two containers can then be given as,

$$q_d(diffusive\ flux) = C_1\mathbf{V}' - C_2\mathbf{V}' = \mathbf{V}'(C_1 - C_2); \quad \left[ML^{-2}T^{-1}\right] \tag{3.5}$$

By multiplying and dividing Eq. (3.5) by the distance between the centers of the two boxes, which is the mean travel distance for all contaminant particles in the two chambers, we can write Eq. (3.5) as,

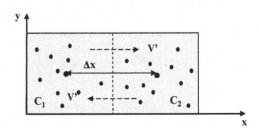

Fig. 3.3 Diffusion process in the absence of advection

$$q_d = -(\mathbf{V'}\Delta x)\frac{\Delta C}{\Delta x} = -D\frac{dC}{dx} \; ; \quad [ML^{-2}T^{-1}] \tag{3.6}$$

where the concentration difference between the containers is $(\Delta C = C_2 - C_1)$, D is identified as the diffusion coefficient and has the units of $[L^2T^{-1}]$. The ratio of the concentration difference ΔC over $\Delta x = (x_2 - x_1)$ can be substituted by the definition of the ordinary derivative as Δx approaches zero. The negative sign is the natural outcome of the definitions given above and implies that the diffusive flux is from high concentration to low concentration $\left(\frac{dC}{dx} = \frac{C_2 - C_1}{x_2 - x_1}\right)$ in an x-coordinate axis direction as indicated in Fig. (3.3). This transport process is identified as the diffusive flux.

Based on the definitions given above, advective flux is associated with the bulk movement of the carrying fluid and diffusive flux is associated with the gradient of concentrations between two points. If there is no ambient advective velocity then advective flux will be zero. If there is no concentration difference between the chambers or between two points then the diffusive flux will be zero or as the concentration differences increase the diffusive flux will increase. This definition of flux is also analogous to the definition of Fick's law of diffusion used in mass flux analysis or Fourier's law of heat conduction used in energy flux analysis.

The overall transport process is the combination of both advective and diffusive transport processes.

$$q = q_a + q_d = C\overrightarrow{\mathbf{V}} - \sum_i D\frac{dC}{dx_i} \; ; \quad i = x, y, z \tag{3.7}$$

where $\left(\overrightarrow{\mathbf{V}} = u\overrightarrow{i} + v\overrightarrow{j} + w\overrightarrow{k}\right)$ is the standard definition of the velocity vector in three dimensions. In the equation given above the discussion is extended to a three dimensional form by the use of the vector notation for velocity and also the use of summation for diffusion terms in x-, y-, z-directions. Thus, flux terms defined above are valid for a three dimensional domain. Other possible transport processes such as settling of particles under gravitational attraction or evaporation from surfaces or other definitions of fluxes can simply be added to the two transport components identified above as additional transport mechanisms.

Diffusive Mass Flux: In the context of the conservation of mass principle, flux of mass across a cross-section area at the boundary of the REV will be considered. As stated earlier, a flux of mass is the quantity of mass passing through a cross-section area per unit area per unit time. For conservation of mass principles, Fick's law can be used to determine the flux of mass across boundaries (cross-section areas) of the REV. This flux is more conveniently expressed in terms of concentration units given the definition above. Fick's law for a concentration species, i, can be given as,

$$F_{i,x_j} = -D_{i,j}\frac{dC_i}{dx_j} \; ; \quad j = 1, 2, 3; \quad i = 1, 2, 3, ..., n; \quad [ML^{-2}T^{-1}] \tag{3.8}$$

where F_{i,x_j} is the diffusive flux of concentration species C_i in the direction x_j, and $D_{i,j}$ is a proportionality coefficient identified as the diffusion coefficient of the concentration species i in the x_j direction (x-, y-, z-coordinates for $j = 1, 2, 3$). In association with the definition given in Eq. (3.6), Eq. (3.8) includes a negative sign to indicate that the flux is negative when the concentration is increasing in the positive x_j-coordinate direction. Since in Eq. (3.8) a gradient term and a directional diffusion coefficient are used, the mass flux in transverse directions can also be defined in terms of the gradients of concentrations in the transverse directions, and the associated transverse diffusion or dispersion coefficient tensor components in terms of that direction. The diffusion coefficient has the units of $[L^2T^{-1}]$ (Eq. (3.6)). The gradient term has the units of $[ML^{-4}]$. Thus the units of diffusive mass flux can be given as $[ML^{-2}T^{-1}]$. The total dispersive mass flux across a surface area $[L^2]$ is the mass flux (Eq. (3.8)) times the cross-section area. Thus the total mass flux across a surface takes on the dimensions $[MT^{-1}]$.

Momentum Flux: The linear momentum of a fluid element is defined as the product of mass m, times its velocity $\vec{V}$, of the entity and is an extensive property.

$$Momentum_{i,j} = m_i v_{i,j}; \quad j = 1, 2, 3; \quad i = 1, 2, 3, ..., n; \quad [MLT^{-1}] \qquad (3.9)$$

The Eq. (3.9) is given for species i and the species velocity in directions x_j(x-, y-, z-coordinates for $j=1, 2, 3$). The momentum per unit volume, an intensive property of the fluid for specified flow conditions, is mass times velocity divided by volume, or simply density times the velocity. Since velocity is a vector quantity, the momentum per unit volume has three components, one for each coordinate direction. Thus momentum of species i, per unit volume of REV can be given as,

$$Momentum_{i,j} = \frac{m_i v_{i,j}}{\cancel{V}} = \rho_i v_{i,j}; \quad j = 1, 2, 3; \quad i = 1, 2, 3, ..., n; \quad [ML^{-2}T^{-1}] \qquad (3.10)$$

where ρ_i and $v_{i,j}$ are the density and the components of the velocity (u, v, w) in three coordinate directions for the species i.

The local rate of momentum transfer in a fluid element is determined in part by the stresses. As in Newton's law a force represents an overall rate of transfer of momentum $\vec{F} = m\vec{a} = d(m\vec{V})/dt$, in which $\vec{a}$ is the acceleration vector. Then stress (force per unit area) represents the flux of momentum. In the context of conservation principles, the flux of momentum across a cross-section area, i.e. the boundaries of the REV, will be considered. The momentum flux across boundaries (cross-sections areas) of the REV per unit cross-section area can then be given as,

$$M_{i,j} = \frac{1}{A} \frac{d(m_i v_{i,j})}{dt}; \quad j = 1, 2, 3; \quad i = 1, 2, 3, ..., n; \quad [ML^{-1}T^{-2}] \qquad (3.11)$$

or in the case of momentum per unit volume,

$$M_{i,j} = \frac{1}{A} \frac{d(\rho_i v_{i,j})}{dt}; \quad j = 1, 2, 3; \quad i = 1, 2, 3, ..., n; \quad [ML^{-4}T^{-2}] \qquad (3.12)$$

where $M_{i,j}$ is the momentum flux of species i associated with the rate of momentum change for the species in the Jth coordinate direction.

The time averaged velocity gradients used in the Eqs. (3.11) and (3.12) can be defined in terms of the time averaged turbulent or laminar shear stresses (force per unit area) parallel to the surface areas of the REV. Using the definition of time averaged turbulent shear stress for the time averaged velocity gradients defined in the equations above, the momentum flux can be represented as,

$$M_{i,j} = \tau_{i,mj} = \mu_v \frac{d\mathbf{v}_{i,j}}{dx_m}; \ j,m=1,2,3 \ (m \neq j); \ i=1,2,3,...,n; \ \left[ML^{-1}T^{-2}\right] \quad (3.13)$$

or representing momentum flux per unit volume one can write,

$$M_{i,j} = \tau_{i,mj} = \frac{\mu_v}{\rho_i} \frac{d(\rho_i \mathbf{v}_{i,j})}{dx_m}; \ j,m=1,2,3 \ (m \neq j); \ i=1,2,3,...,n; \ \left[ML^{-1}T^{-2}\right] \quad (3.14)$$

in the equations above μ_v is a proportionality constant that is identified as the momentum transfer coefficient (time averaged eddy viscosity in the turbulent flow field case) that combines the flow, fluid and ambient medium properties. Given the gradient definition and also the directional momentum transfer coefficient, one can also define the momentum flux of species i in transverse directions by considering the velocity gradients in that direction as well as the momentum transfer coefficient in transverse directions (Weber and DiGiano 1996). The momentum transfer coefficient has the dimensions of $[ML^{-1}T^{-1}]$ since it is defined in terms of fluid and flow properties. The gradient term divided by density has the units of $[T^{-1}]$. Thus the dimensions of momentum flux can be given as $[ML^{-1}T^{-2}]$. The total momentum flux across a surface area $[L^2]$ is momentum flux times the cross-section area. Thus the total momentum flux across a surface takes on the dimensions $[MLT^{-2}]$.

Energy Flux: Energy is a measure of the capacity to perform work. This capacity may occur in nature in mechanical, thermal, chemical, biological, nuclear, solar or electrical energy forms. Energy has the dimensions of $[ML^2T^{-2}]$ and in SI units it is measured in joules (J). By definition a joule is the energy exerted by a force of 1 N (Newton) over a distance of 1 m. Thus the units of joule can be given as kg m^2/s^2 or g m^2/s^2. Units of energy in other systems such as in British units, are given in British Thermal Units (BTU), calorie, erg, foot–pounds or kilowatt–hours.

In the context of conservation principles, flux of energy across a cross-section area will be considered. In the case of conservation of energy principles Fourier's law of heat conduction (notice the change of the term diffusion to conduction in this case) can be used to determine the flux of heat energy across boundaries (cross-sections) of the REV. Fourier's law can be given as,

$$E_{T,x_j} = -k_c \frac{d\widehat{T}}{dx_j}; \ j=1,2,3; \ \left[MT^{-3}\right] \quad (3.15)$$

where E_{T,x_j} is the energy flux of the intensive property $\widehat{T}$ (temperature) which is associated with the gradient of the intensive property and k_c is a proportionality constant that is identified as the energy transfer coefficient. If temperature is the intensive property k_c is the thermal conductivity of the medium. The energy flux per unit volume can be given as,

$$E_{T,x_j} = -\frac{k_c}{\rho_i Q_T^o}\frac{d\left(\rho_i Q_T^o \widehat{T}\right)}{dx_j}; j = 1,2,3; \quad \left[MT^{-3}\right] \tag{3.16}$$

where $\rho_i Q_T^o$ is the product of the density of the medium and the thermal capacity. Given the gradient definition and the directional energy transfer coefficient for heterogeneous domains, one may also define the energy flux of the intensive property $\widehat{T}$ in transverse directions by considering the gradients in that direction (Weber and DiGiano 1996) and also the directional values of the energy transfer coefficient for heterogeneous domains. The energy transfer coefficient, k_c (thermal conductivity), has the dimensions of $\left[\widehat{T}^{-1}MLT^{-3}\right]$ since it is defined in terms of the intensive property dimensions of energy. In this case temperature is considered to be the intensive property. The gradient term in Eq. (3.15) has the dimensions of $\left[\widehat{T}L^{-1}\right]$. Thus the dimension of energy flux across a surface area per unit surface area is $[MT^{-3}]$. The total energy flux across a surface area $[L^2]$ is energy flux per unit area times the cross-section area. Thus, the total energy flux across a surface takes on the dimension $[ML^2T^{-3}]$.

Similarity Between Mass, Momentum and Energy Flux Principles: The first similarity between the mass, momentum and energy flux principles given above may be observed in the mathematical forms of Eqs. (3.8), (3.13) and (3.15). For all three cases the flux terms are defined in terms of the spatial gradient of the intensive property times a proportionality coefficient. The second similarity between these principles can be observed in the dimensions of the proportionality coefficient. If we define these proportionality coefficients as the momentum per unit volume for the momentum transfer coefficient (Eq. (3.14)), as the energy per unit volume for the energy transfer coefficient (Eq. (3.16)), as given in Eq. (3.8) for the mass transfer coefficient (mass per unit volume, concentration) then we observe that the proportionality coefficients for the three flux principles have the dimensions of $[L^2T^{-1}]$ (Weber and DiGiano 1996).

3.2 Conservation Principles

The definitions of fluxes for mass, momentum and energy given above can now be used in developing the conservation principles within a control volume (Fig. 3.4).

Conservation of Mass Principle: We will derive the conservation of mass principle for an unsteady state case in three-dimensions (3D) first. Then we will reduce this form to other special cases such as 2D or 1D and/or steady state cases

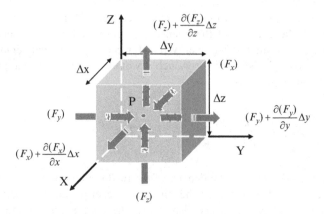

Fig. 3.4 Mass flux across the boundary surfaces of a REV

without going through the details of the derivation of each case. It is stated earlier, as well as in Chapter 1, that the conservation principle is an accounting process of substances in a REV. Thus, the derivation of the conservation of mass principle will be based on mass flux entering and leaving a REV and the time rate of change of mass within the control volume, plus the accounting of the external sources and sinks and transformation of mass species within the control volume. In identifying the mass fluxes entering and leaving the boundaries of the REV relative to each other, we will use the first order Taylor series approximation defined in Chapter 2 (Fig. 3.4). In differential form the accounting process of mass fluxes per unit volume (REV) and the presence of sources, sinks and reactions can be given as,

$$\sum\left(\frac{Mass\ flux\ in}{\cancel{V}}\right)\Delta A - \sum\left(\frac{Mass\ flux\ out}{\cancel{V}}\right)\Delta A = \frac{\partial\left(\frac{Mass\ in\ REV}{\cancel{V}}\right)}{\partial t} \qquad (3.17)$$
$$+ \sum Sources/Sinks/Reactions$$

The source, sink and reaction terms of Eq. (3.17) will be included in the analysis later on. First let's focus on the flux and rate of change of mass terms in the REV, the first two terms of Eq. (3.17) and the first term on the right hand side of Eq. (3.17). Mass flux in and out of the REV (Fig. 3.4) can be represented as the sum of the advective and diffusive fluxes (Eq. (3.7)) in each coordinate direction.

As an extension of the earlier definitions, the advective flux can be given as,

$$F_{a,j} = \overline{nCv_j}; \quad j = 1, 2, 3 \qquad (3.18)$$

and the diffusive flux can be defined as,

$$F_{d,j} = -nD_{H_j}\overline{\frac{\partial C}{\partial x_j}}; \quad j = 1, 2, 3 \qquad (3.19)$$

where n is the porosity associated with the control volume if the control volume is composed of solid particles and pores (soil medium). For example, in a soil medium the contaminants will mainly reside in the pores of the soil which are occupied with fluid under saturated conditions and also with air and fluid in partially saturated conditions. The porosity of a REV which is completely occupied by a fluid would be one. An air pathway or surface water pathway may represent such cases. Thus, in those applications this parameter may be omitted. In Eqs. (3.18) and (3.19) the subscript j represents the coordinate directions (x, y, z for $j = 1, 2, 3$). The bar over the products used in these equations represents the time average values of the terms used in these equations. Given Eqs. (3.18) and (3.19), the total flux per unit area across the boundaries of the REV can be given as,

$$F_j = F_{a,j} + F_{d,j}; \quad j = 1, 2, 3 \tag{3.20}$$

Thus,

$$F_x = \overline{nCu} - \overline{nD_{H_x} \frac{\partial C}{\partial x}}$$
$$F_y = \overline{nCv} - \overline{nD_{H_y} \frac{\partial C}{\partial y}} \tag{3.21}$$
$$F_z = \overline{nCw} - \overline{nD_{H_z} \frac{\partial C}{\partial z}}$$

and the gradients of these fluxes in each coordinate direction can also be defined.

$$\frac{\partial F_x}{\partial x} = \frac{\partial}{\partial x} \left(\overline{nCu} - \overline{nD_{H_x} \frac{\partial C}{\partial x}} \right)$$
$$\frac{\partial F_y}{\partial y} = \frac{\partial}{\partial y} \left(\overline{nCv} - \overline{nD_{H_y} \frac{\partial C}{\partial y}} \right) \tag{3.22}$$
$$\frac{\partial F_z}{\partial z} = \frac{\partial}{\partial z} \left(\overline{nCw} - \overline{nD_{H_z} \frac{\partial C}{\partial z}} \right)$$

Using the Taylor series expansion of the mass flux terms in each coordinate direction, the mass flux in minus the mass flux out per unit volume through the boundaries of the REV ($\Delta y\Delta z, \Delta x\Delta z, \Delta x\Delta y$) in x-, y- and z-directions can be written as shown in Fig. 3.4, respectively.

$$\left(F_x - \left(F_x + \frac{\partial F_x}{\partial x} \Delta x \right) \right) \Delta y \Delta z = -\frac{\partial F_x}{\partial x} \Delta x \Delta y \Delta z = -\frac{\partial}{\partial x} \left(\overline{nCu} - \overline{nD_{H_x} \frac{\partial C}{\partial x}} \right) \Delta x \Delta y \Delta z$$
$$\left(F_y - \left(F_y + \frac{\partial F_y}{\partial y} \Delta y \right) \right) \Delta x \Delta z = -\frac{\partial F_y}{\partial y} \Delta x \Delta y \Delta z = -\frac{\partial}{\partial y} \left(\overline{nCv} - \overline{nD_{H_y} \frac{\partial C}{\partial y}} \right) \Delta x \Delta y \Delta z$$
$$\left(F_z - \left(F_z + \frac{\partial F_z}{\partial z} \Delta z \right) \right) \Delta x \Delta y = -\frac{\partial F_z}{\partial z} \Delta x \Delta y \Delta z = -\frac{\partial}{\partial z} \left(\overline{nCw} - \overline{nD_{H_z} \frac{\partial C}{\partial z}} \right) \Delta x \Delta y \Delta z$$

$$\tag{3.23}$$

The rate of change of mass within REV per unit volume can be written as,

$$\frac{\partial\left(\dfrac{Mass\ in\ REV}{\cancel{V}}\right)}{\partial t}\Delta x \Delta y \Delta z = n\frac{\partial C}{\partial t}\Delta x \Delta y \Delta z \tag{3.24}$$

where it is assumed that porosity and REV are constant with respect to time. To simplify the derivation later on we will further assume that the porosity is also constant spatially, which is true for a homogeneous domain.

Substituting Eqs. (3.23) and (3.24) into Eq. (3.17) and cancelling the volume we obtain the conservation principle in three dimensions.

$$-\frac{\partial}{\partial x}\left(\overline{nCu - nD_{H_x}\frac{\partial C}{\partial x}}\right) - \frac{\partial}{\partial y}\left(\overline{nCv - nD_{H_y}\frac{\partial C}{\partial y}}\right) - \frac{\partial}{\partial z}\left(\overline{nCw - nD_{H_z}\frac{\partial C}{\partial z}}\right) = n\frac{\partial C}{\partial t}$$

$$\tag{3.25}$$

Based on the assumption that porosity are constant both spatially as well as time (its value is 1 for a REV completely filled with fluid) and eliminating the time average representation of the terms above, by assuming that time averaging is implied for those terms identified as time averaged in the equation above, we obtain the final three dimensional conservation mass equation below.

$$\frac{\partial C}{\partial t} + \frac{\partial(Cu)}{\partial x} + \frac{\partial(Cv)}{\partial y} + \frac{\partial(Cw)}{\partial z} = \frac{\partial}{\partial x}\left(D_{H_x}\frac{\partial C}{\partial x}\right) + \frac{\partial}{\partial y}\left(D_{H_y}\frac{\partial C}{\partial y}\right) + \frac{\partial}{\partial z}\left(D_{H_z}\frac{\partial C}{\partial z}\right)$$

$$\tag{3.26}$$

One should recognize that this form of the conservation principle does not include external source or sink terms or the transformation processes that may be present in the REV. The transformation processes within the REV will be identified by chemical or biological reactions, and these contributions will be added to the equation above later on.

Simpler Forms of the Conservation of Mass Principle: Based on Eq. (3.26) it is now straightforward to identify the following reduced forms of the conservation of mass principle again in the absence of source, sink and reaction terms.

Two-dimensional time dependent form:

$$\frac{\partial C}{\partial t} + \frac{\partial(Cu)}{\partial x} + \frac{\partial(Cv)}{\partial y} = \frac{\partial}{\partial x}\left(D_{H_x}\frac{\partial C}{\partial x}\right) + \frac{\partial}{\partial y}\left(D_{H_y}\frac{\partial C}{\partial y}\right) \tag{3.27}$$

One-dimensional time dependent form:

$$\frac{\partial C}{\partial t} + \frac{\partial(Cu)}{\partial x} = \frac{\partial}{\partial x}\left(D_{H_x}\frac{\partial C}{\partial x}\right) \tag{3.28}$$

3D and 2D time dependent form with only longitudinal velocity in x-direction and diffusion in x-, y- and/or z-directions: This is a very common assumption in applications where longitudinal advection is dominant when compared to advection in transverse directions.

$$\frac{\partial C}{\partial t} + \frac{\partial (Cu)}{\partial x} = \frac{\partial}{\partial x}\left(D_{H_x}\frac{\partial C}{\partial x}\right) + \frac{\partial}{\partial y}\left(D_{H_y}\frac{\partial C}{\partial y}\right) + \frac{\partial}{\partial z}\left(D_{H_z}\frac{\partial C}{\partial z}\right)$$

$$\frac{\partial C}{\partial t} + \frac{\partial (Cu)}{\partial x} = \frac{\partial}{\partial x}\left(D_{H_x}\frac{\partial C}{\partial x}\right) + \frac{\partial}{\partial y}\left(D_{H_y}\frac{\partial C}{\partial y}\right)$$

(3.29)

Steady state 3D, 2D and 1D forms:

$$\frac{\partial (Cu)}{\partial x} + \frac{\partial (Cv)}{\partial y} + \frac{\partial (Cw)}{\partial z} = \frac{\partial}{\partial x}\left(D_{H_x}\frac{\partial C}{\partial x}\right) + \frac{\partial}{\partial y}\left(D_{H_y}\frac{\partial C}{\partial y}\right) + \frac{\partial}{\partial z}\left(D_{H_z}\frac{\partial C}{\partial z}\right)$$

$$\frac{\partial (Cu)}{\partial x} + \frac{\partial (Cv)}{\partial y} = \frac{\partial}{\partial x}\left(D_{H_x}\frac{\partial C}{\partial x}\right) + \frac{\partial}{\partial y}\left(D_{H_y}\frac{\partial C}{\partial y}\right)$$

(3.30)

$$\frac{d(Cu)}{dx} = \frac{d}{dx}\left(D_{H_x}\frac{dC}{dx}\right)$$

Given the reduced forms of the 3D equation shown above, it is also possible to deduce other combinations without much difficulty. These cases, which will be used in the following chapters, may include the situations in which the velocities in each direction are constant or one may include different combinations of advective or diffusive fluxes into the equation. The development of these reduced forms will be left to the reader as an exercise.

We can also recognize from the reduced forms given above that if the contaminant concentration in the REV, that is material properties, is constant with respect to independent coordinates (x, y, z) and time, (Eq. (3.26)) will also yield the continuity equation for an incompressible fluid.

$$\frac{\partial (u)}{\partial x} + \frac{\partial (v)}{\partial y} + \frac{\partial (w)}{\partial z} = 0$$

(3.31)

Similarly, the reduced forms of the continuity Eq. (3.31) may also be written. One should also recognize that Eq. (3.31) is true for both steady and unsteady incompressible fluid flow cases.

In the conservation of mass equations given above we observe that the longitudinal advection and diffusion terms (dominant advective and dominant diffusion direction) will be associated, most commonly, with their x-directional components as shown below:

$$u\frac{\partial C}{\partial x} \qquad \text{(dominant advective transport component)}$$

$$D_x\frac{\partial^2 C}{\partial x^2} \qquad \text{(dominant diffusive transport component)}$$

(3.32)

These equations can be represented dimensionally as: vC/L for the advection term and DC/L^2 for the diffusion term. In a transport process, we can now compare the dominance of these processes over each other by evaluating the ratio of their scales.

$$P_e = \frac{(advection)}{(diffusion)} = \frac{vC/L}{DC/L^2} = \frac{vL}{D} \qquad (3.33)$$

This is a dimensionless ratio identified as the Peclet number $P_e = vL/D$. If $P_e \ll 1$ then the advection term is significantly smaller than the diffusion term, and it can be concluded that the transport process is a diffusion dominant process. In this case the spreading of the contaminant is expected to occur symmetrically around the source for homogeneous domains since diffusion from a source will be the same in all directions. If $P_e \gg 1$ then the advection term is significantly larger than the diffusion term. In that case it can be concluded that the transport process is an advection dominant process. The spreading of the contaminant in the domain is expected to occur non-symmetrically, since advection from a source will elongate the plume in the dominant advection direction. Considering these two cases, one may also choose to drop the advective term from the governing equations for $(P_e \ll 1)$ or drop the diffusion term from the governing equations for $(P_e \gg 1)$ as an approximation to simplify the mathematical model used in the analysis. In between these two extreme cases is the transition range in which both mechanisms will influence the transformation and transport process and need to be included in the mathematical model.

Conservation of Energy Principle: Heat sources introduced to the environment by mankind can be considered to be contamination. Thus, changes of temperature in the environment caused by heat sources that may adversely affect the receptors need to be evaluated. Using the basic principles of thermodynamics it is possible to define conservation principles for energy, in this case for heat energy, in a similar way to the conservation principles for mass developed earlier. In differential form the accounting process of energy fluxes per unit volume (REV) and energy sources and sinks can be given as,

$$\sum\left(\frac{\text{Heat flux in}}{\cancel{V}}\right)\Delta A - \sum\left(\frac{\text{Heat flux out}}{\cancel{V}}\right)\Delta A = \frac{\partial\left(\dfrac{\text{Heat in REV}}{\cancel{V}}\right)}{\partial t} + \sum\text{Heat Sources/Sinks} \qquad (3.34)$$

The source and sink terms will be treated in the next section. The heat content of a fluid element will be identified as the internal energy of the fluid. This energy is associated with the molecular energy of the fluid element, i.e. the energy associated with the agitation of the atoms comprising the molecule. Since the agitation of atoms of a molecule increases or decreases as a function of temperature, then the internal energy of a fluid element can be considered to be a function of temperature. The internal energy of a fluid element, which is proportional to temperature can be given as,

$$Int.\ E = mQ_T^o\widehat{T}\ ;\quad \left[ML^2T^{-2}\right] \tag{3.35}$$

where $m\ [M]$ is the mass of the fluid element, Q_T^o is the thermal capacity $\left[\widehat{T}^{-1}L^2T^{-2}\right]$ of the medium and $\widehat{T}$ is the absolute temperature $[\widehat{T}]$. Using the definition of energy flux per unit volume per unit area (Eq. (3.16)),

$$E_{T,x_j} = -\frac{k_c}{\rho_i Q_T^o}\frac{d\left(\rho_i Q_T^o\widehat{T}\right)}{dx_j} = -K_c\frac{d\left(\widehat{T}\right)}{dx_j}\ ;\ j=1,2,3;\quad \left[MT^{-3}\right] \tag{3.36}$$

where for simplicity we replace,

$$K_c = \frac{k_c}{\rho_i Q_T^o}\quad \text{and}\quad \widehat{\overline{T}} = \rho_i Q_T^o\widehat{T} \tag{3.37}$$

We can write the conservation of energy principle as follows.

$$-\frac{\partial}{\partial x}\left(\widehat{\overline{T}}u - K_c\frac{\partial\widehat{\overline{T}}}{\partial x}\right) - \frac{\partial}{\partial y}\left(\widehat{\overline{T}}v - K_c\frac{\partial\widehat{\overline{T}}}{\partial y}\right) - \frac{\partial}{\partial z}\left(\widehat{\overline{T}}w - K_c\frac{\partial\widehat{\overline{T}}}{\partial z}\right) = \frac{\partial\widehat{\overline{T}}}{\partial t} \tag{3.38}$$

The equation above indicates that heat energy may enter or leave the REV by the mechanism of heat conduction according to Fourier law. Heat energy may also enter or leave the REV by the overall fluid motion, which will be identified as convective transport. The heat energy entering and leaving the REV this way is sometimes identified as the sensible heat into or out of the system. The rate of heat energy change over time may be associated with the slowing down of atomic agitation of the molecules, degradation of mechanical energy or conversion of chemical energy into heat. It must be emphasized that in Eq. (3.38) only a restricted form of energy balance is considered. In the energy balance equation above we have not introduced the concepts of kinetic energy, potential energy or external work components of the overall energy conservation principle (Bird et al. 2002). Nevertheless, the simpler energy balance statement given in Eq. (3.38) will be useful in defining and solving a number of heat transfer problems in solids or incompressible fluids which is of interest in this book. If we consider only diffusive processes, Eq. (3.38) yields the familiar heat transfer equation.

$$\frac{\partial}{\partial x}\left(K_c\frac{\partial \widehat{T}}{\partial x}\right) + \frac{\partial}{\partial y}\left(K_c\frac{\partial \widehat{T}}{\partial y}\right) + \frac{\partial}{\partial z}\left(K_c\frac{\partial \widehat{T}}{\partial z}\right) = \frac{\partial \widehat{T}}{\partial t} \tag{3.39}$$

Similar to the conservation of mass principle discussed earlier the reduced forms of the equations above in 2D or 1D in steady and unsteady forms can be obtained with relative ease. The description of these reduced cases will be left as an exercise.

3.3 Sources and Sinks

Given Eq. (3.26), or Eq. (3.39) for that matter, one can now include the external sources and sinks, and the reaction terms of the conservation of mass equation, which will represent the complete transformation and transport processes within the REV. In this section we will treat the source and sink terms using the equations of conservation of mass which will be used extensively in the environmental pathway analysis that will be covered in the following chapters of this book. The analysis of source and sink terms in energy principles can be found in Weber and DiGiano (1996), Bird et al. (2002).

Based on the conservation of mass equation derived earlier (Eq. (3.26)), the source, sink and reaction terms can be included as,

$$\frac{\partial C}{\partial t} + \frac{\partial (Cu)}{\partial x} + \frac{\partial (Cv)}{\partial y} + \frac{\partial (Cw)}{\partial z} = \frac{\partial}{\partial x}\left(D_{H_x}\frac{\partial C}{\partial x}\right) + \frac{\partial}{\partial y}\left(D_{H_y}\frac{\partial C}{\partial y}\right) + \frac{\partial}{\partial z}\left(D_{H_z}\frac{\partial C}{\partial z}\right)$$
$$+ \sum_{w=1}^{N} C_s Q_w \delta(x_w, y_w, z_w) + \sum R_{reaction} \tag{3.40}$$

Here the external point sources/sinks are identified as $C_s Q_w$ where C_s represents the contaminant concentration in the source/sink and $Q_w [L^3 T^{-1}]$ is the strength of the source/sink, $\delta(x_j, y_j, z_j)$ $[L^{-3}]$ is the Dirac-delta function which assumes a value of 1 at the point (x_w, y_w, z_w) and zero elsewhere in the solution domain. In this equation sinks (extraction) will be identified with a positive sign and sources (injection) will be identified with a negative sign as the sign convention. An elaborate treatment of the use of the Dirac-delta function for source and sink terms can be found in Gunduz and Aral (2005). The term $\sum R_{reaction}$ will include all possible reactions that describe the transformation processes within the REV. We will introduce the reaction terms in the next section.

If we focus our attention on the external source/sink terms of the equation above, several observations can be made. First we will rewrite Eq. (3.40) as,

$$\frac{\partial C}{\partial t} + u\frac{\partial C}{\partial x} + v\frac{\partial C}{\partial y} + w\frac{\partial C}{\partial z} + C\left(\frac{\partial u}{\partial x} + \frac{\partial v}{\partial y} + \frac{\partial w}{\partial z}\right)$$
$$= +\frac{\partial}{\partial x}\left(D_{H_x}\frac{\partial C}{\partial x}\right) + \frac{\partial}{\partial y}\left(D_{H_y}\frac{\partial C}{\partial y}\right) + \frac{\partial}{\partial z}\left(D_{H_z}\frac{\partial C}{\partial z}\right) \qquad (3.41)$$
$$+ \sum_{w=1}^{N} C_s Q_w \delta(x_w, y_w, z_w) + \sum R_{reaction}$$

where the continuity principle component of Eq. (3.41), the fifth term in the equation above, is zero everywhere except the singular point where the source/sink is located $\delta(x_w, y_w, z_w)$. At these singularity points and throughout the domain the following is valid.

$$\frac{\partial u}{\partial x} + \frac{\partial v}{\partial y} + \frac{\partial w}{\partial z} = 0 \quad at \quad (x, y, z) \neq \delta(x_w, y_w, z_w) \qquad (3.42)$$

At the source/sink locations (x_w, y_w, z_w) we can replace this term as,

$$C\left(\frac{\partial u}{\partial x} + \frac{\partial v}{\partial y} + \frac{\partial w}{\partial z}\right) = \sum_{w=1}^{N} CQ_w \delta(x_w, y_w, z_w) \quad at \quad (x, y, z) = \delta(x_w, y_w, z_w) \quad (3.43)$$

When the definitions given in Eqs. (3.42) and (3.43) are substituted in Eq. (3.41) we obtain,

$$\frac{\partial C}{\partial t} + u\frac{\partial C}{\partial x} + v\frac{\partial C}{\partial y} + w\frac{\partial C}{\partial z} = \frac{\partial}{\partial x}\left(D_{H_x}\frac{\partial C}{\partial x}\right) + \frac{\partial}{\partial y}\left(D_{H_y}\frac{\partial C}{\partial y}\right) + \frac{\partial}{\partial z}\left(D_{H_z}\frac{\partial C}{\partial z}\right)$$
$$+ \sum_{w=1}^{N} (C_s - C) Q_w \delta(x_w, y_w, z_w) + \sum R_{reaction}$$

$$(3.44)$$

Equation (3.44) is now the proper form of the conservation of mass principle if there are external sources/sinks in the REV. Given this form of the governing equation the following observations can now be made.

If there is a sink in the domain, that is if there is extraction of contaminants from the REV, then $C_s = C$. Thus for this case the governing equation reduces to,

$$\frac{\partial C}{\partial t} + u\frac{\partial C}{\partial x} + v\frac{\partial C}{\partial y} + w\frac{\partial C}{\partial z} = \frac{\partial}{\partial x}\left(D_{H_x}\frac{\partial C}{\partial x}\right) + \frac{\partial}{\partial y}\left(D_{H_y}\frac{\partial C}{\partial y}\right)$$
$$+ \frac{\partial}{\partial z}\left(D_{H_z}\frac{\partial C}{\partial z}\right) + \sum R_{reaction}$$

$$(3.45)$$

If we are injecting clean fluids into an otherwise contaminated REV, then $C_s = 0$. Thus for this case the governing equation reduces to,

$$\frac{\partial C}{\partial t} + u\frac{\partial C}{\partial x} + v\frac{\partial C}{\partial y} + w\frac{\partial C}{\partial z} = \frac{\partial}{\partial x}\left(D_{H_x}\frac{\partial C}{\partial x}\right) + \frac{\partial}{\partial y}\left(D_{H_y}\frac{\partial C}{\partial y}\right) + \frac{\partial}{\partial z}\left(D_{H_z}\frac{\partial C}{\partial z}\right)$$
$$+ \sum_{w=1}^{N} CQ_w\delta(x_w, y_w, z_w) + \sum R_{reaction}$$

$$(3.46)$$

One should notice that Q_w is negative for injection. If we are injecting contaminants into an otherwise clean REV, than at that point $C = 0$ and $C_s \neq 0$. For this case the governing equation can be written as,

$$\frac{\partial C}{\partial t} + u\frac{\partial C}{\partial x} + v\frac{\partial C}{\partial y} + w\frac{\partial C}{\partial z} = \frac{\partial}{\partial x}\left(D_{H_x}\frac{\partial C}{\partial x}\right) + \frac{\partial}{\partial y}\left(D_{H_y}\frac{\partial C}{\partial y}\right) + \frac{\partial}{\partial z}\left(D_{H_z}\frac{\partial C}{\partial z}\right)$$
$$- \sum_{w=1}^{N} (C_s)Q_w\delta(x_w, y_w, z_w) + \sum R_{reaction}$$

$$(3.47)$$

These are all important outcomes of the presence of sources and sinks in the solution domain and will require attention to interpret the physical problem considered properly (Galeati and Gambolati 1989).

As given earlier, the reduced dimensional forms of these equations can be deduced with relative ease. In these cases the definition of the sources and sinks also change from volumetric source or sink to line source or sink as expected. In the one dimensional form the source or sink terms cannot be defined. The description of these reduced cases will be left as an exercise.

3.4 Reactions

We can now add the definitions of the reaction terms to the conservation of mass equation given in Eq. (3.40). This will yield the final form of the conservation of mass principle within the REV. The reaction terms in the conservation of mass principle can be represented using the following general form,

$$\sum R_{reaction} = \left(\frac{dC}{dt}\right) + \left(\frac{\rho_B}{n}\frac{dC^*}{dt}\right)$$

$$(3.48)$$

The first term in Eq. (3.48) represents the behavior of the solute over time within the solution and C is the concentration of the chemical. This form of the reaction equation is commonly used to represent biological or chemical reactions of the solute, which are identified as homogeneous reactions. Typically this type of reaction is associated with the degradation or decay of the solute. The second term in Eq. (3.48) is used to represent the behavior of the solute in the ambient environment, that is the interaction of the solute with the ambient environment where ρ_B is the bulk density of the solid media and n is the porosity of the ambient environment. Here the term C^* is the mass of chemical interacting with the ambient environment. Depending on the type of the ambient environment considered in the REV these type of reactions are commonly identified with processes such as sorption, desorption, precipitation, and transfer of solute from liquid phase to solid phase or vice versa. Since only solute is considered in the first term, the reaction is identified as homogeneous. Since the interaction of solute and the ambient environment are considered in the second term, these reactions will be identified as heterogeneous reactions. This categorization leads to six types of reactions as shown in Fig. 3.5 (Rubin 1983; Fetter 1993). The task now is to identify the proper mathematical forms of these six reactions to represent them in the conservation of mass principle (Eq. (3.40)).

The identifiers "sufficiently fast" and "insufficiently fast" used in Fig. 3.5 refer to the reaction rate relative to the solute advection time frame within a REV. A chemical reaction can be considered to be "sufficiently fast" if the rate of reaction is faster than the velocity of particles moving within and through the REV. This reference time is

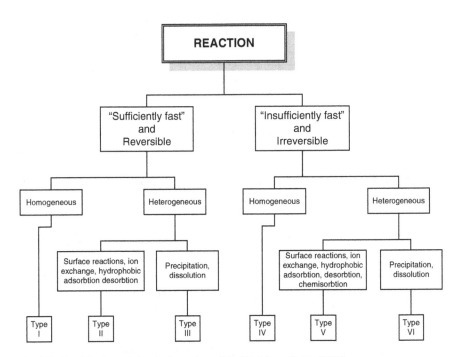

Fig. 3.5 Classification of chemical reactions (Modified from Rubin 1983)

also associated with the residence time of the solute within the control volume. Thus in this case the particles would experience the full reaction while in the REV. Otherwise the reaction can be identified as "insufficiently fast".

Homogeneous Reactions: Homogeneous reactions are reactions that are associated with the chemical behavior of the constituent independent of the interaction of the solute with the surroundings within the REV. In these reactions, which are categorized as Type I and Type IV reactions in Fig. 3.5, the question is what model is best to represent the first term on the right hand side of Eq. (3.48). If a reaction is sufficiently fast and reversible (Type I), then equilibrium conditions would be considered and the time derivative of concentration would be zero. If a reaction is slow or irreversible (Type IV), local equilibrium conditions cannot be considered and a kinetic process model will be used to describe the reaction. The general form of this reaction can be given as,

$$\frac{dC}{dt} = -KC^a \qquad (3.49)$$

where K is the reaction rate constant and "a" is the reaction order. If the reaction is a "zero" order reaction ($a = 0$), than the reaction is represented as,

$$\frac{dC}{dt} = -K \qquad (3.50)$$

This case is associated with a linear decay where the negative sign implies the loss of substance due to decay. The solution to Eq. (3.50) can also be given as,

$$\left. \begin{aligned} \int dC &= -\int K dt \\ C &= C_o - Kt \end{aligned} \right\} \qquad (3.51)$$

which indicates that the loss of substance is a linear function of time at a rate K, and C_o is the initial concentration.

A first order reaction ($a = 1$) is a more common form of this reaction type (Type IV). In this reaction, the rate of loss of a substance is proportional to the amount of substance present in the REV, which can be given as:

$$\frac{\partial C}{\partial t} = -KC \qquad (3.52)$$

where K is the first order reaction rate constant, C is the concentration and the negative sign again implies the loss of the substance. This is a single phase reaction and yields the most commonly used definition of decay in a chemical reaction process. The solution of this differential equation can be given as,

$$C = C_o \exp(-Kt) \qquad (3.53)$$

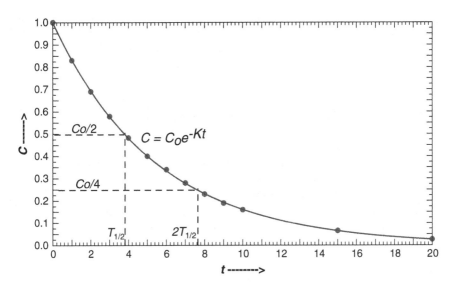

Fig. 3.6 The solution of the first order reaction equation

where C_o is the initial concentration in the REV. This is an exponential decay function as shown in Fig. 3.6.

This solution can be used to define the half-life of a chemical, which is defined as the time required for the chemical to reduce to half its original concentration. Thus the half-life of a substance is $T_{1/2}$ as its concentration reaches $(C_o/2)$. Accordingly, from Eq. (3.53) we can write,

$$\frac{C_o}{2} = C_o \exp\left(-KT_{1/2}\right) \tag{3.54}$$

which implies,

$$T_{1/2} = \frac{\ln 2}{K} \tag{3.55}$$

or,

$$K = \frac{\ln 2}{T_{1/2}} \tag{3.56}$$

This gives us the definition of the decay constant in terms of the half-life of a chemical, which is considered to be a standard property of chemicals under constant temperature and pressure conditions.

In analogy to the discussion above it is also possible to define higher order reactions where a > 1. For these cases the contaminant fate and transport equation becomes a nonlinear equation due to the presence of the power of concentration in the reaction term.

A common form of a higher order reaction that is used to describe biological reactions is the Monod kinetics form, in which the biodegradation rate is defined as:

$$\frac{dC}{dt} = -\mu_{max} \frac{C}{K_m + C} \tag{3.57}$$

in which μ_{max} is the maximum rate of increase and K_m is the half saturation constant which is also identified as the rate limiting constant. As the definition of the rate constants indicates, this reaction is more commonly used in subsurface pathway analysis (Cherry et al. 1984; Charbeneau 2000).

Using the two more common and simpler cases represented in Eqs. (3.50) and (3.52) the fate and transport equation given in Eq. (3.44) can be described as,

$$\frac{\partial C}{\partial t} + u\frac{\partial C}{\partial x} + v\frac{\partial C}{\partial y} + w\frac{\partial C}{\partial z} = \frac{\partial}{\partial x}\left(D_{H_x}\frac{\partial C}{\partial x}\right) + \frac{\partial}{\partial y}\left(D_{H_y}\frac{\partial C}{\partial y}\right) + \frac{\partial}{\partial z}\left(D_{H_z}\frac{\partial C}{\partial z}\right)$$
$$+ \sum_{w=1}^{N}(C_s - C)Q_w\delta(x_w, y_w, z_w) \pm K_1 \pm K_2C \tag{3.58}$$

where K_1 and K_2 are the rate constants for the zero and first order reactions respectively.

Heterogeneous Reactions: Heterogeneous reactions are associated with the chemical constituent behavior in interaction with its surroundings. These reactions involve the interaction of the chemical constituent in the dissolved phase and the solid phase that may exist in the REV such as granular soil particles in the groundwater pathway, sediments in the surface water pathway or dust particles in the air pathway. Sorption processes that belong to this category may include adsorbtion, chemisorption, absorption and ion exchange. It is this type of heterogeneous reaction that leads to the definition of the retardation coefficient in the advection-diffusion equation, as the reaction terms are interpreted mathematically in terms of sorption desorption rates in groundwater pathway applications. Thus, these types of reaction terms are more commonly used in subsurface pathway analysis in the form of a retardation coefficient due to the presence of a soil granular matrix where the adsorption and desorption is taking place, and a pore space where the advective transport is occurring. We will review these reactions from the subsurface pathway perspective, although these processes may also become important in surface water applications where sediments are involved or air shed applications where dust particles are involved.

In these reactions, which are categorized as Types II, III and Types V, VI reactions in Fig. 3.5, the question centers around what model is best to represent the second term on the right hand side of Eq. (3.48). If a reaction is sufficiently fast and reversible (Types II and III) than equilibrium conditions can be considered. If a reaction is slow or irreversible (Types V and VI) local equilibrium conditions cannot be considered and a kinetic process model will be used to describe the reaction.

The kinetic processes used in these cases are based on laboratory batch experiments conducted under constant temperatures. The simplest batch experiment can be the placement of solute at a certain concentration in a container which contains solid material or soil. Measurements are made over time to establish the amount of concentration in the liquid phase C (mass per unit volume, mg/L), and the amount of concentration in the solid phase, C^* (mass per unit weight solids, mg/kg). Since the batch experiment is done under constant temperature conditions the term isotherm is commonly associated with the interpretation of the outcome. Depending on the complexity of the relationship between C and C^*, linear sorbtion, Freundlich sorbtion and Langmuir sorbtion isotherms can be defined as shown in Fig. 3.7. The difference between these isotherms is the rate limited nature of the final representation and of course the mathematical form of the relationship used to describe this rate limited reaction. In the linear sorbtion isotherm the rate of sorbtion is not rate limited. In this reaction it is assumed that as the concentration in the solution increases the amount absorbed to the solid phase will increase indefinitely. Of course, this unending growth is not a realistic condition since there is a limit to the absorbtion that can take place given the amount of solid media in a mixture. The other two isotherms are formulated to compensate for this physical impossibility. In the Freundlich sorbtion isotherm the relationship between C and C^* is nonlinear, but the rate limited nature of the reaction is still not fully captured (Fig. 3.7). In the Langmuir sorbtion isotherm the rate limited nature of the reaction is captured.

In the linear sorbtion isotherm it is assumed that the relationship between the amount sorbed on to the soil and the concentration in the solution is linear,

$$\left.\begin{array}{c} C^* = K_d C \\ \dfrac{dC^*}{dt} = -K_d \dfrac{dC}{dt} \end{array}\right\} \tag{3.59}$$

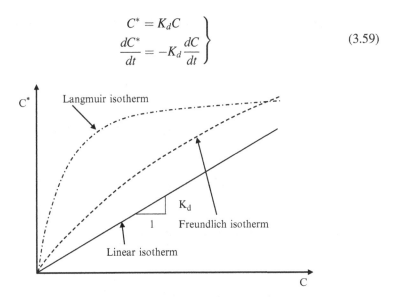

Fig. 3.7 Linear, Freundlich and Langmuir isotherms

where K_d is identified as the partition coefficient (volume per unit weight of solids, L/kg). If the rate equation above is substituted into the second term of Eq. (3.48) we get,

$$R_{reaction} = \left(\frac{\rho_B}{n}\frac{dC^*}{dt}\right) = -\left(\frac{\rho_B K_d}{n}\frac{dC}{dt}\right) \tag{3.60}$$

When this reaction is used in Eq. (3.44) the following is the outcome:

$$\frac{\partial C}{\partial t} + u\frac{\partial C}{\partial x} + v\frac{\partial C}{\partial y} + w\frac{\partial C}{\partial z} = \frac{\partial}{\partial x}\left(D_{H_x}\frac{\partial C}{\partial x}\right) + \frac{\partial}{\partial y}\left(D_{H_y}\frac{\partial C}{\partial y}\right) + \frac{\partial}{\partial z}\left(D_{H_z}\frac{\partial C}{\partial z}\right)$$
$$+ \sum_{w=1}^{N}(C_s - C)Q_w\delta(x_w, y_w, z_w) - \frac{\rho_B K_d}{n}\frac{\partial C}{\partial t} \tag{3.61}$$

or

$$\left(1 + \frac{\rho_B K_d}{n}\right)\frac{\partial C}{\partial t} + u\frac{\partial C}{\partial x} + v\frac{\partial C}{\partial y} + w\frac{\partial C}{\partial z}$$
$$= \frac{\partial}{\partial x}\left(D_{H_x}\frac{\partial C}{\partial x}\right) + \frac{\partial}{\partial y}\left(D_{H_y}\frac{\partial C}{\partial y}\right) + \frac{\partial}{\partial z}\left(D_{H_z}\frac{\partial C}{\partial z}\right) + \sum_{w=1}^{N}(C_s - C)Q_w\delta(x_w, y_w, z_w) \tag{3.62}$$

This leads to the definition of the dimensionless number identified as the retardation coefficient that is commonly used in subsurface analysis.

$$R = \left(1 + \frac{\rho_B K_d}{n}\right) \tag{3.63}$$

For the Freundlich isotherm the relationship between C and C^* is defined as,

$$\left.\begin{array}{l} C^* = KC^N \\ \ln C^* = \ln K + N\ln C \end{array}\right\} \tag{3.64}$$

where the constants K and N can be interpreted from the batch experiment outcome as shown in Fig. 3.8. If the rate equation above is substituted into the second term of Eq. (3.48) we get,

$$R_{reaction} = \left(\frac{\rho_B}{n}\frac{dC^*}{dt}\right) = -\left(\frac{\rho_B KN}{n}C^{N-1}\frac{dC}{dt}\right) \tag{3.65}$$

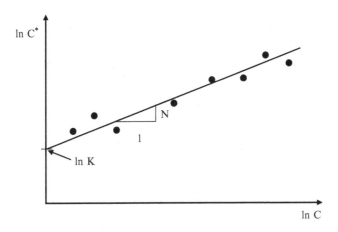

Fig. 3.8 Freundlich isotherm

When this reaction is used in Eq. (3.44) the following is the outcome which leads to a nonlinear transformation and transport equation:

$$\frac{\partial C}{\partial t} + u\frac{\partial C}{\partial x} + v\frac{\partial C}{\partial y} + w\frac{\partial C}{\partial z} = \frac{\partial}{\partial x}\left(D_{H_x}\frac{\partial C}{\partial x}\right) + \frac{\partial}{\partial y}\left(D_{H_y}\frac{\partial C}{\partial y}\right) + \frac{\partial}{\partial z}\left(D_{H_z}\frac{\partial C}{\partial z}\right)$$
$$+ \sum_{w=1}^{N}(C_s - C)Q_w\delta(x_w, y_w, z_w) - \frac{\rho_B KN}{n}C^{N-1}\frac{\partial C}{\partial t}$$

$$(3.66)$$

or

$$\left(1 + \frac{\rho_B KN}{n}C^{N-1}\right)\frac{\partial C}{\partial t} + u\frac{\partial C}{\partial x} + v\frac{\partial C}{\partial y} + w\frac{\partial C}{\partial z}$$
$$= \frac{\partial}{\partial x}\left(D_{H_x}\frac{\partial C}{\partial x}\right) + \frac{\partial}{\partial y}\left(D_{H_y}\frac{\partial C}{\partial y}\right) + \frac{\partial}{\partial z}\left(D_{H_z}\frac{\partial C}{\partial z}\right) + \sum_{w=1}^{N}(C_s - C)Q_w\delta(x_w, y_w, z_w)$$

$$(3.67)$$

For the Langmuir isotherm the rate limited relationship between C and C^* is defined as,

$$\left.\begin{aligned}\frac{C}{C^*} &= \frac{1}{\alpha\beta} + \frac{C}{\beta} \\ C^* &= \frac{\alpha\beta C}{1 + \alpha C} \\ \frac{dC^*}{dC} &= \frac{\alpha\beta}{(1 + \alpha C)^2}\end{aligned}\right\}$$

$$(3.68)$$

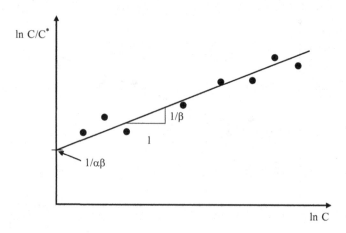

Fig. 3.9 Langmuir isotherm

where the constant α is the absorbtion constant related to the binding energy (volume per unit mass) and β is the maximum amount of solute that can be absorbed by the solid (mass per unit weight of solids). The experimental data in this case can be interpreted as shown in Fig. 3.9. If the rate equation above is substituted into the second term of Eq. (3.48) we get,

$$R_{reaction} = \left(\frac{\rho_B}{n}\frac{dC^*}{dt}\right) = -\left(\frac{\rho_B\alpha\beta}{n(1+\alpha C)^2}\frac{dC}{dt}\right) \tag{3.69}$$

When this reaction is used in Eq. (3.44) the following is the outcome which leads to a nonlinear transformation and transport equation:

$$\frac{\partial C}{\partial t} + u\frac{\partial C}{\partial x} + v\frac{\partial C}{\partial y} + w\frac{\partial C}{\partial z} = \frac{\partial}{\partial x}\left(D_{H_x}\frac{\partial C}{\partial x}\right) + \frac{\partial}{\partial y}\left(D_{H_y}\frac{\partial C}{\partial y}\right) + \frac{\partial}{\partial z}\left(D_{H_z}\frac{\partial C}{\partial z}\right)$$
$$+ \sum_{w=1}^{N}(C_s - C)Q_w\delta(x_w, y_w, z_w) - \left(\frac{\rho_B\alpha\beta}{n(1+\alpha C)^2}\frac{dC}{dt}\right) \tag{3.70}$$

or

$$\left(1 + \frac{\rho_B\alpha\beta}{n(1+\alpha C)^2}\right)\frac{\partial C}{\partial t} + u\frac{\partial C}{\partial x} + v\frac{\partial C}{\partial y} + w\frac{\partial C}{\partial z}$$
$$= \frac{\partial}{\partial x}\left(D_{H_x}\frac{\partial C}{\partial x}\right) + \frac{\partial}{\partial y}\left(D_{H_y}\frac{\partial C}{\partial y}\right) + \frac{\partial}{\partial z}\left(D_{H_z}\frac{\partial C}{\partial z}\right) + \sum_{w=1}^{N}(C_s - C)Q_w\delta(x_w, y_w, z_w) \tag{3.71}$$

In the discussion above we have introduced various standard forms for the irreversible reactions that are commonly used in environmental analysis. In this context it is also possible to define mathematical models for the reversible linear and nonlinear reactions. These reactions will yield more complex forms of the transformation and transport equation which will not be treated in this book. The reader is referred to the following references where the treatment of these cases can be found (Bear 1972; Rubin 1983; Schnoor 1996; Weber and DiGiano 1996; Charbeneau 2000; Hemond and Fechner-Levy 2000).

3.5 Boundary and Initial Conditions

The advection–diffusion-reaction equation that is used in the analysis of contaminant migration in three environmental pathways is a second order parabolic partial differential equation. In this equation the independent variables are (x, y, z, t). Thus the concentration distribution in the solution domain is a function of these independent variables $C(x, y, z, t)$. For the closure of the models that utilize the general Eq. (3.44), the boundary conditions at the boundaries of the solution domain and the initial condition that describes the contaminant distribution in the solution domain at the start of the solution are necessary.

The initial condition in the solution domain refers to the contaminant distribution at time zero $C(x, y, z, 0)$. Based on some boundary conditions the solution will build on this initial distribution. Mathematically the initial condition is described as,

$$C(x, y, z, 0) = C_o(x, y, z) \tag{3.72}$$

where $C_o(x, y, z)$ is a given function. If the solution is starting from an uncontaminated condition in the domain then $C_o(x, y, z) = 0$.

The boundary conditions of the advection-diffusion-reaction equation are more complex. Three different boundary conditions can be used to represent the physical conditions that may exist at the boundaries of the solution domain. These are either concentration based or flux based representations. Concentration based boundary condition is identified as Dirichlet condition in the literature, which can be given as:

$$C(x_b, y_b, z_b, t) = C_1(x_b, y_b, z_b, t) \quad \forall (x_b, y_b, z_b) \in \Omega_{b,D}; \quad t > 0 \tag{3.73}$$

where $\Omega_{b,D}$ is the segment of the boundary on which a Dirichlet boundary condition is defined. Here, (x_b, y_b, z_b) are the x-, y-, z-coordinates of the Dirichlet boundary segment, $C_1(x_b, y_b, z_b, t)$ is a given function defined on this boundary segment. This boundary condition is usually used to characterize the source term. For a constant source at a point or a segment on the boundary the $C_1(x_b, y_b, z_b, t)$ can be selected to be a constant. This boundary condition can also be defined as a function of time which may be used to define changing concentration values at the boundary.

The concentration flux based boundary conditions can be used to model two different physical conditions that may exist at the boundaries of the solution domain. In either case the concentration flux is defined as the concentration flux normal to the boundary. The first flux condition is usually identified as the Neuman condition, for which the concentration flux at the boundary is defined in terms of a given function. The mathematical model for this boundary condition can be given as:

$$D_{x_{b,n}} \frac{\partial C}{\partial n} = C_2(x_b, y_b, z_b, t) \quad \forall (x_b, y_b, z_b) \in \Omega_{b,N}; \quad t > 0 \qquad (3.74)$$

where $\Omega_{b,N}$ is the segment of the boundary on which a Neuman boundary condition is defined, where (x_b, y_b, z_b) are the x-, y-, z-coordinates of the Neuman boundary segment, $C_2(x_b, y_b, z_b, t)$ is a given function on this boundary segment, $D_{x_{b,n}}$ is the diffusion coefficient at the boundary and n is the normal direction to the boundary. This boundary condition is usually used to characterize the boundaries at which the diffusive escape or entry of the contaminant concentration is defined. For a constant diffusive flux at a segment on the boundary, the $C_2(x_b, y_b, z_b, t)$ can be selected to be a constant. A special case of this condition occurs when $C_2(x_b, y_b, z_b, t)$ is selected as zero, which implies an impervious boundary through which the concentration flux is zero. This condition may also be used on downstream boundaries of the solution domain as a boundary condition for which the contaminant concentration in the domain is not expected to reach the boundary. This boundary condition can also be defined as a function of time which may be used to define time dependent concentration flux values at the boundary.

The second flux boundary condition is identified as the Cauchy condition, for which the advective–diffusive contaminant concentration flux at the boundary is defined in terms of a given function which is also a function of the concentration at the boundary. The mathematical model for this boundary condition can be given as:

$$v_n C - D_{x_{b,n}} \frac{\partial C}{\partial n} = C_3(C, x_b, y_b, z_b, t) \quad \forall (x_b, y_b, z_b) \in \Omega_{b,C}; \quad t > 0 \qquad (3.75)$$

where $\Omega_{b,C}$ is the segment of the boundary on which a Neuman boundary condition is defined, (x_b, y_b, z_b) are the x-, y-, z-coordinates of the Cauchy boundary segment, $C_3(C, x_b, y_b, z_b, t)$ is a given function, v_n is the advective velocity at the boundary and n is the normal direction to the boundary. This boundary condition is usually used to characterize the boundaries where the contaminant flux escaping or entering the solution domain happens to be a function of the contaminant concentration at the boundary. This boundary condition can also be defined as a function of time which may be used to define changing concentration flux values at the boundary.

These three types of boundary conditions constitute the most common boundary conditions employed in the solution of the advection–dispersion-reaction equation. Specific forms of these boundary conditions will be employed throughout the remainder of this book. Given the emphasis on analytical solutions in the three environmental pathways discussed in this book, the complex forms of the boundary conditions defined above will be further simplified using the pathway specific characteristics of the problem analyzed.

3.6 Multi-pathway and Inter-pathway Mass Transport

Environmental pathways are continuous systems. Contaminants in one pathway are usually transferred from one pathway to the other through the interfaces of the boundaries. The contaminants that start in one pathway may pass to the next pathway and continue to migrate to the next pathways based on the advective dispersive properties of the next pathway. Although it is a simpler process to analyze the contaminant migration in any one pathway, it is a very difficult task to link the pathways and analyze the contaminant migration in all linked pathways. The fugacity analysis approach yields the simplest technique to solve multi-pathway problems since fugacity, which is a measure of the escaping tendency of a chemical from one medium to the other, is a constant at the interface. This property of fugacity models makes the multi-pathway analysis much simpler. The examples of transformation and transport applications that use this approach can be found in (Mackay 2001; Kilic and Aral 2008, 2009), which are beyond the scope of this book.

There are specific models that can be used in inter-pathway analysis, such as the models that are used to describe water–air interface conditions. These cases will be treated as specific models in various chapters of this book.

Short of using the fugacity approach, multi-pathway analysis using concentration as the unknown variable is very complex. This can only be achieved if the solution obtained from one pathway can be used as the time dependent boundary condition of the next pathway. A continuous solution obtained in this manner is the proper solution of these types of problems. This solution can only be achieved in an iterative manner. When analytical solutions are utilized for the solution of the advection-dispersion-reaction equation, it is very difficult to accomplish iterative solutions mainly due to the restrictions of the analytical models. For simpler cases this can be accomplished relatively easily as will be discussed in the following chapters. For more complicated multi-pathways analysis numerical methods need to be used which provide a more flexible computational environment.

References

Abdel-Magid IS, Mohammed A-WH et al (1997) Modeling methods for environmental engineers. CRC Lewis, Boca Raton, FL

Anderson MP, Woessner WW (1992) Applied groundwater modeling: simulation of flow and advective transport. Academic, San Diego, CA

Bear J (1972) Dynamics of fluids in porous media. American Elsevier Pub. Co., New York

Bird RB, Stewart WE et al (2002) Transport phenomena. Wiley, New York

Charbeneau RJ (2000) Groundwater hydraulics and pollutant transport. Prentice Hall, Upper Saddle River, NJ

Cheng RTS (1976) Modeling of hydraulic systems. Adv Hydrosci 11:278–284

Cherry JA, Gillham RW et al (1984) Contaminants in groundwater – chemical processes. National Academy Press, Washington, D.C

Clark MM (1996) Transport modeling for environmental engineers and scientists. Wiley, New York

Fetter CW (1993) Contaminant hydrogeology. Prentice Hall, Upper Saddle River, NJ

Galeati G, Gambolati G (1989) On boundary-conditions and point sources in the finite-element integration of the transport-equation. Water Resour Res 25(5):847–856

Gunduz O, Aral MM (2005) A Dirac-delta function notation for source/sink terms in groundwater flow. ASCE J Hydrol Eng 10(5):420–427

Hemond HF, Fechner-Levy EJ (2000) Chemical fate and transport in the environment. Academic, San Diego, CA

Kilic SG, Aral MM (2008) Probabilistic fugacity analysis of Lake Pontchartrain pollution after Hurricane Katrina. J Environ Manage 88(3):448–457

Kilic SG, Aral MM (2009) A fugacity based continuous and dynamic fate and transport model for river networks and its application to Altamaha River. J Sci Total Environ 407(12):855–3866

Mackay D (2001) Multimedia environmental models: the fugacity approach. Lewis, Boca Raton, FL

Nirmalakhandan N (2002) Modeling tools for environmental engineers and scientists. CRC, Boca Raton, FL

Rubin J (1983) Transport of reacting solutes in porous-media – relation between mathematical nature of problem formulation and chemical nature of reactions. Water Resour Res 19 (5):1231–1252

Schnoor JL (1996) Environmental modeling: fate and transport of pollutants in water, air, and soil. Wiley, New York

Sun N-Z (1996) Mathematical modelling of groundwater pollution. Springer, New York

Weber WJ, DiGiano FA (1996) Process dynamics in environmental systems. Wiley, New York

Zheng C, Bennett GD (1995) Applied contaminant transport modeling: theory and practice. Van Nostrand Reinhold, New York

Chapter 4
Air Pathway Analysis

You can never solve a problem at the level it was created.
Albert Einstein

As our societies become more centralized in and around large population centers and as the demand for energy, food, water and technological need increases proportional to an exponentially increasing population of the world, air pollution and its adverse health effects outcome will continue to be an important concern. Relatively speaking, the pollution of environmental media such as water, soil and plants are not as critical as the pollution of air, which we need to breathe regularly. This is because the other environmental media such as water, soil and plants can be processed, remediated or treated before we come into contact with them. On the other hand, air has to be clean anywhere and everywhere that we go, and it cannot be isolated, other than probably indoor air, if it needs to be treated. Thus, source control is the most effective remedy for the control of air pollution. In the United States the regulatory branches of the government started addressing air pollution problems during the early 1960s with the enactment of the Clean Air Act followed by its subsequent amendments in 1963, 1966, 1970, 1977 and 1990. Early regulations focused on point sources such as emissions from smoke stacks originating from industrial, commercial or power plants. Later on these regulations were extended to cover distributed sources originating from roads and highways and indoor air pollution which could originate from natural diffusion of environmental contaminants, uncirculated indoor conditions or circulation based dispersion of contamination sources. Whatever is the source, indoor and outdoor air pollution is a major health concern and we need to understand and evaluate contaminant migration patterns in this environmental pathway.

In the United States the air quality standards and emission standards are regulated by US EPA under the umbrella of National Ambient Air Quality Standard (NAAQS) rules. Under this regulatory umbrella not only are the emissions from industrial sources regulated but also the air we breathe is under regulatory control. For toxic contaminants very low levels are allowed to be present in the air around us. There are other contaminants which are not immediately harmful but which can

M.M. Aral, *Environmental Modeling and Health Risk Analysis (ACTS/RISK)*,
DOI 10.1007/978-90-481-8608-2_4, © Springer Science+Business Media B.V. 2010

be harmful if their concentrations are high or if the exposure duration to these contamination is long. The contaminants of the latter type are identified as "Criteria Pollutants" as they are characterized below.

Carbon monoxide (CO) is a colorless, odorless gas that is formed when carbon in fuel is not burned completely. It is a component of motor vehicle exhaust, which contributes about 56% of all CO emissions in the US. Carbon monoxide poisoning is the most common type of fatal air poisoning in many countries. When inhaled, it combines with hemoglobin to produce carboxyhemoglobin, which is ineffective at delivering oxygen to bodily tissues. This condition is known as anoxemia. The most common symptoms of carbon monoxide poisoning may resemble other types of poisonings and infections (such as the flu), including headache, nausea, vomiting, dizziness, lethargy and a feeling of weakness. Infants may be irritable and feed poorly. Neurological signs include confusion, disorientation, visual disturbance, syncope and seizures.

Nitrogen oxides (NO_x) is the generic term for a group of highly reactive gases, all of which contain nitrogen and oxygen in varying amounts. Many of the nitrogen oxides are colorless and odorless. However, the common pollutant, nitrogen dioxide (NO_2) interacts with particles in the air, and can often be seen and recognized as a reddish-brown layer over many urban areas. Nitrogen oxides form when fuel is burned at high temperatures, as in a combustion process. The primary manmade sources of NO_x are motor vehicles, electric utilities, and other industrial, commercial, and residential sources that burn fuels. NO_x may also be formed naturally. NO_x react with volatile organic compounds in the presence of heat and sunlight to form Ozone. Ozone may cause adverse effects such as damage to lung tissue and reduction in lung function as described below. NO_x (especially NO_2) destroys the ozone layer. This layer absorbs ultraviolet light, which is potentially damaging to life on earth.

Sulfur dioxide (SO_2) belongs to the family of sulfur oxide gases (SO_x). These gases dissolve easily in water. Sulfur is prevalent in all raw materials, including crude oil, coal, and ore that contains common metals like aluminum, copper, zinc, lead, and iron. SO_x gases are formed when fuel containing sulfur, such as coal and oil, is burned, and when gasoline is extracted from oil or metals are extracted from ore. SO_2 dissolves in water vapor to form acid, and interacts with other gases and particles in the air to form sulfates and other products that can be harmful to people and also the environment. Inhaling sulfur dioxide is associated with increased respiratory symptoms and disease, difficulty in breathing, and premature death.

Ozone (O_3) is a gas composed of three oxygen atoms. It is not usually emitted directly into the air, but at lower altitudes it is created by a chemical reaction between oxides of nitrogen (NO_x) and volatile organic compounds (VOC) in the presence of sunlight. Ozone has the same chemical structure whether it occurs miles above the earth or at ground-level and can be "good" or "bad" for the environment depending on its location in the atmosphere. Ground level ozone is an air pollutant with harmful effects on the respiratory systems. Although ozone was present at ground level before the industrialization of societies, peak concentrations are now far higher than the pre-industrial levels. The ozone layer in the upper atmosphere filters potentially damaging ultraviolet light from reaching the Earth's surface, which is a necessary and important function.

Lead (Pb) is a metal found naturally in the environment as well as in manufactured products. The major sources of lead emissions have historically been motor vehicles (such as cars and trucks) and industrial or commercial sources. As a result of US EPA's regulatory efforts to remove lead from gasoline, emissions of lead from the transportation sector dramatically declined by 95% between 1980 and 1999. Parallel to this reduction, levels of lead in the air decreased by 94% between 1980 and 1999. Today, the highest levels of lead in air are usually found near lead smelters. Other stationary sources are waste incinerators, utilities, and lead-acid battery manufacturers. Lead is a poisonous metal that can damage nervous systems, especially in young children and may cause blood and brain disorders. Long-term exposure to lead or its salts, especially soluble salts or the strong oxidant PbO_2, can cause nephropathy, and colic-like abdominal pains. The effects of lead are the same whether it enters the body through breathing or swallowing. Lead can affect almost every organ and system in the body. However, the main target for lead toxicity is the nervous system, both in adults and children. Lead exposure may also increase blood pressure, particularly in middle-aged and older people and can cause anemia. Exposure to high lead levels can severely damage the brain and kidneys in adults or children and ultimately cause death. In pregnant women, high levels of exposure to lead may cause miscarriage.

Particulate matter (PM) is a complex mixture of extremely small particles and liquid droplets. Particulate pollution is made up of a number of components, including acids (such as nitrates and sulfates), organic chemicals, metals, and soil or dust particles. The size of particles is directly linked to their potential health effects. The effects of inhaling particulate matter may lead to asthma, lung cancer, cardiovascular issues, and premature death in humans. The size of the particle is the main determinant of where in the respiratory tract the particle will come to rest when inhaled. Depending on the size of the particle, it can penetrate the deepest part of the lungs. Larger particles are generally filtered in the nose and throat, but particulate matter smaller than about 10 μm, referred to as PM_{10}, can settle in the bronchi and lungs and may cause health problems. The 10 μm size does not represent a strict boundary between reparable and non-reparable particles, but has been agreed upon for monitoring of airborne particulate matter by most regulatory agencies. Similarly, particles smaller than 2.5 μm, $PM_{2.5}$, tend to penetrate into the gas-exchange regions of the lung, and the very small particles (<100 nm) may pass through the lungs to affect other organs. US EPA is concerned about particles that are in the range 10–2.5 μm in diameter or smaller because those are the particles that generally pass through the throat and nose and enter the lungs.

Exposure to these "Criteria Pollutants" pollutants is associated with numerous human health effects, including increased respiratory symptoms, hospitalization for heart or lung diseases, and even premature death. Some of these health effects and the regulatory levels are given in Table 4.1.

Table 4.1 Air pollution standards for criteria pollutants and potential health effects

Pollutant	Exposure duration	NAAQS	Health and environmental outcome
CO	1 h	35 ppm	Headaches, asphyxiation
	8 h	9 ppm	Angina, pectoris
NO_2	1 year	0.053 ppm	Respiratory disease
SO_2	3 h	0.50 ppm	Shortness of breath,
	1 day	0.14 ppm	Odor, acid precipitation
	1 year	0.03 ppm	
O_3	1 h	0.12 ppm	Eye irritation, breathing damage,
	8 h	0.075 ppm	Bronchitis, heart attack
Pb	3 months	1.5 $\mu g/m^3$	Blood poisoning
$PM_{2.5}$	24 h	35 $\mu g/m^3$	Lung damage
	1 year	15 $\mu g/m^3$	
PM_{10}	24 h	150 $\mu g/m^3$	Respiratory disease, visibility

4.1 Lapse Rate and Atmospheric Stability

To understand air diffusion and air circulation processes we need to have an understanding of the processes that create or inhibit atmospheric circulation. The concepts and definitions that are necessary in this analysis are the lapse rate and the latent heat of condensation. Atmospheric stability conditions that are important in the characterization of air circulation are described in terms of, or are tied to the definitions of lapse rate and latent heat of condensation concepts. Lapse rate is associated to the manner in which the temperature in an air packet changes with altitude or elevation. A positive lapse rate implies a decrease in temperature with increasing elevation. There are several definitions that can be used in the description of the lapse rate. First, there is the environmental lapse rate (ELR) which refers to the actual variation of temperature with altitude at a certain geographic location and time. This implies that ELR is not constant and may change over time and with geographic location. The other lapse rate definition is the adiabatic lapse rate (ALR). Here, the use of the term adiabatic implies that in this case it is assumed that heat neither enters nor leaves the air pocket or the system under consideration. Thus, heat transfer with the environment is not considered in this analysis. An adiabatic process is in contrast to a diabatic process in which heat is added or subtracted from the system, e.g., solar heating, radiation cooling. ALR is commonly divided into two categories, the dry adiabatic lapse rate (DALR) and the saturated adiabatic lapse rate (SALR). Dry adiabatic lapse rate refers to the rate at which a non-saturated air parcel cools as it rises. This rate is 9.8°C/km. This rate is constant until the ascending air parcel becomes saturated, that is until it reaches its dew point temperature. Once the dew point is reached (water saturation) latent heat is released as an outcome of condensation and the lapse rate drops. The SALR is variable since it depends on how much latent heat is made available as the condensation occurs for a saturated air parcel. At lower elevations SALR varies between 3.9°C/km and 7.2°C/km when the ambient temperatures are in the range of 26°C to −10°C.

Stability or instability of the atmosphere can be defined in terms of the reaction that a parcel of air has to initial vertical upward or downward displacement. It is a parcel's resistance to further movement or the enhancement of its movement in the direction of initial displacement, or the tendency of the parcel to return to the original position. The stability condition of the atmosphere determines the likelihood of further convective activity, likelihood of atmospheric turbulence or even the cloud types that may be formed due to the displacement process. Accordingly, in analogy to mechanical instability conditions, we can define three stability criteria (Fig. 4.1a). As shown in this figure, similar to the mechanical stability concepts, an air parcel may behave in one of the following three stability cases. A displaced parcel will change its temperature (if adiabatic) at the adiabatic lapse rate (ALR). The case in Fig. 4.1b is for an atmosphere that is stable for a dry process with no condensation. If the parcel of air is displaced upwards (or downwards) it will cool (or warm) at the ALR. That is it will become cooler (warmer) than its surroundings and therefore denser (lighter), and thus it will tend to return to its original position. This condition would represent a stable atmosphere. However, if the ELR is greater

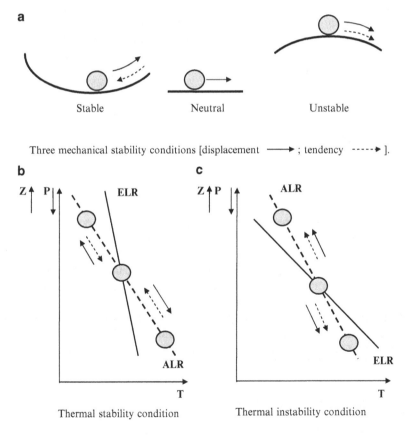

Three mechanical stability conditions [displacement ⟶ ; tendency ----▶].

Thermal stability condition   Thermal instability condition

Fig. 4.1 Atmospheric stability conditions

than the ALR, the displaced parcel will keep moving in the direction of the initial displacement due to density effects. This condition would represent an unstable atmosphere (Fig. 4.1c) (Cole 1970; Dunnivant and Anders 2006; Hemond and Fechner-Levy 2000).

Based on the definitions and the description given above, the following stability conditions can be defined which will lead to the expectation of the conditions that might occur after the initial vertical displacement of an air pocket.

Absolute instability occurs when the ELR is greater than the DALR. As we know the DALR is 9.8°C/km, so we can conclude that absolute instability exists when ELR is equal to or greater than 9.8°C/km. This condition is sometimes identified as a "super-adiabatic lapse rate," since the heat loss is very rapid.

Natural instability occurs when the ELR and DALR are equal. In this term the word "natural" refers to the fact that thermal momentum is not going to be accelerated or decelerated.

Conditional instability occurs when the ELR is less than the DALR but more than the SALR. The SALR is usually considered to be in the range 3.9–7.2°C/km. The use of the word "conditional" is associated with the criteria that instability is expected to occur only when the thermal becomes saturated and not before.

Absolute stability occurs when the ELR is less than the SALR.

Potential instability occurs when air is moist at lower elevations but dry at higher elevations. The potential for instability is only realized when the thermal ascends and reaches saturation.

4.2   Principles of Atmospheric Stability

When compared to liquids, gases are more compressible. Thus, the common simplifying incompressibility assumption that is made for most problems that involve liquids cannot be made for the analysis of problems that involve gases. This is also the case for atmospheric studies. In the analysis of atmospheric stability principles one has to consider the change of pressure with elevation $P(z)$, the change of density with elevation $\rho(z)$ and also the change of temperature with elevation $T(z)$. To derive the governing equations of the stability conditions described above we need to analyze the behavior of an air pocket as shown in Fig. 4.2 under the effect of pressure and gravitational forces.

Using Newton's second law, vertical and horizontal equilibrium between these forces can be analyzed. The interest here is the equilibrium in the vertical direction since the horizontal equilibrium outcome is trivial. In the z-direction,

$$P_3 - P_4 - W = 0 \qquad (4.1)$$

where P_i; $i = 1, 2, 3, 4$ are pressure forces and W is the weight of the air parcel in the control volume. For an infinitesimal volume,

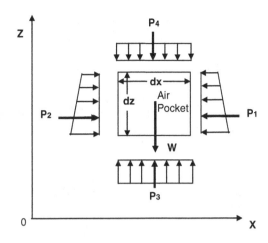

Fig. 4.2 Forces acting on an air pocket control volume

$$p(z + dz) - p(z) = -\rho g dz \qquad (4.2)$$

or,

$$\frac{dp}{dz} = -\rho g = -\gamma \qquad (4.3)$$

which is the hydrostatic condition, where p is the pressure $[ML^{-1}T^{-2}]$, is the density $[ML^{-3}]$, γ is the specific weight $[ML^{-2}T^{-2}]$ of air, g is the gravitational acceleration $[LT^{-2}]$, W is the weight force $[MLT^{-2}]$ and z is the elevation $[L]$. As expected from hydrostatic conditions, the change in pressure in the vertical direction is negatively proportional to the specific weight of the fluid within the control volume.

In the case of gases the relationship between pressure, density and temperature can be expressed in terms of the ideal gas law.

$$p = \frac{\overline{R}}{M} \rho \widehat{T} \qquad (4.4)$$

where $\overline{R}$ is the ideal gas constant, 8.314 J/mol K, $\widehat{T}$ is temperature measured in terms of absolute temperature kelvin ($^\circ$C + 273.15) and M is the molar mass. If we assume that air is mostly composed of nitrogen and oxygen, although various other gases are in the mixture, and if we assume that air can be treated as an ideal gas Eq. (4.4) can be used to define the relationship between pressure, density and temperature in which $R = \overline{R}/M = 287$ J/kg K $= 287$ m^2/s^2 K. Here we have assumed that the molecular weight of air to be in between the molecular weights of nitrogen and oxygen.

Equation (4.4) will also yield the differential relationship between these three variables if we differentiate Eq. (4.4) with respect to elevation, z.

$$\frac{dp}{dz} = R\frac{d\rho}{dz}\widehat{T} + R\rho\frac{d\widehat{T}}{dz} \tag{4.5}$$

Given these two equations, if an air parcel moves from elevation z to $(z + dz)$ its pressure will decrease according to Eq. (4.3) and its density and temperature or both will decrease according to Eq. (4.5). A decrease in density is associated with an increase in volume if the number of molecules is kept constant within the air pocket. The work done in expanding the air pocket $(pd\mathcal{V})$ is balanced by the internal energy loss $\left(mC_v d\widehat{T}\right)$. Thus,

$$mC_v d\widehat{T} = -pd\mathcal{V} \tag{4.6}$$

where m is the air pocket mass and C_v is the specific heat capacity $[J/mol\ K]$. Given the mass, density and volume relationship, Eq. (4.6) can also be written as,

$$C_v d\widehat{T} = -pd\left(\frac{1}{\rho}\right) \tag{4.7}$$

or considering the elevation change,

$$C_v \frac{d\widehat{T}}{dz} = \frac{p}{\rho^2}\frac{d\rho}{dz} \tag{4.8}$$

Equations (4.1), (4.5) and (4.8) are three independent equations which may be used for the solution of the three gradient terms for pressure, temperature and density $\left\{\frac{dp}{dz}; \frac{d\widehat{T}}{dz}; \frac{d\rho}{dz}\right\}$. This will yield the solution for the temperature gradient as,

$$(C_v + R)\frac{d\widehat{T}}{dz} = -g \tag{4.9}$$

According to Eq. (4.9) the temperature decreases with elevation at a constant rate,

$$\frac{d\widehat{T}}{dz} = -\frac{g}{C_p} \tag{4.10}$$

where $C_p = C_v + R$. The constant gradient $(\Gamma = g/C_p)$ is the adiabatic lapse rate (ALR) we have defined earlier. The adiabatic lapse rate is approximately $1°$ for every 100 m. This outcome is consistent with the observations we make in nature; it gets cold as we move up the mountain and the temperatures are freezing outside of a plane at high altitudes.

Now let's reanalyze the stability conditions using these relationships. Consider an air pocket at elevation z at temperature $\widehat{T}(z)$ which is slightly displaced upward an incremental distance dz. At its new location it will be under lower pressure, thus the air pocket will expand acquiring lower density and at the same time losing temperature in doing so. The pressure drop is $(dp = -\gamma dz)$ and the temperature drop is $\left(d\widehat{T} = -\Gamma dz \right)$. If the atmosphere is not in a neutral state, the ambient temperature $\left(\widehat{T}_a = \widehat{T}(z + dz) \right)$ is not going to be equal to the new temperature of the air pocket at its new position $\left(\widehat{T}_p = \widehat{T}(z) + d\widehat{T} \right)$. This is because the air parcel moving upward has adjusted its temperature according to the adiabatic lapse rate but that rate may not correspond to the rate of decrease in temperature as a function of elevation change for ambient conditions. The difference between the two temperatures can be calculated.

$$\widehat{T}_a - \widehat{T}_p = \widehat{T}(z + dz) - \widehat{T}(z) + \frac{g}{C_p} dz$$

$$\approx \left(\frac{d\widehat{T}}{dz} + \frac{g}{C_p} \right) dz \qquad (4.11)$$

Because of this temperature change the displaced air parcel will experience a net buoyancy force that is not equal to its weight, which would result in a net upward force.

$$
\begin{aligned}
F_{net} &= F_{bouyancy} - W \\
&= \rho_a \cancel{V} g - \rho_p \cancel{V} g \\
&= \left(\frac{p}{R\widehat{T}_a} - \frac{p}{R\widehat{T}_p} \right) \cancel{V} g = \frac{p}{R\widehat{T}_a \widehat{T}_p} \left(\widehat{T}_p - \widehat{T}_a \right) \cancel{V} g \\
&= \rho_a \frac{\left(\widehat{T}_p - \widehat{T}_a \right)}{\widehat{T}_p} \cancel{V} g
\end{aligned}
\qquad (4.12)
$$

Again using Newton's second law:

$$\rho_a \frac{\left(\widehat{T}_p - \widehat{T}_a \right)}{\widehat{T}_p} \cancel{V} g = \rho_a \cancel{V} \frac{d^2(dz)}{dt^2} \qquad (4.13)$$

Accordingly, the acceleration of the air pocket can be given as:

$$\frac{d^2(dz)}{dt^2} = \frac{\left(\widehat{T}_p - \widehat{T}_a \right)}{\widehat{T}_p} g = -\frac{g}{\widehat{T}_p} \left(\left(\frac{d\widehat{T}}{dz} \right)_a + \Gamma \right) dz \qquad (4.14)$$

As can be seen from Eq. (4.14) $\left(d\widehat{T}/dz\right)$ is always negative and $\left(\Gamma = g/C_p\right)$ is always positive. Thus, the acceleration of the disturbed air pocket is a function of the difference between ELR and ALR, and this difference can be positive or negative depending on the magnitudes of these two lapse rates which gives rise to the stability conditions defined earlier (Fig. 4.3).

Based on Eq. (4.14) the following stability conditions can now be defined in reference to adiabatic condition. Let,

$$\Phi^2 = \frac{g}{\widehat{T}_p}\left(\left(\frac{d\widehat{T}}{dz}\right)_a + \Gamma\right) \qquad (4.15)$$

which is one form of the Richardson number that is used in meteorology. Then,

$$\left|\left(\frac{d\widehat{T}}{dz}\right)_a\right| > \Gamma \quad \Rightarrow \quad \Phi^2 < 0 \quad \Rightarrow \quad \textit{Unstable atmosphere}$$

$$\left|\left(\frac{d\widehat{T}}{dz}\right)_a\right| = \Gamma \quad \Rightarrow \quad \Phi^2 = 0 \quad \Rightarrow \quad \textit{Neutral atmosphere} \qquad (4.16)$$

$$\left|\left(\frac{d\widehat{T}}{dz}\right)_a\right| < \Gamma \quad \Rightarrow \quad \Phi^2 > 0 \quad \Rightarrow \quad \textit{Stable atmosphere}$$

For very limited cases the stable atmospheric condition defined for the last case in Eq. (4.16) can be split into two. As expected there is the possibility of.

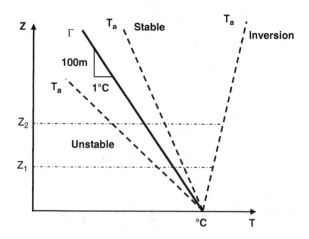

Fig. 4.3 Lapse rate and stability conditions

$$\left(\frac{d\widehat{T}}{dz}\right)_a \leq 0 \quad or \quad \left(\frac{d\widehat{T}}{dz}\right)_a > 0 \quad while \quad \left|\left(\frac{d\widehat{T}}{dz}\right)_a\right| < \Gamma \qquad (4.17)$$

For this case, the condition $\left(d\widehat{T}/dz\right)_a > 0$ is referred to as inversion and attributed to very stable atmospheric conditions.

In summary, stability is the tendency to resist vertical motion or to suppress existing turbulence, and the stability of air in the atmosphere depends on the temperature of rising air relative to the ambient air temperature that it passes through, which as discussed above varies from place to place and changes according to atmospheric conditions. When a pocket of air near the Earth's surface is heated it rises, as it is lighter than the surrounding air. Whether or not this air packet will continue to rise will depend on how the temperature in the ambient air changes with elevation. The rising pocket of air will lose heat because it expands as atmospheric pressure falls, and its temperature drops. If the temperature of the surrounding air does not fall as quickly with increasing altitude as the ambient air temperature, the air pocket will quickly become colder than the surrounding air and lose its buoyancy, and will sink back to its original position. In this case the atmosphere is said to be stable. If the temperature of the surrounding air falls more quickly with increasing altitude, the pocket of air will continue to rise. The atmosphere in this circumstance is said to be unstable. This tendency directly influences the ability of the atmosphere to disperse pollutants emitted into it. As a consequence, when the stability is low, vertical motion is not suppressed and pollutants may be dispersed higher from the ground surface.

The stability conditions described above are going to be used extensively in the air dispersion models included in the ACTS software. As we will discuss later in this chapter, knowing the atmospheric stability, category is very important in modeling plume dispersion in the atmosphere. If sufficient data is available for the site under consideration, users of these models can compute the temperature gradient and choose the stability category accurately. However, more often than not these data are not available and the decisions on stability conditions must be made based on observations. The most widely used procedure for this purpose is based on the method developed by Pasquill, hence the name Pasquill Stability Criteria (Pasquill 1961, 1976).

A simple approach to estimate some of the parameters of the atmospheric dispersion models, which we will discuss later on, is the employment of atmospheric stability categories based on meteorological conditions. The commonly used Pasquill–Gifford stability categories are developed from correlations found at a particular geographic location in Britain (Gifford 1976; Pasquill 1961, 1976). In Table 4.2 the Pasquill–Gifford stability categories are given as a function of insolation (solar heat input) and wind speed. In this table, category A corresponds to conditions under which atmospheric mixing is augmented by instability during periods of intense sunlight due to solar heating of the ground surface and overlying air. Category D corresponds to an atmosphere of neutral stability, while categories E and F correspond to increasingly stable conditions associated with atmospheric

Table 4.2 Pasquill–Gifford stability categories

Surface wind speed (m/s)	Insolation			Night	
	Strong	Moderate	Slight	Thinly overcast or $\geq$4/8 low cloud	$\leq$3/8 Cloud
<2	A	A–B	B	–	–
2–3	A–B	B	C	E	F
3–5	B	B-C	C	D	E
5–6	C	C-D	D	D	D
>6	C	D	D	D	D

For A–B, the average of values for A and B are taken, and similarly for other cases.

Strong insolation corresponds to sunny midday in midsummer in England; slight insolation to similar conditions in midwinter.

Night refers to the period from 1 h before sunset to 1 h after sunrise.

The neutral category D should also be used, regardless of wind speed, for overcast conditions during day or night and for any sky conditions during the hour preceding or following night as defined above.

inversion. There are significant limitations to using atmospheric stability categories. Errors may result if the user applies them to settings that differ in local topography or climatic conditions (Cole 1970; Hemond and Fechner-Levy 2000). The Pasquill–Gifford stability categories can be used to choose the appropriate parameters for the Gaussian plume models as a function of downwind distance from the source, as will be discussed later. As expected, at a given downwind distance, the less stable categories will correspond to more mixing, with an increase of one or more orders of magnitude in the actual vertical or horizontal width of the plume (Cole 1970; Hemond and Fechner-Levy 2000).

The other meteorological factor affecting the concentration of air pollutants is the wind speed and wind direction. Wind speed or wind velocity is influenced by topography near the earth's surface. Movement of air near the earth's surface is retarded by friction effects proportional to surface roughness. Thus the wind speed effect will be greater farther from the ground surface, since it is expected that friction will reduce the wind speed near the ground surface.

Stability and wind speed are related in that when air near the earth surface is pushed down because of greater stability, the wind speed increases. The effects of higher wind speed may at first seem counterintuitive; higher wind speeds, which cause more atmospheric turbulence, shift the classification in the direction of higher stability categories that would seem to result in less mixing. This occurs because the Pasquill–Gifford stability categories are used to determine the width and height of a pollutant plume at a particular downwind distance. They are not used to estimate a Fickian mixing coefficient. Higher wind velocity may actually decrease the absolute amount of spreading a pollutant plume may undergo before reaching a fixed downwind distance. This may override the effects of more intense mixing, as there is less time for mixing to occur, and therefore the latter effect may predominate (Cole 1970; Hemond and Fechner-Levy 2000).

Based on these stability conditions, the plume dispersion coefficients for the Gaussian models that are used in the ACTS software can be obtained from the charts shown in Figs. 4.4 and 4.5. While using the ACTS software, the user does not

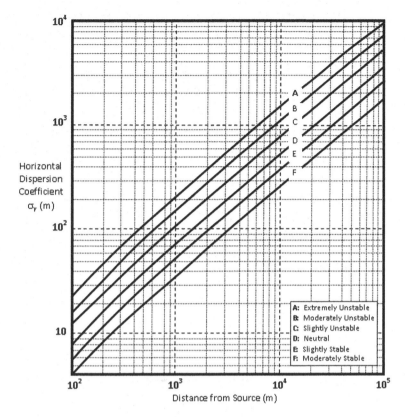

Fig. 4.4 Lateral dispersion coefficients (NRC 1982; Turner 1994)

have to obtain these dispersion coefficients and insert them manually. The selection of the appropriate stability conditions for the problem analyzed obtained from Table 4.2 results in automatic selection and use of the appropriate dispersion coefficients in the Gaussian models of the ACTS software. This process, which uses Figs. 4.4 and 4.5, is transparent to the user.

4.3 Air Pathway Models

The models that are included in the air pathway module of the ACTS software (Figs. 4.6 and 4.7) are based on chemical properties of the contaminants. For this purpose, a chemical database is included to this module, which can be updated and customized by the user. Based on this database, the models included in the air pathway module are divided into two subgroups, the emission models and air dispersion models. Air pollution is always associated with an emission source. The emission source may be an emission from a factory stack, which can be identified as

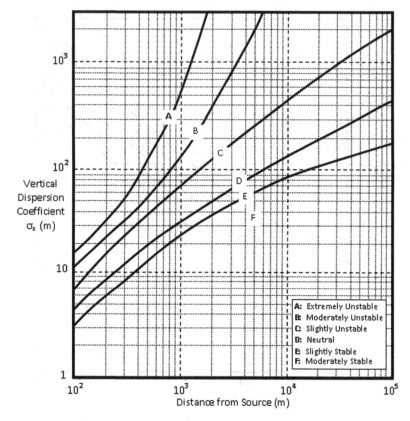

Fig. 4.5 Vertical dispersion coefficients (NRC 1982; Turner 1994)

a point source or emissions from vehicles on a highway, which may be interpreted as a line source, if one is interested in analyzing air pollution associated with highways and roads. If indoor pollution is of concern emissions from soil entering indoors through cracks in the foundation and crawl spaces must be considered. Thus, the first step in air pollution analysis is the estimation of the emission rate. The next step in the analysis is the evaluation of the spread of the emission source indoors or outdoors as the case may be.

In the emission model subgroup, six models are considered: the Farmers emission model, the Thibodeaux-Hwang emission model, the Cowherd particulate emission model, the Jury unsaturated zone emission model, the landfill gas emission model and the volatilization from water surfaces model. These models can be used to estimate emissions from land and water based contaminant sources. The theoretical background, the data entry and the output analysis procedures for these models will be discussed in detail in this chapter. Emission rates are one of the required input data for the air dispersion models. The emission models of the ACTS software are dynamically linked to air dispersion models, which may be considered

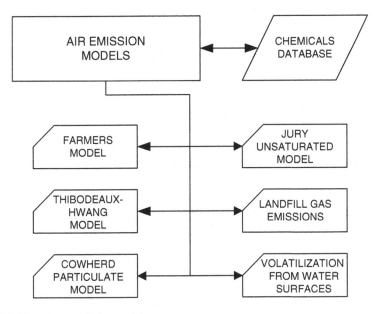

Fig. 4.6 Air pathway emission models

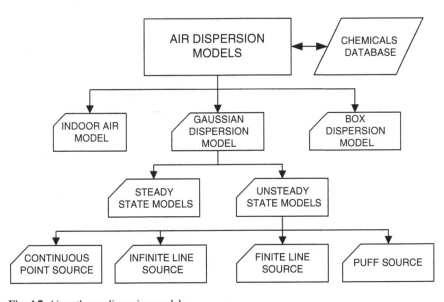

Fig. 4.7 Air pathway dispersion models

to be the second stage of analysis in air pollution. Thus, if the user selects to use these emission rates as input data to air dispersion models, the emission rates generated in the emission models module can be directly transferred to air dispersion models. If this direct transfer is not desired, the emission models and air dispersion models can be used independently to analyze site-specific problems. In that case the user should enter emission rates as an external input data in the air dispersion models.

In the air dispersion module there are three models which can be used to evaluate dispersion of contaminants in the air pathway. These are the Box air dispersion model, Gaussian air dispersion models and the indoor air dispersion model. The Gaussian air dispersion models include steady state and unsteady state models. Each of the Gaussian models has several subcategory models, which can be used to model various source and atmospheric conditions to provide further site specific options to the user (Fig. 4.7).

Similar to emission models, the air dispersion models require chemical specific databases. For this purpose, the air pathway module also includes one generic and two editable chemical databases which contain chemical properties of several contaminants. These databases can be directly linked to all models through the "Preferences" menu button on the opening window of the ACTS software (see Appendix 3). The generic chemical database identified as (CHEMICAL.MDB) is a master reference chemical database file which cannot be edited. The purpose of this uneditable database is to make a database available to the user which is error free and shows all the proper data categories, which are necessary to run the models included in the air pathway module. When the user selects a chemical to work with from this database, the appropriate data categories available in this database will be dynamically linked to all other models the user selects to use in a specific application. The other two chemical databases (CHEM1.MDB and CHEM2.MDB) are copies of the generic (CHEMICAL.MDB) database, which are editable. The User may work with these files and develop his or her own databases to use in applications. Editing can be done after selecting one of the editable databases as the default database in the preferences menu and opening this database from the "Chemicals!" menu button in the air pathway module and selecting the "Chemicals" pull down menu button. This button will be in in-active mode if the user has selected the generic un-editable chemical database. If the default database selected is editable, new chemical databases can be added, undesired chemical data can be deleted or data entered for a chemical can be edited using the pull down menu options. It is important to note that when a new chemical is entered all data categories with proper units must be entered for the new chemical that is added to the database. Otherwise, the user will observe errors as the air emission or air dispersion models chosen by the user utilize these data internally, which is transparent to the user. Given a selected model, the code will internally attempt to access the data necessary for the chemical in question to run the model. If the data is not available, an error message will appear. Using the editing option appropriately, the user may generate custom databases for his or her specific purposes. After saving the edited database as ".MDB" file the database will be available for use in all air emission and air dispersion models.

The air emission and air dispersion models are also directly linked to the Monte Carlo simulation package, which provides uncertainty analysis for most of the parameters used in these models. The generic graphics package developed for the ACTS software is also available in this module for viewing the results of the analysis or preparing report ready hard copy or standard computer file figures. WINDOWS™ based file editors may be used to view the input and output data using the menu options. Hard copy of the numerical results obtained may also be printed using the file editor functions, or these files can be opened using other software to perform the functions of that specific software.

All air pathway models used in the ACTS software are generic models. Thus, their application in site-specific cases requires knowledge of the assumptions and limitations inherent in these models. In this chapter, a review of the models used to estimate the air emission from contaminated land and water based sources, and the dispersion of these emissions in the air pathway are reviewed.

To start the air pathway module application from the opening ACTS window, the user may select the air pathway icon or the "Air Path" option under the "Pathways" pull down menu. The module will start with the window shown in Fig. 4.8. In this window there are four options that are available to the user. The "File" menu option allows the user to create a "New" model data file, "Open" an old model data file, "Edit" an existing open data file, "Close" an open model data file and "Exit" the air pathway module. When a new data entry option is selected, the user has the option of starting an "Emissions Models" or "Dispersion Models" data preparation option. When either of these two options is selected, the user is given further options to go into the specific model type as described above (Figs. 4.6 and 4.7) (see Appendix 3). The "Chemicals!" menu option allows the user to select chemicals for the site-specific application. In the air pathway module a chemical or a set of chemicals must be selected before the "File" option is selected to start a new project, since the computations in the air module will require chemical properties data. Once a chemical or a set of chemicals is selected, this database will be

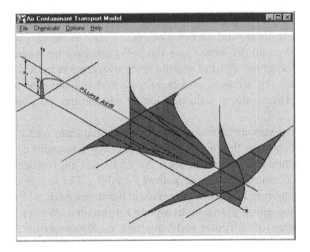

Fig. 4.8 Opening window for the air pathway models

automatically linked to the computational modules of all air pathway models included in the software (Figs. 4.6 and 4.7). The selection of a chemical in this window is done by pressing the "control" button and clicking the left mouse button when the mouse pointer is on the selector column on the left of the table. When on this column, the mouse pointer indicator turns to a solid right pointing arrow symbol indicating that the mouse pointer is on the selection column. Using this operation, the user may select several chemicals to be used in air pathway models. Once they are selected, and the user moves on to other options in the air pathway modules, the selected chemicals become a characteristic database selection for that analysis. These options cannot be modified elsewhere in the module. During another session, if the user opens the input data file prepared in an earlier session, the chemical selections made in the data file will always be linked to the input data. Chemicals can be unselected using the same operations. "Options" menu allows the user to change the text editor, default chemical database, temporary directory paths, and input data directory path preferences. The user must save the preferences selected by clicking on the "Save Preferences" button for the selections to become default options during the following round of sessions of the use of the ACTS software. The "Help" menu accesses segments of this book as context and search sensitive help. All menu operations follow the standard menu operation characteristics of WINDOWS^{TM} environment and can be easily mastered (see Appendix 3).

4.4   Air Emission Models

Air pollutants emitted from point and distributed sources are transported and dispersed by meteorological and topographical conditions that characterize wind speed and stability. The air pollution cycle is usually initiated with the emission of the pollutants, followed by their transport and diffusion through the atmosphere. The cycle is completed when the pollutants are deposited on vegetation, soil, livestock, water surfaces, and on other objects. They can be also washed out of the atmosphere by rain. In some cases the pollutants may be reinserted into the atmosphere by the action of wind erosion or by evaporation from water surfaces. During their path the airborne pollutants may undergo physical and chemical transformation. The results of such transformations and transport can be harmful or beneficial.

Emissions from various land and water based contaminant sources may contribute to harmful exposures through a number of pathways, including exposure due to inhalation, ingestion and dermal contact. Air emissions are considered to be the source of these chemicals in the air pathway module. The source emissions are affected by advection and diffusion processes in the atmosphere, which will diffuse the effluent as the entire plume is transported downwind. We will identify the combined influence of diffusion and advection as dispersion or air dispersion in the air pathway module. In order to perform an air dispersion analysis, air emission rates of chemicals from land or water surface sources must be known.

These emission rates, calculated from emission models, may then be used as input data for the air dispersion models to evaluate the dispersion of chemicals in the air pathway. Alternatively, if field data are available, air emission rates can be directly entered into air dispersion models as input data, bypassing the emission rate calculation step.

This chapter provides an introduction to atmospheric emission and air dispersion models that are included in the ACTS software. The reader is recommended to review the literature cited throughout the text for more detailed information on these models. These references are referenced throughout this document to provide the users' of the ACTS software the source of these technical documents. In particular, the more important literature on which this section is based includes: (Abdel-Magid et al. 1997; Bird et al. 2002; Briggs 1975; Carslaw and Jaeger 1959; Clark 1996; Cole 1970; Cowherd 1983; Cowherd et al. 1985; Csanady 1973; Domenico and Schwartz 1990; Draxler 1979a, b; Farmer et al. 1980, 1978; Fletcher and Dotson 1971; Gifford 1976; Günther 1961; Heinsohn and Kabel 1999; Hemond and Fechner-Levy 2000; Iman and Helton 1988; Kaiser 1979; Louvar and Louvar 1998; Lyons and Scott 1990; Masters 1991; Milton and Stegun 1964; Nirmalakhandan 2002; NRC 1982; Pasquill 1961, 1976; Philp 1995; Schnelle and Dey 2000; Schnoor 1996; Stern 1976; Stern et al. 1984; Thibodeaux 1979, 1982; Thibodeaux and Hwang 1982; Turner 1994; USEPA 1984, 1985, 1997; Vesilind et al. 1994; Viegle and Head 1978; Weber and DiGiano 1996; Zheng and Bennett 1995). This comprehensive reference list is included here to direct the user of the ACTS software to the sources of the models that are included in this software.

4.4.1   Farmer's Model

The Farmer's model can be used to estimate volatile emissions of a buried contaminant source below the soil surface. Figure 4.9 shows a schematic diagram for a typical subsurface contaminant source placement scenario adopted in the Farmer's model (Farmer et al. 1978). As shown in Fig. 4.9, the contaminated soil source is

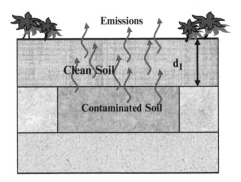

Fig. 4.9 Definition sketch of the Farmer's model

located at a depth d_1 (m) measured from the ground surface. The volatile gases are originating from this source.

In this model, vapor emission from the soil is treated as a diffusion-controlled process. Based on the assumption that this diffusion process can be represented by Fick's law, the steady state emission rate at the soil surface can be estimated using Eq. (4.18),

$$E = AD_e \frac{(C_{vs} - C_a)}{d_1} \left(10^2\right) \tag{4.18}$$

where E is the steady state emission rate of the gases (g/s), A is the surface area of the contamination source in the soil (m^2), D_e is the effective diffusion coefficient of the contaminant for air (cm^2/s), C_{vs} is the vapor phase concentration of the contaminant (g/cm^3), C_a is the air concentration of the contaminant at soil surface (g/cm^3). In this case the air concentration C_a is usually assumed to be equal to zero, d_1 is the depth of soil cover (m) above the contaminant source and 10^2 is the conversion factor from meters to centimeters for the parameters A and d_1.

Given these definitions, the user should note that this model, as it is implemented in the ACTS software uses metric units and that the input values for the parameters used in this equation should be entered in the units given above for the conversion factor that is used in the equation to be correct. This is typical of all air pathway models, and it stems from the fact that the chemical properties used in these models are all entered into the chemical database module in metric units that are linked to all air emission and dispersion models in the ACTS software. Thus, when the user updates the chemical database, the new data should always be entered in metric units to maintain consistency between the data and the models used in this module. The user should also note that the emission rate output is presented in the units of kg/year in the output window grid. The conversion from g/s to kg/year is carried out internally in the ACTS model.

In this model, the soil vapor concentration, C_{vs} (g/cm^3) is given by,

$$C_{vs} = H'C_w \tag{4.19}$$

where H' is the dimensionless Henry's constant (mg/L)/(mg/L) and is defined as,

$$H' = \frac{H}{R\widehat{T}} \tag{4.20}$$

where H is the Henry's law constant (atm-m^3/mol), R is the universal gas constant (8.21E-5 atm-m^3/K) and $\widehat{T}$ is the absolute temperature (K). The aqueous phase concentration, C_w (g/cm^3), is calculated by,

$$C_w = \frac{C_T(\rho_b + \theta_w\rho_w)}{(\theta_T - \theta_w)H' + \theta_w + \rho_bK_d} \tag{4.21}$$

where C_T is the contaminant concentration in the soil (g-contaminant/g-wet soil), θ_T is the soil porosity (dimensionless), ρ_b is the soil bulk density (g-dry soil/cm³-wet soil), ρ_w is the density of water (g/cm³), θ_w is the volumetric water content (dimensionless) and K_d is the soil–water partition coefficient, which is a chemical and soil property dependent parameter ((g/g)/(g/cm³)), and can be given as,

$$K_d = K_{oc} f_{oc} \qquad (4.22)$$

where K_{oc} is the organic carbon partition coefficient ((g/g)/(g/cm³)) and f_{oc} is the fractional organic carbon content of the soil. The user should note that in Eq. (4.21) the soil porosity value entered should be less than the volumetric water content, $\theta_w < \theta_T$.

The effective diffusion coefficient D_e is computed by the relationship given below (Millington and Quirk 1961),

$$D_e = D_{air} \left(\frac{\theta_a^{3.33}}{\theta_T^2} \right) \qquad (4.23)$$

where D_{air} is the diffusion coefficient for the chemical in air (cm³/s), θ_a is the air filled porosity of soil (cm³-air/cm³ soil) and θ_T is the total porosity of soil (cm³-voids/cm³-soil). Again in Eq. (4.23) the air filled porosity should be less than the total porosity of the soil, $\theta_a < \theta_T$.

In the emission rate calculations, when the Farmer's model is used, a temperature correction is also made to the effective diffusion coefficient using the equation below (Lyman et al. 1990),

$$D_{T_2} = D_{T_1} \left(\frac{\widehat{T}_2}{\widehat{T}_1} \right)^{1.75} \qquad (4.24)$$

where $\widehat{T}_1$ is the air temperature at which the diffusion coefficient is known (K), $\widehat{T}_2$ is the air temperature at which the diffusion coefficient is estimated (K) and $D_{\widehat{T}_1}$ and $D_{\widehat{T}_2}$ are the diffusion coefficients of the chemical (cm³/s) at temperatures $\widehat{T}_1$ and $\widehat{T}_2$ respectively. The sequence of calculations described above is automatically executed in sequence when the user implements Farmer's model, thus these steps are all transparent to the user.

Farmer's Model Menu Options: As is the case with all emission models included in the ACTS software, all of the calculations given above are executed sequentially once a site-specific data set is entered that characterizes the application. The output from this model is the emission rate in kg/year for the contaminant in question at the soil surface.

Using this model, the user has the option to calculate emission rates for several contaminants. This may be accomplished by using the "Chemicals!" menu as described in Appendix 3. Once the chemicals database window is entered, several chemicals can be selected. This operation creates a list of chemicals to be linked to the emission models. Once this is accomplished, this list will automatically appear in the Farmer's model output window whenever Farmer's model is selected. In the calculation stage, it is assumed that the input data for soil properties and the contaminant source depth data will be the same for all chemicals in the selected list. If that is not the case, a separate input database has to be generated for each chemical using the "New" menu option in the air pathway window, which creates a new problem for each contaminant with different soil source characteristics. The emission rate computation starts with the first chemical selected in the list. When this calculation is completed, the user may go to the next chemical emission rate calculation by clicking the chemical's name in the output window of the Farmer's model. When the next chemical is selected, the user should note that the "Total Soil Concentration, C_T" input box for this chemical is empty, the chemical parameter(s) are updated to the new chemical data automatically using the chemical database assigned to the problem, and all other input boxes carry the previous site specific data entered into the model. This allows the user to input another source concentration value for the new chemical as input to the model. Once this is accomplished, clicking the "Calculate!" button yields the emission rate for the second chemical. In this manner, emission rates for all chemicals in the list can be calculated. Once this task is completed, the emission rates for all chemicals will be available for use in the air dispersion module as input data.

The Monte Carlo analysis option is available for this model as it is available for all models of the ACTS software. Using the "Monte Carlo" menu option on the menu bar, the user may choose to conduct an uncertainty analysis for most of the parameters of this model using the standard probability distributions imbedded into the ACTS software and the Monte Carlo analysis procedures. A review of Monte Carlo analysis is described in Chapter 7 and menu input operations are described in Appendix 3.

A typical input window for Farmer's model is shown in Fig. 4.10. The menu options on this window are the same as the other emission model input window options. The functions of these menus are described in more detail in Appendix 3.

Assumptions and Limitations of Farmer's Model: The following are the assumptions and limitations of Farmer's model. The assumptions listed below tend to overestimate the emission rate calculated.

    i. In Farmer's model it is assumed that the source concentration of contaminants does not decrease as the emissions occur. Also decay of the contaminant source is not considered. This implies that the amount of contaminant mass in the soil is infinite.

    ii. Adsorption of the chemical to the soil is considered.

    iii. The location of the contaminant source is fixed at a depth d_1 below the surface of the soil.

    iv. Emissions from the soil originating from the contaminant source at depth d_1 are in steady state.

    v. The concentration of the chemical in air at the soil surface is negligible as compared to the vapor concentration within the soil.

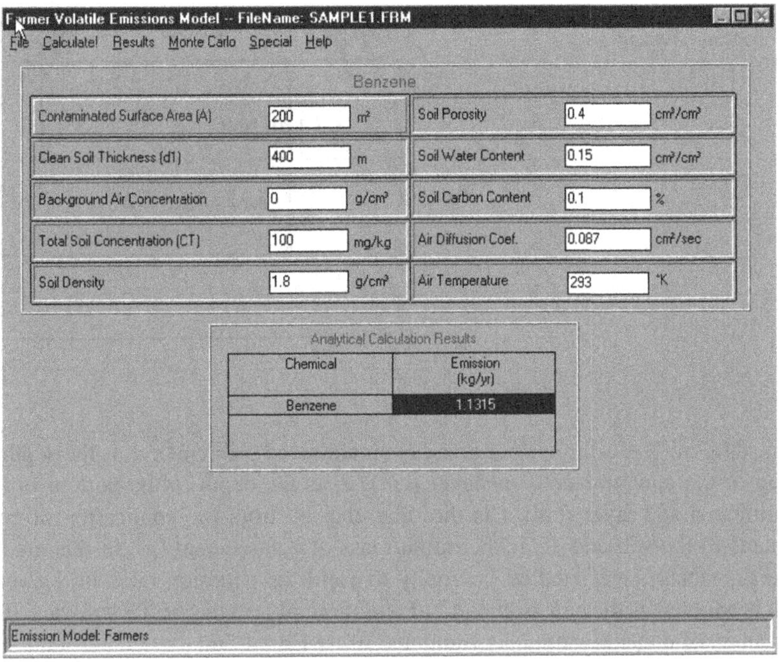

Fig. 4.10 Farmer model input window

4.4.2 Thibodeaux–Hwang Model

The Thibodeaux–Hwang model may be used to estimate time dependent emissions of volatile contaminants that are buried below the soil surface. This model was initially developed to estimate the time dependent emissions of volatile chemicals from petroleum land farming operations based on the analysis described in Thibodeaux and Hwang (1982). A modified version of this emission calculation is also presented in the superfund exposure assessment manual (USEPA 1988). The model may also be used for surface application of contaminants or it may be used in cases of buried chemicals where the zone of contaminated soil is covered by a layer of clean soil. A definition sketch of this model is illustrated in Fig. 4.11.

According to the Thibodeaux–Hwang model, instantaneous emissions originating from volatile chemical sources within the soil can be estimated using Eq. (4.25),

$$E(t) = \frac{D_e C_{vs}}{\sqrt{d_1^2 + \frac{2 D_e A t (d_2 - d_1) C_{vs}}{m_o}}} \tag{4.25}$$

where $E(t)$ is the volatile gas emission rate (g/cm^2 s), D_e is the effective diffusion coefficient of the chemical in air (cm^2/s), C_{vs} is the vapor phase concentration of the

Fig. 4.11 Definition sketch of the Thibodeaux–Hwang model

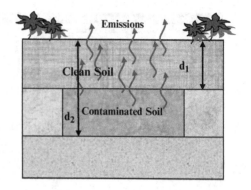

chemical (g/cm³), A is the surface area of contaminated soil (cm²), d_1 is the depth to the top of the contaminated soil layer (cm), d_2 is the depth to the bottom of the contaminated soil layer (cm), t is the time elapsed from the application of contaminants to the soil and m_o is the initial mass of contaminant (g). In this model, conversion factors are handled internally to yield the emission rates in (kg/year) which is adopted as the consistent unit of emission rates in the ACTS software. The initial mass of contaminant m_o is computed using Eq. (4.26),

$$m_o = (d_2 - d_1)AC_b \qquad (4.26)$$

where C_b is the bulk contaminant concentration in the soil (g/cm³) and is computed by,

$$C_b = C_T(\rho_b + \rho_w\theta_w) \qquad (4.27)$$

where C_T is the soil contaminant concentration (g of chemical/g of wet soil), ρ_b is the bulk density of the soil (g of dry soil/cm³ of wet soil), ρ_w is the density of water (g/cm³), and θ_w is the volumetric water content (cm³ of water/cm³ of wet soil). Parameters D_e and C_{vs} are estimated using equations presented in Section 4.4.1.

Based on Eq. (4.25), the average emission rate can be computed by integrating the instantaneous emission rate equation over a time period Δt,

$$E'(\Delta t) = \frac{1}{\Delta t}\int_0^{\Delta t} E(t)dt \qquad (4.28)$$

where, Δt is the averaging period and $E'(\Delta t)$ in g/s is the average emission rate of the chemical over time Δt (s). Substituting $E(t)$ in the equation above one may obtain,

$$E'(t_{max}) = \frac{2D_eC_{vs}A}{d_1 + \sqrt{\frac{2D_eC_{vs}t_{max}}{C_b} + d_1^2}} \qquad (4.29)$$

where the maximum time t_{max} is defined as the evaporation diffusion lifetime, $t_{max} = t_d$. After the evaporation diffusion lifetime, it is assumed that the emission rate is zero.

In the Thibodeaux–Hwang model (Thibodeaux and Hwang 1982) the evaporation diffusion lifetime, t_d, of an initial mass of contaminants that is placed in the subsurface is defined as the time it would take for the entire contaminant mass to volatize. This volatilization time can be estimated as,

$$t_d = \frac{(d_2 + d_1)m_o}{2D_e AC_{vs}} = \left[\frac{d_2^2 - d_1^2}{2D_e}\right]\frac{C_b}{C_{vs}} \tag{4.30}$$

Therefore the average emission over depletion time can be given by $E'(t_{max} = t_d)$.

For risk assessment purposes, the average emission rate needs to be estimated for the exposure period, t_e, which may be longer than t_d. For exposure duration $t_e < t_d$ the average emission rate is $E'(t_e)$. For an exposure duration where $t_e > t_d$, the average emission rate is estimated from,

$$E'(t_e) = \frac{t_d}{t_e}E'(t_e) \tag{4.31}$$

where $E'(t_e)$ (g/s) is the average emission rate over the exposure duration t_e (s) and $E'(t_d)$ (g/s) is the average emission over the duration t_d (s).

Thibodeaux–Hwang Model Menu Options: As is the case with all emission models included into the ACTS software, all of the sequential calculations given above are executed simultaneously once the site-specific data is entered for the application. This model will produce three outputs. In the sequence they appear in the output window grid of the Thibodeaux–Hwang model. These outputs are, "average emission over exposure period (kg/year)," "average emission over depletion time (kg/year)," and "instantaneous emission (kg/year)" for the chemical selected. Any of these emission rates may be selected as the emission rate to be used in the air dispersion models during the second stage of the air dispersion analysis.

Similar to Farmer's model, the user has the option of calculating emission rates for several chemicals. This may be accomplished by following the procedure described in Section 4.4.1. The Monte Carlo analysis option is also available to this model. Using the "Monte Carlo" menu option on the menu bar, the user may choose to conduct an uncertainty analysis for most of the parameters of this model using standard probability distributions imbedded in the ACST software and the Monte Carlo methods. A review of Monte Carlo analysis is described in Chapter 7 and the description of menu operations of this module is given in Appendix 3. A typical input window for the Thibodeaux–Hwang model is shown in Fig. 4.12. The menu options on this window are the same as the other emission model input window options. The functions of these menus are described in more detail in Appendix 3.

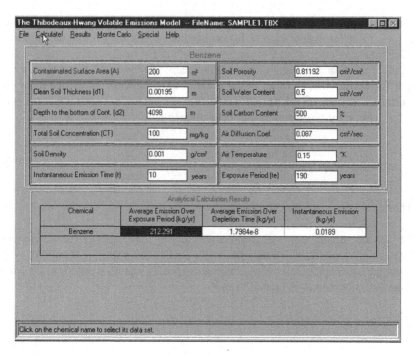

Fig. 4.12 Thibodeaux–Hwang model input window

Assumptions and Limitations of the Thibodeaux–Hwang Model: The following lists the assumptions and limitations of the Thibodeaux–Hwang model:

i. Contaminant mass is distributed as a uniform concentration between the depths d_1 and d_2 that define the zone of contaminant source below the soil surface. The thickness of the contaminated zone is given as $(d_2 - d_1)$.

ii. Contaminant release occurs by molecular diffusion represented by Fick's law and the peeling away of successive layers from the top of the contaminated zone. In other words, the concentration within the contaminated layer is assumed to remain constant, but the thickness of the layer decreases over time. This assumption tends to underestimate the duration of the release, t_d.

iii. The concentration of the chemical in the air zone at the soil surface is assumed to be zero or negligible relative to the soil vapor concentration within the soil.

iv. In the Thibodeaux–Hwang model it is assumed that the entire contaminant mass is volatilized and that none leads to the water table or degrades. This tends to overestimate the emission rate.

4.4.3 Cowherd Particulate Emission Model

The Cowherd particulate emission model estimates the emission rate of respirable soil particles, i.e., those particles with a diameter of 10 μm or less. This model is adopted and described in the US EPA Rapid Assessment of Exposure to Particulate Emissions from Surface Contaminated Sites manual (USEPA 1985) and referred to as the Cowherd model. In the Cowherd model it is assumed that there is a limited reservoir of soil available for erosion. The quantitative model was derived empirically based on wind tunnel experiments conducted for mining soils.

The emission rate using the Cowherd Particulate model is estimated by,

$$E_{10} = \frac{0.83 fAP(u^*)(1-F)}{\left(\frac{PE}{50}\right)^2} \qquad (4.32)$$

where E_{10} is the annual average emission rate of particles less than 10 μm in diameter (PM_{10}) (mg/h), f is the frequency of disturbance per month (month^{-1}), A is the surface area of contaminated soil (m^2), F is the fraction of vegetative cover (dimensionless), and PE is the Thronthwaite's precipitation evaporation index used as a measure of soil moisture content (dimensionless). In this equation, $P(u^*)$ (g/m^2) is defined as $P(u^*) = 6.7(u^+ - u^t)$, where u^+ is the fastest wind speed (m/s) and u^t is the erosion threshold wind speed at 7 m height (m/s). The fastest wind speed may be obtained from the climatologic data station nearest to the site under investigation. The erosion threshold wind speed is related to the soil particle size distribution, which is a measure that quantifies the erosion potential and can be obtained from references on soil erosion (USEPA 1985). Thus $P(u^*)$ is the erosion potential, i.e., a measure of the quantity of particles present on the surface prior to the onset of wind erosion that can be eroded after the application of the wind. Thornthwaite's PE index quantifies average surface soil moisture. A map showing the distribution of this index in the US is given in Fig. 4.13. Estimation of various parameters in Eq. (4.32) is discussed in greater detail in the USEPA manual (USEPA 1985).

In this model, a disturbance is defined as an action which results in the exposure of the soil surface material and occurs whenever soil material is added to the surface or removed from the old surface. For example, breaking the crust of soil due to vehicular traffic may expose erodible material and would be considered a disturbance in certain applications.

The emission rate of chemicals due to wind erosion is computed as the product of the concentration of chemicals in the soil and the E_{10} soil particle emission rate as,

$$E = \frac{E_{10}S}{3600} \qquad (4.33)$$

where E is the annual average emission rate of the chemical (mg/s), S is the particulate contaminant concentration (mg/kg) and 3,600 is the conversion factor from hours to seconds.

Cowherd Particulate Emissions Model Menu Options: As is the case with all emission models that are included in the ACTS software, all of the above calculations are executed sequentially once the site-specific data is entered for an

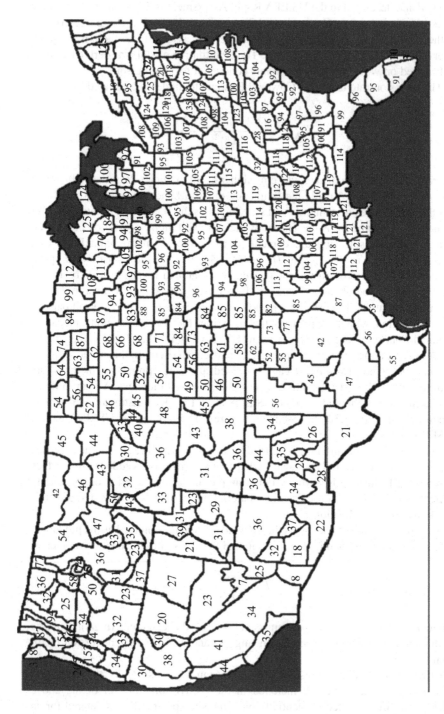

Fig. 4.13 Thornthwaite's PE index for average surface soil moisture

application. Based on the equations given above, the Cowherd particulate emissions model yields the particulate emission rate in (kg/year) in the output box for the selected chemical or chemicals.

Using this model, the user has the option to calculate emission rates for several chemicals. This may be accomplished by using the procedure described in Section 4.4.1. The Monte Carlo analysis option is also available for this model. Using the "Monte Carlo" menu option on the menu bar, the user may choose to conduct an uncertainty analysis for most of the parameters of this model using standard probability distributions imbedded in the ACTS software and the Monte Carlo methods. The details of implementing this calculation sequence are described in Chapter 7 and the menu operations are discussed in Appendix 3. The input window for the Cowherd particulate emissions model is shown in Fig. 4.14. The menu options on this window are the same as the other emission model input window options. The functions of these menus are also described in Appendix 3.

The following lists the assumptions and limitations of the Cowherd Particulate Emissions Model:

  i. The Cowherd model can be used for estimating respirable particulate emissions from soil surfaces due to wind erosion.
 ii. The model assumes a limited soil reservoir surface, with surface erosion potential restored after each disturbance.

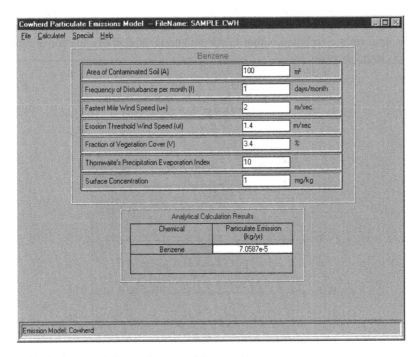

Fig. 4.14 Cowherd particulate emission model input window

iii. The model was developed based on field measurements using a portable wind tunnel, thus it is an empirical model.

iv. The model uses the Thornwaite precipitation–evaporation (*PE*) index as a useful indicator of the average soil surface moisture.

4.4.4 Jury Unsaturated Zone Emission Model

The Jury model is a screening level model that can be used to estimate the gaseous contaminants volatilizing from the soil and the time dependent concentration profile resulting from this volatilization within the unsaturated zone. A definition sketch of the soil profile used in the Jury model is shown in Fig. 4.15.

The Jury model is based on the analytical solution of the advection diffusion Eq. (4.34) (see also Chapter 3) along with certain boundary and initial conditions as discussed below (Jury et al. 1990),

$$\frac{\partial C_T}{\partial t} + v_p \frac{\partial C_T}{\partial z} = D_E \frac{\partial^2 C_T}{\partial z^2} - \lambda C_T \tag{4.34}$$

where C_T is the soil concentration, t is the time (day), λ is the first order decay rate, D_E is the effective diffusion coefficient, z is the depth from the soil surface measured positive downward, v_p is the pore Darcy velocity in z-direction.

The initial conditions for the contaminant source, which imply that the contaminant is uniformly distributed within the depth and the soil above the contaminant zone is clean, are given as,

$$C_T(D<z<D+L, t=0) = C_o \tag{4.35}$$

$$C_T(z \leqslant D, t=0) = 0 \tag{4.36}$$

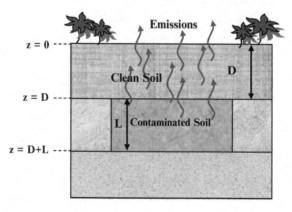

Fig. 4.15 Definition sketch of the Jury model

For the case of a contaminant source buried under a clean fill, the solution is obtained by superposition. This is the configuration shown in Fig. 4.15. The boundary condition at the soil surface is given as:

$$-D_E \frac{\partial C_T}{\partial z} + v_p C_T = -H_E C_T \quad at \quad z = 0 \tag{4.37}$$

which represents a concentration dependent flux boundary condition. This model can be used to simulate the volatilization of chemical vapor to the atmosphere through a stagnant air boundary layer. Above the soil surface it is assumed that the chemical concentration in air is zero. The lower boundary condition is,

$$C_T(z = \infty, t) = 0 \tag{4.38}$$

In the equations given above, C_o is the initial contaminant concentration in the soil, L is the depth of contaminated soil, and H_E is the mass transfer coefficient through the stagnant boundary layer.

Jury Unsaturated Zone Emissions Model Menu Options: The input window for the Jury model is shown in Fig. 4.16. The menu options on this window are the same as the other emission model input window options. The functions of these menus are described in Appendix 3.

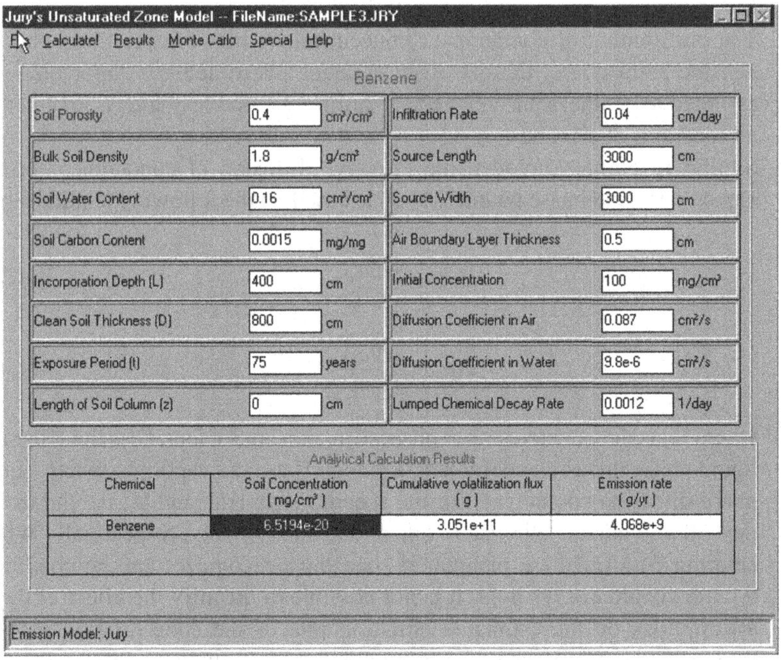

Jury's Unsaturated Zone Model -- FileName:SAMPLE3.JRY

Calculate! Results Monte Carlo Special Help

Benzene

Soil Porosity	0.4	cm³/cm³	Infiltration Rate	0.04	cm/day
Bulk Soil Density	1.8	g/cm³	Source Length	3000	cm
Soil Water Content	0.16	cm³/cm³	Source Width	3000	cm
Soil Carbon Content	0.0015	mg/mg	Air Boundary Layer Thickness	0.5	cm
Incorporation Depth (L)	400	cm	Initial Concentration	100	mg/cm³
Clean Soil Thickness (D)	800	cm	Diffusion Coefficient in Air	0.087	cm²/s
Exposure Period (t)	75	years	Diffusion Coefficient in Water	9.8e-6	cm²/s
Length of Soil Column (z)	0	cm	Lumped Chemical Decay Rate	0.0012	1/day

Analytical Calculation Results

Chemical	Soil Concentration (mg/cm³)	Cumulative volatilization flux (g)	Emission rate (g/yr)
Benzene	6.5194e-20	3.051e+11	4.068e+9

Emission Model: Jury

Fig. 4.16 Jury emission model input window

As is the case with all emission models that are included in the ACTS software, all of the calculations described above are sequentially executed once the site-specific data is entered for an application. Based on the equations given above, the Jury emissions model yields the emission rate in (kg/year) in the output box for the selected chemical or chemicals. Using this model, the user has the option to calculate emission rates for several chemicals. This may be accomplished by using the procedure described in Section 4.4.1. The Monte Carlo analysis option is also available for this model. Using the "Monte Carlo" menu option on the menu bar, the user may choose to conduct an uncertainty analysis for most of the parameters of this model using standard probability distributions imbedded in the ACTS software and the Monte Carlo methods. Details of implementing this calculation sequence are described in Chapter 7. The menu options on this window are the same as the other emission model input window options. The functions of these menus are described in Appendix 3.

Assumptions and Limitations of the Jury model: The following are the assumptions and limitations of the Jury model.

i. The soil column is assumed to be homogeneous and isotropic.
ii. The infiltration rate is assumed to be uniform and steady.
iii. The contamination is initially incorporated uniformly from the top of the soil column to a depth L (cm) below the surface. When the contaminant is incorporated below a depth D (cm), or when the contaminated soil is buried below a clean layer of soil, the contaminant is analyzed using the principle of superposition.
iv. Contaminant decay is assumed to follow a first order decay rate.
v. The partitioning of contaminant concentrations between the three phases, i.e., the solid phase, the dissolved aqueous phase and the vapor phase is assumed to be linear. Instantaneous equilibrium among phases is assumed at all locations at all times.
vi. Similar to Farmer's model, the effective diffusion of contaminant in the vapor and liquid phase within the soil is based on the following relationships:

$$D_g = D_g^a \frac{\theta_a^{3.33}}{\theta_l^2} \tag{4.39}$$

$$D_l = D_l^a \frac{\theta_w^{3.33}}{\theta_l^2} \tag{4.40}$$

where D_g is the effective vapor phase diffusion coefficient, D_l is the effective liquid phase diffusion coefficient, D_l^a and D_g^a are the vapor phase and liquid phase diffusion coefficients for the specific chemical, and θ_a, θ_w and θ_l are the soil air content, soil water content and soil porosity, respectively. When entering data for these parameters, one must recognize that the condition $\theta_w < \theta_l$ should always hold. It is not possible to quantify the effect of these assumptions on the estimated emission rate or the concentration profile. Depending on the degree of departure between the field conditions and the assumptions given above, the Jury model may overestimate or, in some cases, underestimate the emission rate.

4.4.5 Landfill Gas Emissions Model

Municipal solid waste (MSW) landfills generate significant volumes of various gases during their active life and for a period of time after their closure. As such, they are considered to be the largest U.S. anthropogenic source of gases. Worldwide, methane emissions from landfills and open dumps have been estimated to produce approximately 30 teragrams (Tg) year^{-1} or 6% of total global methane emissions (Thorneloe et al. 1993, 1994). Most of the gas generated in landfill emissions is methane and carbon dioxide with smaller amounts of volatile organic compounds (VOCs). The gas is emitted into the atmosphere and can also travel long distances in the porous space of the soil medium. Landfill gases, VOCs in particular, contribute to air pollution and are considered to be one of the important sources of ground-level ozone. Methane is a colorless and odorless gas. It is highly explosive at concentrations of 5–15% in air and can accumulate to dangerous levels virtually undetected. Methane and other emissions from landfills are important contributors to environmental degradation and of concern for their health effects consequences. Therefore, at MSW landfills it is necessary to monitor the migration of methane gas to ensure the safety of both on-site and off-site structures, and to ensure the safety and protection of personnel and populations.

Air emissions from landfills come from landfill gas that is generated by the decomposition of refuse in the landfill. The Landfill Gas Emissions module of the ACTS Software is based on the USEPA model LandGEM (USEPA 1991, 1998). In this model the landfill gas is assumed to be roughly half methane and half carbon dioxide, with additional, relatively low concentrations of other air pollutants. The estimation method used by the model is based on a simple first-order decay equation and requires limited input data such as: (i) the design capacity of the landfill; (ii) the amount of refuse in place in the landfill, or the annual refuse acceptance rate for the landfill; (iii) the methane generation rate; (iv) the potential methane generation capacity; (v) the concentration of total nonmethane organic compounds (NMOC) and speciated NMOC found in the landfill gas; (vi) the years the landfill has been in operation; and, (vii) whether the landfill has been used for disposal of hazardous waste (co-disposal). Because the data available on the quantity, age and composition of the refuse in the landfill are limited, using a more sophisticated calculation method was not attempted in this model (USEPA 1991). The Landfill Gas Emissions Model estimates emissions of methane, carbon dioxide, nonmethane organic compounds, and selected air pollutants. Information on the assumptions used in this model can be found in the document (USEPA 1991).

The following mathematical model is used to estimate gas emissions if the actual year-to-year solid waste acceptance rate to the landfill is known.

$$M_{NMOC} = 2 \sum_{i=1}^{n} kL_oM_iC_{NMOC}\left(e^{-kt_i}\right)\left(3.6 \times 10^{-9}\right) \qquad (4.41)$$

where M_{NMOC} is the total landfill NMOC emission rate, (Mg/year); k is the methane generation rate constant, (year^{-1}); L_o is the methane generation potential, (m^3/Mg of waste); M_i is the mass of solid waste in the ith section of the landfill, (Mg); t_i is the age of the ith section, (years); C_{NMOC} is the concentration of NMOC, (parts per million by volume as hexane) and the constant used in the equation is the conversion factor to render the outcome in (Mg/year). The mass of non-degradable solid waste may be subtracted from the total mass of solid waste in a particular section of the landfill when calculating the value for M_i, if documentation of the nature and amount of such wastes is maintained.

The following mathematical model is used if the actual year-to-year solid waste acceptance rate is unknown.

$$M_{NMOC} = 2RL_oC_{NMOC}\left(e^{-kc} - e^{-kt}\right)\left(3.6 \times 10^{-9}\right) \tag{4.42}$$

where, M_{NMOC}; k; L_o; C_{NMOC} have the same definitions as above and t is the age of the landfill, (years); R is the average annual acceptance rate, (Mg/year) and c is the time since closure, (years). For active landfills it is assumed that $c = 0$.

In the equations given above, the methane generation rate constant, k, reflects the rate of generation of methane for each submass of refuse in the landfill. The higher the value of k, the faster the methane generation rate increases and then decays over time. The value of k is a function of the following factors: (i) refuse moisture content; (ii) availability of the nutrients for methanogens; (iii) pH; and, (iv) the temperature. The k values obtained from the field data collected range from 0.003 to 0.21. The value for the potential methane generation capacity of refuse L_o depends only on the type of refuse present in the landfill. The higher the cellulose content of the refuse, the higher the value of L_o. The values of theoretical and field data for L_o range from 6.2 to 270 m^3/Mg refuse. In Table 4.3 values of the methane generation rate and methane generation capacity are given for typical landfill conditions (USEPA 1991).

Landfill Gas Emissions Model Menu Options: A typical input window for the Landfill Emissions model is shown in Fig. 4.17. The menu options on this window are the same as the other emission model input window options. The functions of these menus are described in Appendix 3. For this model, the user does not need to select a chemical from the chemicals database to determine the emissions, as they are only evaluated for the $NMOC$.

Table 4.3 Values for the methane generation rate, k and potential methane generation capacity, L_o

Emission type	Landfill type	k (year^{-1})	L_o (m^3/Mg)
CAA	Conventional	0.05 (Default)	170 (Default)
CAA	Arid area	0.02	170
Inventory	Conventional	0.04	100
Inventory	Arid area	0.02	100
Inventory	Wet (bioreactor)	0.7	96

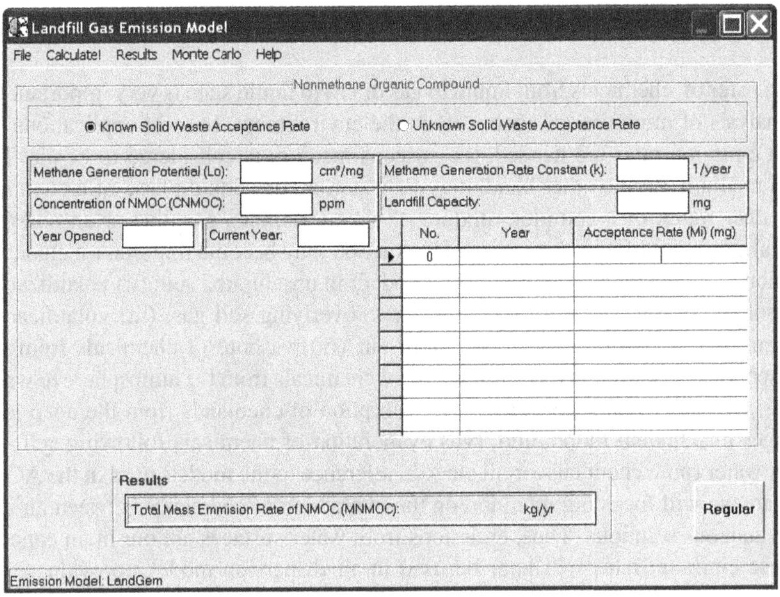

Fig. 4.17 Landfill emissions model input window

As is the case with all emission models that are included in the ACTS software, all of these calculations are sequentially executed once the site-specific data are entered for an application. Based on the equations given above, the landfill emissions model yields the emission rate in kg/year in the output box for the selected chemical *NMOC*. Using this model, the user does not have the option to calculate emission rates for several other chemicals. The Monte Carlo analysis option is also available for this model. Using the "Monte Carlo" menu option on the menu bar, the user may choose to conduct uncertainty analysis for most of the parameters of this model using standard probability distributions imbedded in the ACTS software and the Monte Carlo methods. Details of implementing this calculation sequence are described in Chapter 7. The menu options on this window are the same as the other emission model input window options. The functions of these menus are described in Appendix 3.

Assumptions and Limitations of the Landfill Emissions model: The following are the assumptions and limitations of the Landfill Emissions model.

i. The methane generation process is based on a first order decay mechanism.
ii. Parameters of the model are empirical variables.
iii. Some important physical conditions of a landfill which may affect the methane generation are not considered, or these effects are incorporated into the model based on empirical variables.
iv. These limitations may tend to underestimate or overestimate the methane generation at a site.

4.4.6 Volatilization from Water Bodies

The transfer of chemicals from liquid to gas or gas to liquid state is very important for the analysis of migration of chemicals in the environment. In most applications for which emission rates are needed, the gaseous medium is considered to be air. The liquid medium, on the other hand, may be water, a pure liquid phase of a chemical other than water, or a complex mixture of chemicals, e.g., gasoline or an oil spill. Typical examples of where gas–liquid interaction may become important include: (i) Partitioning of chemicals between water and air in unsaturated soil; (ii) volatilization of gasoline constituents from groundwater to overlying soil gas; (iii) volatilization of chemicals from groundwater to indoor air; (iv) washout of chemicals from the atmosphere to rain droplets; (v) absorption of chemicals from the atmosphere to water bodies, e.g., the Great Lakes case; (vi) absorption of chemicals from the deep lung passages into human blood; and, (vii) evaporation of chemicals following spills to soil or water (pure chemicals or oil etc.). In reference to the models used in the ACTS software we will focus our attention on the interaction of chemicals between air and dilute aqueous solutions. Thus, emissions from water surfaces are our main concern as these emission rates will later be used in air dispersion models to evaluate the spread of gaseous emissions from water bodies.

To understand this mass transfer process in terms of the basic principles of thermodynamics a review of the equilibrium partitioning principle (Henry's law) and the kinetics of the gas-liquid mass transfer concept will be helpful.

Henry's Law: The vapor phase of a substance can be defined as an air dispersion of molecules of that substance which is a liquid or solid phase in its normal state under standard temperature and pressure. The vapor pressure of a substance can also be defined as the pressure characteristic of the substance at any given temperature of a vapor that is in equilibrium with its liquid or solid form. The vapor pressure of a pure gas is 1 atm at standard temperature and pressure. Vapor density, on the other hand, is the mass concentration of the substance in air, with the saturated vapor density being equal to the vapor pressure, representing the maximum concentration of that substance in air.

When a liquid and a gas are in contact, the weight of the gas that dissolves in a given quantity of liquid is proportional to the pressure of the gas which is formed above the liquid. Henry's law applies to chemicals dissolved in dilute aqueous solutions that have reached equilibrium between the aqueous and adjacent vapor phase. At equilibrium, for a fixed temperature and chemical, the ratio of the chemical concentration in the vapor phase to the chemical concentration in water or liquid phase is a constant. This proportionality is referred to as the Henry's law constant. At equilibrium, Henry's law can be given as,

$$H = \frac{C_a}{C_w} \tag{4.43}$$

where C_a is the concentration of the chemical in the air or vapor phase, C_w is the concentration of the chemical in the aqueous phase and H is the Henry's law constant. It is conventional to define H in terms of gas concentrations in atmospheres and liquid concentrations in mol/m³. Thus, the most typical unit for H is atm-m³/mol. The Henry's law constant is an important parameter that is required for estimating the equilibrium distribution of chemicals between the two phases and it is also important in estimating the rate of gas–liquid mass transfer rates for the chemical in question.

It is often easier to work with the "dimensionless" Henry's law constant H' which can be obtained by converting gas concentrations from atmospheres to mol/m³. To do this we can use the ideal gas law:

$$\frac{n}{V} = \frac{P}{R\hat{T}} \tag{4.44}$$

where n is the number of moles of gas, V is the volume of the gas, P is the absolute pressure of the gas, $\hat{T}$ is the absolute temperature (K) and R is the universal gas constant ($8.314472 \text{ J mol}^{-1} \text{ K}^{-1}$ or $8.2 \times 10^{-5} \text{ m}^3$ gas-atm/mol-K). From this relationship we can conclude that the dimensionless Henry's constant H' can be given as:

$$H' = \frac{H}{R\hat{T}} \tag{4.45}$$

Note that H' really has units of $\text{mol/m}_{\text{gas}}^3/\text{mol/m}_{\text{liq}}^3$ or $\text{m}_{\text{liq}}^3/\text{m}_{\text{gas}}^3$. We should also acknowledge that the Henry's law constant is a function of chemical structure and the temperature.

Two Film Theory of Gas–Liquid Mass Transfer: Volatilization of chemicals from water surfaces is commonly described by the two-film model (Bird et al. 2002). In this model we assume a uniformly mixed water and air phases that are separated by two thin films of air and water through which mass transfer occurs (Fig. 4.18). It is further assumed that this mass transfer is governed by molecular diffusion only. Mass transfer coefficients are commonly identified as the liquid-film and the gas-film coefficients.

Empirical evidence indicates that in the two-film model the relative importance of the water and air resistances for the transfer of a specific volatile organic compound to either phase depends on the Henry's law constant. However, early

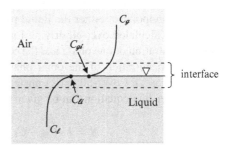

Fig. 4.18 Two-film theory definition sketch and concentration gradients

Table 4.4 Processes affecting the mass transfer between phases

Chemical properties	Water properties	Atmospheric properties
Aqueous solubility	Depth	Wind speed
Vapor pressure	Flow rate	Turbulence
Henry's law constant	Turbulence	Temperature
Diffusivity	Temperature	
	Ionic strength	

experimental work also indicates that water film resistance is very important. In any case, both of these resistances play an important role in defining the mass transfer rate between phases. In actuality the mass transfer rate process between phases is very complex, non-linear and transient, and depends on the chemical and physical properties of the chemical, the water body, and the atmosphere as given in Table 4.4.

Based on Fick's law, the mass transfer flux between the two phases can be given as,

$$N = k_g(C_g - C_{gi}) = k_\ell(C_{\ell i} - C_\ell) \qquad (4.46)$$

In the equation above,

$$k_\ell = \frac{D_\ell}{\delta_\ell}; \quad k_g = \frac{D_g}{\delta_g} \qquad (4.47)$$

where δ_ℓ and δ_g are the liquid and gas-phase film thickness and D_ℓ and D_g are the liquid and gas-phase diffusion coefficients, respectively; N is the mass transfer flux; k_g is the gas-phase exchange coefficient (cm/s); C_g is the concentration in gas phase at the outer edge of the film (g/cm^3); C_{gi} is the concentration in gas phase at the interface (g/cm^3); k_ℓ is the liquid-phase exchange coefficient (cm/s); $C_{\ell i}$ is the concentration in the liquid phase at interface (g/cm^3); and, C_ℓ is the concentration in the liquid phase at the outer edge of the film (g/cm^3).

The concentrations of the diffusing material in the two phases immediately adjacent to the interface are generally unequal, but are usually assumed to be related to each other by the laws of thermodynamic equilibrium.

Experimental determination of the coefficients k_ℓ and k_g is difficult. When the Henry's absorption isotherm is linear, over-all coefficients, which are more easily determined by an experiment, can be used. Over-all coefficients can be defined from the standpoint of either the liquid phase or the gas phase. Each coefficient is based on a calculated over-all driving force, defined as the difference between the bulk concentration of one phase and the equilibrium concentration corresponding to the bulk concentration of the other phase. When the controlling resistance is in the liquid phase, the over-all mass transfer coefficient K_L is generally used. Then the mass transfer flux equation can be given as,

$$N = k_\ell(C_{\ell i} - C_\ell) = K_L(C_L^* - C_\ell) \qquad (4.48)$$

where C_L^* is the liquid concentration in equilibrium with the bulk gas concentration.

The non-dimensional Henry's law constant H' relates the concentration of a compound in the gas phase to its concentration in the liquid phase (g/cm³). Thus,

$$H' = \frac{C_{gi}}{C_{\ell i}} = \frac{C_G^*}{C_\ell} = \frac{C_g}{C_L^*} \tag{4.49}$$

Using Eq. (4.49) the relationship between the overall mass transfer coefficient K_L and the gas-phase exchange coefficient k_g and the liquid-phase exchange coefficient k_ℓ can be obtained,

$$\frac{1}{K_L} = \frac{1}{k_\ell} + \frac{1}{k_g H'} \tag{4.50}$$

or

$$K_L = \frac{k_\ell}{1 + \frac{k_\ell}{k_g H'}} \tag{4.51}$$

Based on two-film theory, the overall mass transfer resistance can be conceptualized as the sum of the resistance on the liquid and gas sides of the air–water interface. Note that the first term on the right-hand-side of Eq. (4.46) corresponds to the gas-phase resistance to mass transfer while the second term corresponds to liquid-phase resistance to mass transfer. Thus, the ratio of k_ℓ/k_g to H' is important in identifying the source of the major controlling resistance. When $k_\ell << k_g H'$ the liquid side resistance dominates and $K_L = k_\ell$. This is usually true for oxygen transfer, but may not be true for volatilization of organic compounds. The overall mass transfer coefficient K_L is defined and is valid for systems where $k_\ell \cong k_g H'$. However, for this case the over-all mass transfer coefficient is no longer a function of only the liquid phase parameters, but also of the gas phase parameters. For most typical environmental systems in nature the ratio of k_g to k_ℓ is greater than 10. Thus, for chemicals with Henry's law constants that are much lower than 0.1, the gas-phase resistance to mass transfer can dominate the overall mass transfer process. Conversely, for chemicals with large values of H', the mass transfer is typically dominated by the liquid phase and the second term on the right-hand side can be neglected.

Volatilization from Water Surfaces Model Menu Options: As is the case with all emission models that are included in the ACTS software, all of the calculations given above are sequentially executed once the site-specific data are entered for an application. Based on the model given above, the volatilization from water surfaces module yields the emission rate in (kg/year) in the output box for the selected chemical or chemicals. Using this model, the user has the option to calculate emission rates for several chemicals. This may be accomplished by using the procedure described in Section 4.4.1. The Monte Carlo analysis option is also available for

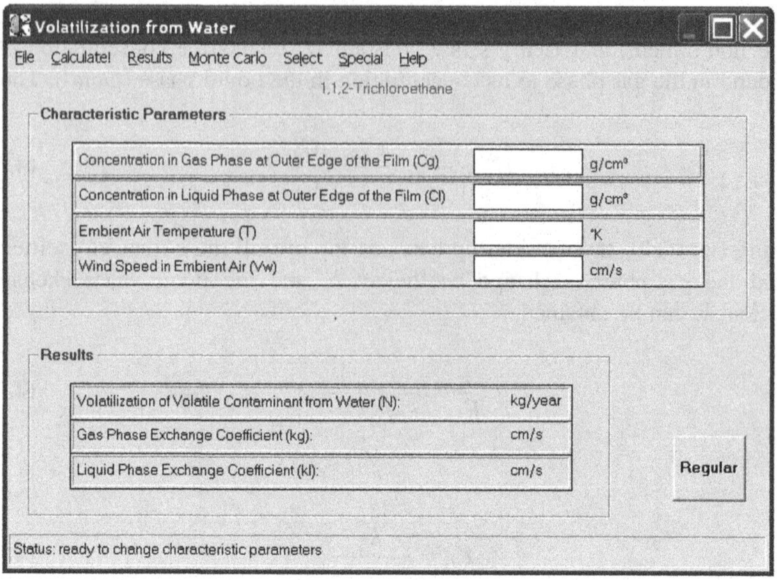

Fig. 4.19 Volatilization from Water surfaces model input window

this model. Using the "Monte Carlo" menu option on the menu bar, the user may choose to conduct uncertainty analysis for most of the parameters of this model using standard probability distributions imbedded in the ACTS software and the Monte Carlo methods. Details of Monte Carlo analysis are described in Chapter 7. The menu options on this window are the same as the other emission model input window options (Fig. 4.19). The functions of these menus are described in Appendix 3.

Assumptions and Limitations of the Volatilization from Water Surfaces Model: The model is restricted to the two-film analogy which may underestimate the fluxes between phases.

4.4.7 Air Dispersion Models

The purpose of air dispersion modeling is the evaluation of the impacts of the emissions sources in the vicinity of the emission study area. Several factors impact the fate and transport of emission sources in the atmosphere including meteorological conditions, site configuration, emission release characteristics, and surrounding terrain among others. In the air dispersion analysis of gaseous plumes the plume source is most commonly associated with an emission source from landfills, contaminated soil or water bodies. Those emission sources that can be used in the air dispersion models are discussed in the previous section. These emission calculations are included in the ACTS software as the first step of the analysis. After a chemical or a set of chemicals are selected from the "Chemicals" data base and the emission rate is calculated using one of the methods described in the previous section, the next step is the calculation of the transformation and dispersion of the air plume generated. This step of the analysis can be based on simple mass balance analysis or it can be based on the solution of the advection-diffusion equation. In the ACTS software both of these approaches are used. In air dispersion models, emissions from stacks are also considered in addition to the emission rate estimates described in the previous section.

In the absence of transport with an average wind velocity, u $[LT^{-1}]$, a chemical will be dispersed in air according to Fick's second law,

$$\frac{\partial C}{\partial t} = D_x \frac{\partial^2 C}{\partial x^2} \tag{4.52}$$

where D_x is the longitudinal diffusion coefficient $[L^2T^{-1}]$, C is the concentration $[ML^{-3}]$ (see also Chapter 3) and x and t are the spatial and temporal coordinates. This is an expression, which describes the rate of change of concentration over time relative to the rate of change of the gradient of the concentration profile with respect to distance. If in addition to dispersion, the concentration plume is advected with a wind velocity u, the following transport equation also applies,

$$\frac{\partial C}{\partial t} = -u \frac{\partial C}{\partial x} \tag{4.53}$$

Combination of Eqs. (4.52) and (4.53) yields the well-known one-dimensional advection–diffusion equation (see Chapter 3).

$$\frac{\partial C}{\partial t} + u \frac{\partial C}{\partial x} = D_x \frac{\partial^2 C}{\partial x^2} \tag{4.54}$$

The advection–diffusion equation (4.54) has a solution that is based on the equation of Gaussian normal distribution,

$$C(x, t) = \frac{M}{(4\pi D_x t)^{1/2}} \exp\left(\frac{-(x - ut)^2}{4D_x t}\right) \tag{4.55}$$

where M is the mass of contaminant released per unit area perpendicular to the air flow direction. Taylor was the first to establish the basis for this one-dimensional analytical solution (Taylor 1953). Similarly, two or three-dimensional forms of the advection-diffusion equation with decay and a unidirectional velocity component can be given as follows,

$$\frac{\partial C}{\partial t} + u \frac{\partial C}{\partial x} = D_x \frac{\partial^2 C}{\partial x^2} + D_y \frac{\partial^2 C}{\partial y^2} + D_z \frac{\partial^2 C}{\partial z^2} + \lambda C \tag{4.56}$$

where D_x, D_y, D_z are diffusion coefficients in the x-, y- and z-coordinate directions respectively and λ is the first order decay coefficient (see Chapter 3).

In this chapter we provide an introduction to atmospheric dispersion models that are included in the ACTS software, which are either based on simplified mass balance equations or analytical solutions of the advection–diffusion equations given above. It is recommended that the reader review the literature cited throughout the text and also Chapter 3 for more detailed information on the models and boundary conditions used in these solutions.

4.4.8 Box Air Dispersion Model

The Box dispersion model can be used to estimate concentrations in the air near an emission source and it is based on the mass balance principle. In this model, it is assumed that the steady-state contaminant emissions originating from the source are uniformly mixed within a fixed volume of air inside the selected "Box." As shown in Fig. 4.20, the "Box" is a bounded mixing zone above the soil surface, and a steady flow of wind passes across this box. Emissions originating from a contaminated soil layer or other emission sources enter the box in a perpendicular direction relative to the wind velocity direction. Concentrations are then assumed to be proportional to the rate of source emission and inversely proportional to the average residence time of air and the inversion height. Based on these assumptions, the average air concentration of the chemical in the "Box" can be calculated by the mass balance relationship,

$$C_{air} = \frac{10^3 E}{vWH} \tag{4.57}$$

where C_{air} is the concentration of chemical in air (mg/m^3), E is the average volatile chemical emission rate for the exposure period (g/s) which is calculated based on

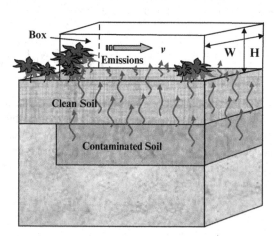

Fig. 4.20 Definition sketch of the Box model

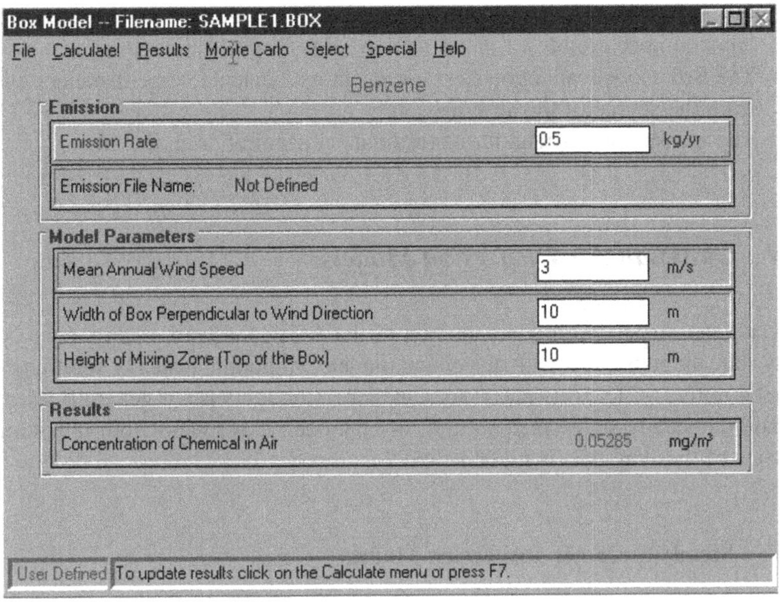

Fig. 4.21 Box model input window

the emission models discussed in the previous section, v is the mean annual wind speed (m/s) in the predominant wind direction, W is the width of the box perpendicular to the predominant wind direction (m), H is the height of the mixing zone (m) and 10^3 is the conversion factor for grams to milligrams conversion (mg/g).

Box Dispersion Model Menu Options: A typical input window for the Box model is shown in Fig. 4.21. Using this model, the user has the option to calculate emission rates for several chemicals. These calculations may be made by using the procedure described in Section 4.4.1. The Monte Carlo analysis option is also available for this model. Using the "Monte Carlo" menu option on the menu bar, the user may choose to conduct an uncertainty analysis for most of the parameters of this model using standard probability distributions imbedded in the ACTS software and the Monte Carlo methods. Details of implementing this calculation sequence are described in Chapter 7. The menu options on this window are the same as the other emission model input window options. The functions of these menus are described in Appendix 3.

Assumptions and Limitations of the Box Dispersion Model: The following lists the assumptions and limitations of the Box model:

i. The Box model does not account for the decrease in concentration with distance in the wind direction or over the height from the soil surface. The Box model is an equilibrium model and does not yield spatial distribution of concentrations within the box. This model may be used as a screening model, and best applications may be for onsite exposure analysis scenarios.

ii. In the Box model, it is assumed that all of the volatile emissions enter the box and none are blown in a direction away from the receptor.
iii. The Box model cannot be used for receptors located at large distances away from the source of the air emission.
iv. These assumptions tend to overestimate concentration in the air.

4.4.9 Gaussian Air Dispersion Models

The Gaussian dispersion models module of the ACTS software incorporates both steady state and unsteady state dispersion models. Although unsteady solutions are provided in the ACTS software, Gaussian dispersion models are generally used to estimate the steady-state concentration of chemicals downwind from the source. A review of these models is given below.

4.4.9.1 Steady State Air Dispersion Models

The steady state models will focus on the emission patterns from stacks or other sources during a short time period (hours and days), as opposed to global balances, which may be averaged over several years. Estimation or predictions of the atmospheric concentration of volatile chemicals resulting from emissions from point sources is complicated and involves a great deal of uncertainty. The uncertainty is mostly due to wide variations in geographical and meteorological conditions that may change in a short period of time, such as terrain, wind speed, turbulence, and temperature. In such cases, the Monte Carlo analysis mode of the ACTS software may be used to quantify the uncertainty in the results.

The atmospheric conditions in the range of (0–1,000 m) above the ground surface significantly affect the dispersion of emissions originating near the earth's surface. This effect can best be observed in the behavior of the emissions discharging from a smoke stack. Depending on the stability conditions near the surface a plume may be dispersed in several different geometries as shown in Fig. 4.22.

If the atmosphere is in a neutral state (Fig. 4.22a) the plume will gradually expand in the vertical direction both above and below the smoke stack exit elevation symmetrically. The plume will also expand in the lateral direction due to transverse dispersion effects (Fig. 4.23). Since the plume expands in the shape of a cone, this condition is identified as coning. If the atmosphere is stably stratified near the elevation of the smoke stack exit elevation (Fig. 4.22b), the turbulence in the atmosphere will be minimal. This would prevent the expansion of the plume in vertical and lateral directions, which would result in a much narrower plume when compared to the coning condition. The plume still expands as it is advected in the downwind direction, but this expansion is very shallow. This condition of the plume is identified as fanning. A combination of stable and unstable conditions is also possible above or below the smoke stack exit elevation. This condition gives rise to

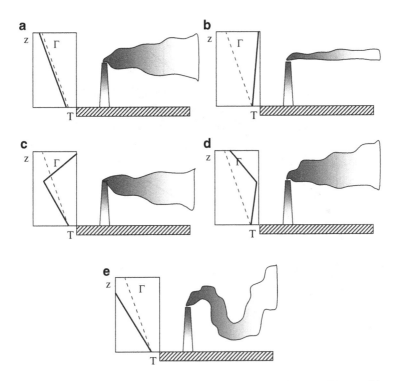

Fig. 4.22 Most commonly observed plume configurations and associated stability conditions: (**a**) neutral, coning; (**b**) stable, fanning; (**c**) natural below and stable above, fumigation; (**d**) stable below and natural aloft, lofting; (**e**) unstable looping

the plume shapes shown in Figs. 4.22c and d. In this case the stable side of the plume will not expand as much as the unstable side of the plume. These conditions are identified as fumigation (Fig. 4.22c) and lofting (Fig. 4.22d). Between these two cases, the fumigation condition would be more critical since the plume will affect the near surface exposure points more significantly than the lofting case in which the plume will be carried away from the near surface exposure points. If the atmosphere is unstable (Fig. 4.22e) the active convective forces tend to push the warmer air upward as the colder air sinks. This condition results in the formation of a looping shape of the plume. This case is identified as the looping plume condition.

The plume that exits a smoke stack is also characterized by its initial rise. The dynamic condition is characterized by buoyancy and exit momentum effects. The buoyancy force and upward acceleration can be written in terms of Newton's second law as,

$$F = V\left(\gamma_a - \gamma_s\right) = m_s a_b \tag{4.58}$$

where F is the buoyancy force $[MLT^{-2}]$, V is the air parcel volume $[L^3]$, γ_a and γ_s are the specific weight of the ambient atmosphere and the air parcel leaving the stack

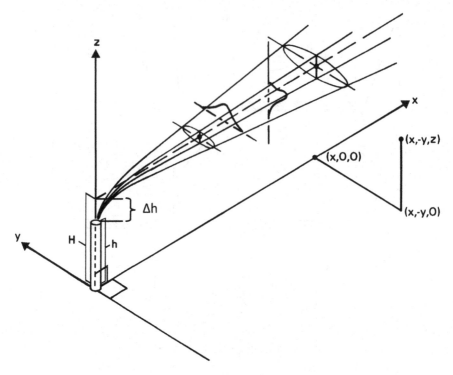

Fig. 4.23 The coordinate system of the Gaussian plume model (Adapted from Slade 1968)

respectively $(\gamma = \rho g)[ML^{-2}T^{-2}]$, m_s is the mass of air parcel exiting the smoke stack $[M]$ and a_b is the buoyancy acceleration of the air parcel $[LT^{-2}]$. Equation (4.58) can be rearranged to yield the buoyancy acceleration in terms of densities of the local atmosphere and the air parcel,

$$a_b = \frac{g(\rho_a - \rho_s)}{\rho_s} \tag{4.59}$$

At the stack exit elevation if we assume that the pressure of the air parcel exiting the smoke stack is the same as the pressure of the local atmosphere the ideal gas law, $\rho = P/R\widehat{T}$, can be introduced to define the buoyancy acceleration in terms of temperature differences.

$$a_b = \frac{g\left(\widehat{T}_s - \widehat{T}_a\right)}{\widehat{T}_a} \tag{4.60}$$

Thus, the initial upward plume acceleration is a function of the temperature difference between the local atmospheric conditions at the stack elevation. In this case higher air parcel temperatures will yield higher initial upward acceleration.

As the air plume rise the plume temperature will decrease due to mixing with the ambient atmosphere. After this initial buoyancy rise the plume spread in the downwind direction will depend on the local stability conditions depicted in Fig. 4.22.

Sutton was one of the first to derive the dispersion equation for estimating concentrations of pollutants in air parcels (Sutton 1932). Later, Cramer derived a diffusion equation which incorporated the standard deviation of the Gaussian distributions in both vertical and horizontal directions to associate the plume spread with the material in the plume (Cramer 1959). This appears to be the method of choice, and most diffusion experiments are now being reported in terms of the standard deviations of plume spread. For the equations that will be discussed in this section, the work-book of (Turner 1994) will serve as the primary reference and can be consulted if more detailed information is desired. The U.S. EPA publication (USEPA 1987) serves as the basis of atmospheric stability classifications and the calculation of relevant standard deviation parameters. The coordinate system used in these calculations is shown in Fig. 4.23.

Based on the governing equations described earlier, the analytical solution for the concentration of a chemical at a point (x, y, z) from a continuous point source discharging from a stack with an effective plume height H (m) can be given as,

$$C(x,y,z,H) = \frac{\Psi Q_s}{2\pi\sigma_y\sigma_z v} \exp\left(-\frac{1}{2}\left(\frac{y}{\sigma_y}\right)^2\right)$$

$$\times \left\{\exp\left(-\frac{1}{2}\left(\frac{z-H}{\sigma_z}\right)^2\right) + \exp\left(-\frac{1}{2}\left(\frac{z+H}{\sigma_z}\right)^2\right)\right\} \quad (4.61)$$

where C is the pollutant concentration (mg/l), Q_s is the stack emission rate (g/s), v is the wind velocity (m/s) at the stack elevation, σ_y and σ_z (m) are the standard deviations of the Gaussian distribution at location (x, y, z). In Eq. (4.61), Ψ is an exponential decay term used to account for transformation and degradation of the chemical.

$$\Psi = \exp\left(-\lambda\frac{x}{v}\right) \quad (4.62)$$

where λ is a first order decay rate (s^{-1}). In this solution, it is assumed that the plume has a Gaussian distribution in both the vertical and the horizontal directions with standard deviations of σ_y and σ_z. Standard deviations of the plume are important parameters that need to be evaluated. Obviously, they will vary greatly with the intensity of turbulence of the wind. These parameters are calculated in the ACTS software based on the stability criteria defined earlier (USEPA 1987) and in other references as indicated in this chapter.

A simple approach for estimating σ_y and σ_z, which requires no direct wind variability measurements, is the employment of atmospheric stability categories based on meteorological conditions. The commonly used Pasquill–Gifford stability categories (Gifford 1976) are developed from correlations found at a particular geographic location in Britain. The Pasquill–Gifford stability categories are given in Table 4.2 as a function of insolation (solar heat input) and wind speed. The categories are related to the dispersion parameters σ_y and σ_z as a function of down-wind distance and are incorporated in to the ACTS software. In this table, category A corresponds to conditions under which atmospheric mixing is augmented by instability during periods of intense sunlight due to solar heating of the ground surface and overlying air. Category D corresponds to an atmosphere of neutral stability, while categories E and F correspond to increasingly stable conditions associated with atmospheric inversion. There are significant limitations to using atmospheric stability categories. Errors may result if the user applies them to settings that differ in local topography or climatic conditions, (Hemond and Fechner-Levy 2000). The Pasquill–Gifford stability categories (Gifford 1976) can be used to choose the appropriate σ_y and σ_z as a function of the downwind distance from the source. As expected, at a given downwind distance x, the less stable categories correspond to more mixing, with an increase of one or more orders of magnitude in the actual vertical or horizontal width of the plume, (Hemond and Fechner-Levy 2000; Stern 1976).

The effects of higher wind speed may at first seem counterintuitive; higher wind speeds, which cause more atmospheric turbulence, shift the classification in the direction of higher stability categories that would seem to result in less mixing. This shift occurs because the Pasquill–Gifford stability categories are used to determine the width and height of a pollutant plume at a particular downwind distance. They are not used to estimate the Fickian mixing coefficient. Higher wind velocity may actually decrease the absolute amount of spreading a pollutant plume may undergo before reaching a fixed downwind distance. This spreading may override the effects of more intense mixing, as there is less time for mixing to occur, and therefore the latter effect may predominate, (Hemond and Fechner-Levy 2000; Stern et al. 1984).

To evaluate Eq. (4.61), the effective plume height H (m) first has to be calculated where $(H = h + \Delta h)$ and Δh is the plume rise. In the ACTS software, the effective height of a source is considered to be the sum of the physical height of the stack release point above the ground level plus any plume rise that might occur due to buoyancy effects. The rise of plumes above their initial point of release is a significant contributing factor to the reduction of ground level concentrations. Under common atmospheric conditions, for typical stacks, the plume rise will often be two or three times the physical height of the stack. Downwind ground level concentrations from a typical stack are often a quarter to a tenth of what they would be if there were no plume rise. The effective height can be associated with either final effective rise or transitional rise while the plume is still rising. The emphasis in the ACTS software is on the final rise and not the transition or partial rise that may occur immediately after the plume leaves the stack.

Over the past several decades, numerous plume rise formulas have been proposed. During the last decade, increased understanding of the physics of plume rise and analyses of the data available from generally larger sources, have led to the acceptance of formulas proposed by Briggs (1975). For the most part these formulas require only readily available input data and are used routinely in models approved for regulatory use.

Plumes rise due to momentum and buoyancy, both as discussed above. Wind velocity tends to bend the plume over in the downwind direction. Under calm, no wind conditions plumes would rise vertically. Because of high elevation of such plumes, the resulting ground level concentrations tend to be negligible. Rise under such conditions is generally of little interest to air dispersion modelers. In general, it is the shape and dimensions of bent over plumes that are of interest in air dispersion modeling. For the most part this interest is in the final effective height, which is largely empirical, as the resulting equations were developed based on observations.

Stack gases that are forcibly ejected vertically from a stack carry the momentum imparted to these gases upward into the air, where this momentum is eventually dissipated by friction and mixing with the ambient air. This vertical momentum is destroyed if a small roof (rain hat) is placed over the outlet. Also, discharges that are horizontal will not have any momentum plume rise. Further, if the density of the effluent is less than that of the ambient air, then the plume will rise due to buoyancy effects. As Eq. (4.60) indicates, excess temperature is almost always the cause of buoyancy. Therefore, buoyancy flux is a function of the difference between stack gas and ambient air temperature. Temperatures in excess of ambient on the order of about 10°C will usually result in buoyant plume rise higher than that due to momentum effects. The dissipation of buoyancy by the mixing of cooler ambient air with the effluent will usually proceed at a slow rate. Thus, the effect of buoyancy will persist on the order of 3–5 min for large power plant stacks when compared to the momentum effect, which dissipates in considerably less time. The actual dissipation of the temperature excess causing the buoyancy will depend upon the ambient turbulence structure at plume level. Upon entering the atmosphere, the buoyant plume will be acted upon by the wind, resulting in the plume's bending over eventuall to become horizontal, or nearly so. With stronger winds the bending of the plume will be more pronounced and the final plume rise will be less.

Buoyancy Flux Effect Rise: Several empirical equations have been formulated to determine the plume rise under buoyant conditions based on the analysis and observation of plume data (Hanna 1989; Turner 1994). First, it is necessary to determine the buoyancy flux based on Eq. (4.63),

$$F_b = \frac{gQ_H}{\pi C_p \rho_s \widehat{T}_s} = \frac{gQ_s(\widehat{T}_s - \widehat{T}_a)}{\pi \widehat{T}_s} = \frac{gv_s d_s^2(\widehat{T}_s - \widehat{T}_a)}{4\widehat{T}_s} \tag{4.63}$$

where F_b is the buoyancy flux (m^4/s^3), g is the gravitational acceleration (9.81 m/s^2), Q_H is the source heat release (Cal/s – 252 Cal = 1 BTU), C_p is the specific heat of air (0.24 Cal/g K), ρ_{atm} is the density of air (1,205 g/m^3 at mean sea level), Q_s is the volumetric flow rate at stack conditions (m^3/s), $\widehat{T}_a$ is the ambient air temperature (K), $\widehat{T}_s$ is the stack gas temperature (K), v_s is the stack gas exit velocity (m/s) and

d_s is the inside diameter of the stack exit (m). The form of the equation on the far right is commonly used to determine buoyancy flux from stacks.

Defining $(\partial\bar{\theta}/\partial z)$ as the potential temperature change with elevation, then for unstable and neutral conditions, when $(\partial\bar{\theta}/\partial z)$ is zero or negative, Eq. (4.64) gives the plume rise as a function of downwind distance, x. The potential temperature change can be approximated as follows $(\partial\bar{\theta}/\partial z = (\partial\widehat{T}/\partial z)_a + \Gamma)$, that is the change of temperature with elevation plus the adiabatic lapse rate. As discussed earlier, the adiabatic lapse rate is 0.0098 K/m, usually rounded to 0.01 K/m. For example, if $(\partial\widehat{T}/\partial z)_a$ is 0.014 K/m, then $(\partial\bar{\theta}/\partial z)$ would be (0.014 + 0.01 = 0.024 K/m). This would indicate a stable atmosphere. The plume rise equation for unstable and natural conditions is,

$$\Delta h = \frac{1.6F_b^{1/3}x^{2/3}}{v} \tag{4.64}$$

where F_b is the buoyancy flux, v is the wind speed at the top of the stack and x is the final rise distance. The height of final rise and the distance where it occurs are also of interest in making dispersion estimates. In many of the data sets used in formulating plume rise equations, the pollutant plume still appears to be rising at its farthest distance of measurement. Therefore, it has been difficult to estimate the height of final rise and its distance. The two equations that are in use depend on the value of the buoyancy flux.

For final rise and the distance to final rise under unstable conditions, the following equation can be used,

For $F_b < 55$,

$$H = h + 21.425\frac{F_b^{3/4}}{v} \tag{4.65}$$

$$x_f = 0.049F_b^{5/8} \tag{4.66}$$

For $F_b \geqslant 55$

$$H = h + 38.71\frac{F_b^{3/5}}{v} \tag{4.67}$$

$$x_f = 0.119F_b^{2/5} \tag{4.68}$$

The coefficients 0.049 and 0.119 in Eqs. (4.66) and (4.68) will give the distance to final rise in km. For x, in meters, the coefficients are 49 and 119 respectively.

For stable conditions, positive $(\partial\bar{\theta}/\partial z)$, the stability parameter, s must first be evaluated. For this purpose the following equation may be used,

$$s = g\left(\partial\bar{\theta}/\partial z\right)/\hat{T} \tag{4.69}$$

Then, the height of the final rise and the distance to final rise can be calculated from,

$$H = h + 2.6\left(\frac{F_b}{vs}\right)^{1/3} \tag{4.70}$$

$$x_f = 0.00207vs^{-1/2} \tag{4.71}$$

In the equations above F_b is the buoyancy flux, h is the physical stack height (m), v is the wind speed at the top of the stack, x_f is the final rise distance (km or m as indicated) and s is the stability parameter.

Momentum Flux Effect Rise: The momentum flux of a gas pocket exiting a stack can be given as,

$$F_m = \frac{Q_s v_s}{\pi} = \frac{d^2 v_s^2}{4} \tag{4.72}$$

where F_m is the momentum flux (m⁴/s²), Q_s is the volumetric flow rate (m³/s), which can be calculated as $\left(\pi d_s^2 v_s/4\right)$, v_s is the gas exit velocity (m/s) and d_s is the stack inside diameter of the top of the stack (m). This flux can be adjusted for the density of the stack gases relative to ambient air, which can arise through molecular weight or temperature differences,

$$F_m = \frac{d_s^2 v_s^2 \rho_s}{4\rho_a} = \frac{d_s^2 v_s^2 \hat{T}_a}{4\hat{T}_s} \tag{4.73}$$

where ρ_s is the density of stack gas effluent, ρ_a is the density of ambient air. Based on these parameters, the final plume rise due to momentum flux, which is expected to take place very quickly after the release, can be computed using Eqs. (4.74) and (4.75),

For unstable conditions,

$$H = h + 3d_s\frac{v_s}{v} \tag{4.74}$$

For stable conditions,

$$H = h + 1.5\left(\frac{F_m}{v}\right)^{1/3}s^{-1/6} \tag{4.75}$$

Given the discussion above, we also need to identify conditions in which the buoyancy flux or the momentum flux effects will be dominant in determining the effective stack height. This choice is a function of the stack plume temperature. For any plume for a specified stability condition there is a critical temperature, below which the plume rise is momentum flux dominated and above which it is buoyancy flux dominated. If the stack temperature is close to the critical temperature, both momentum and buoyancy fluxes will affect the plume rise, although one of the approaches discussed above will be used to calculate the plume rise. Thus, the plume rise for discharges at or near this critical temperature will underestimate the rise. The empirical formulas given above for the buoyancy flux plume rise are based on observed stack plume temperatures in the range of 150–300 K above ambient temperatures. Similarly, momentum flux plume rise formulas are based on plumes at ambient temperature. As a general rule of thumb, it is agreed that if the stack gas temperature is on the order of 10°C higher than the ambient temperature then the buoyancy flux effects will dominate the plume rise.

In the analytical solution given by Eq. (4.61), it is assumed that the emissions are at a uniform rate Q_s, and that there is no deposition or reaction with the earth's surface conditions. As is pointed out in (Turner 1994), this equation is valid where the diffusion effects in the x-direction can be neglected. This is appropriate if the release is continuous or if the duration of the release is equal to or greater than the travel time (x/v). Other forms of this equation, as used in the ACTS software, can be given as follows:

For concentrations at ground level $(z = 0)$, one may use,

$$C(x, y, 0; H) = \frac{\Psi Q_s}{\pi \sigma_y \sigma_z v} \exp\left(-\frac{1}{2}\left(\frac{y}{\sigma_y}\right)^2\right) \exp\left(-\frac{1}{2}\left(\frac{H}{\sigma_z}\right)^2\right) \qquad (4.76)$$

If the concentration distribution is to be calculated along the centerline of the plume$(y = 0)$, one may use,

$$C(x, 0, 0; H) = \frac{\Psi Q_s}{\pi \sigma_y \sigma_z v} \exp\left(-\frac{1}{2}\left(\frac{H}{\sigma_z}\right)^2\right) \qquad (4.77)$$

For a ground level source with no effective plume rise $(H = 0)$, Eq. (4.77) takes the form,

$$C(x, 0, 0; 0) = \frac{\Psi Q_s}{\pi \sigma_y \sigma_z v} \qquad (4.78)$$

There are cases when the source of pollution can be described more appropriately as a line source rather than a point source. For example, the time averaged emissions over an hour from a roadway, or time averaged emissions over an hour from all aircraft using a particular taxiway may be identified as such sources. Steady state analytic solutions for infinite line and finite line source problems, in which the wind is blowing perpendicular across the line can be given as,

$$C(x, 0, 0, H) = \frac{\Psi q}{\sqrt{2\pi}\sigma_z v} \exp\left(-\frac{1}{2}\left(\frac{H}{\sigma_z}\right)^2\right) \qquad (4.79)$$

for the infinite line source, where q is the source strength (g/s m) is different from the point source, in that it is defined as mass per unit time per unit length of line source. For the finite line source the Eq. (4.79) above can be integrated at the source over a finite length to yield the solution for this case.

The analytical solutions discussed above are provided in the ACTS software as the primary steady state models to evaluate concentration distributions in the air pathway. The ACTS software uses a sector-averaged form of the Gaussian dispersion model described above, based on Hanna et al. (1982); Stern (1976).

4.4.9.2 Unsteady State Air Dispersion Models

Up to this point we have been considering only the continuous release cases. It is possible to develop an equation that will allow us to estimate the concentration from an instantaneous or puff release. Of course, one should understand that no source is ever instantaneous. A release always takes some time, but for purposes of modeling, releases that take place over two orders of magnitude shorter than plume migration time can be considered to be instantaneous. For this case the analytical solution can be given as,

$$C = \frac{Q_T}{(2\pi)^{3/2}\sigma_x\sigma_y\sigma_z} \exp\left(-\frac{1}{2}\left(\frac{x - vt}{\sigma_x}\right)^2 - \frac{1}{2}\left(\frac{y}{\sigma_y}\right)^2\right)$$
$$\times \left\{\exp\left(-\frac{1}{2}\left(\frac{z - H}{\sigma_z}\right)^2\right) + \exp\left(-\frac{1}{2}\left(\frac{z + H}{\sigma_z}\right)^2\right)\right\} \qquad (4.80)$$

where all parameters are as described earlier. In this case, instead of an emission rate, one must use the mass of the total release, Q_T, usually expressed in kilograms. In the ACTS software, if the emission rate is transferred from emission models, then Q_T will be estimated as an emission over a short emission period of time. This period will be calculated based on the maximum and minimum time limits specified by the user. Otherwise, the user must enter Q_T in units of kilograms.

In the solution given above, the first exponential term accounts for how far the downwind distance, x, is from the center of the puff source which is at a downwind distance, vt, at t seconds after the release. Obviously, maximum concentrations occur at the center of the puff, that is when $(x = vt)$. It is assumed that the downwind spreading can be estimated by the normal distribution using the dispersion parameter σ_x.

Previously, we considered the time average concentrations from a continuously emitted source and used dispersion parameters that would be used to simulate the spreading for that time average. These included some degree of horizontal meander due to minute-to-minute wind direction changes. For a puff release, minute-to-minute wind direction changes will not aid in dispersion but will only change the direction of transport of the puff affecting the trajectory or path of travel of the puff. Consequentially, the parameters we have been using are not appropriate for puff releases. For this case Slade (1968) suggested the use of a set of parameters for short-term releases based on analysis of photographs of plume segments released over 30 s. These are much more close to puff releases than the time averaged measurements that form the basis for the Pasquill–Gifford dispersion parameters. In the ACTS software Slade parameters are used in the equation given above.

4.4.10   Air Dispersion Model Assumptions and Limitations

Gaussian Dispersion Model Menu Options: Using this model, the user has the option to calculate emission rates for several chemicals. This may be accomplished by using the procedure described in Section 4.4.1. The Monte Carlo analysis option is also available for this model. Using the "Monte Carlo" menu option on the menu bar, the user may choose to conduct an uncertainty analysis for most of the parameters of this model using standard probability distributions imbedded in the ACTS software and the Monte Carlo methods. Details of implementing this calculation sequence are described in Chapter 7. The menu options on this window are the same as the other model input window options (Fig. 4.24). The functions of these menus are described in Appendix 3.

Assumptions and Limitations: The following lists the assumptions and limitations of the Gaussian-dispersion model employed in the ACTS software.

    i. The chemical emissions at the source are steady and continuous except for the puff source model.

    ii. In the continuous source model, it is assumed that a point source emission is present at the source, i.e. at a relatively small source area. Thus, the continuous source Gaussian model is more appropriate when the distance to the receptor is large relative to the size of the source.

    iii. The distribution of chemicals within the plume is Gaussian in the vertical and crosswind directions.

    iv. Longitudinal (downwind) dispersion is negligible.

    v. Wind speed is steady in a constant direction; short-term fluctuations in wind are not accounted for.

    vi. Atmospheric dispersion can be characterized by six stability classes that are used to estimate the dispersivity values.

    vii. No deposition of chemicals or particles occurs during transport.

    viii. The model assumes a flat terrain.

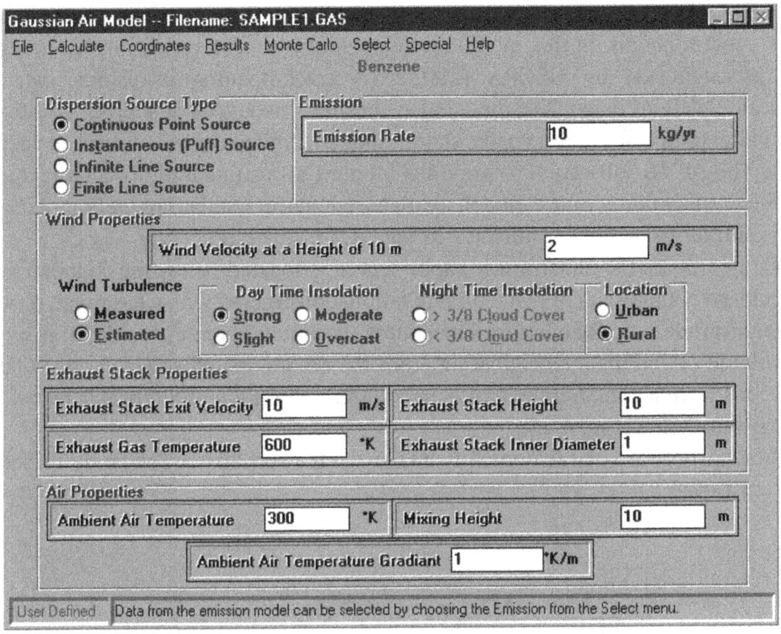

Fig. 4.24 Gaussian plume dispersion model input window

4.4.11   Indoor Air Dispersion Models

Volatilization from contaminated soil or from groundwater contaminant plumes and the subsequent transport of these volatile vapors into building or family dwellings constitute a potential inhalation exposure pathway. Considerable attention has been paid to adverse health and safety effects from this potential indoor inhalation exposure pathway since Johnson and Ettinger's proposed heuristic model for estimating the intrusion rate of subsurface contaminant vapor into buildings (Johnson and Ettinger 1991). Based on recent research results and investigation into this exposure pathway and its related health problems, EPA has developed spreadsheets for evaluating subsurface vapor intrusion into buildings (USEPA 2003) using Microsoft Excel. The spreadsheet may be used to estimate the concentrations in building basements originating from subsurface contaminant vapor and to assess the risk exposure to the contaminants.

The ACTS software incorporates the indoor vapor intrusion model in the air pathway family of models and provides a user-friendly interface for using this model to estimate the contaminant concentration from subsurface vapor intrusion into the building, similar to the standard computational platform utilized for other pathways. The ACTS software considers four types of contamination sources: (i) soil contamination without the presence of a residual phase; (ii) soil contamination with residual phase; (iii) soil gas contamination; and, (iv) emissions from ground-water contamination. In this module two types of intrusion possibility are considered as suggested by USEPA (2003): (i) crack/opening intrusion; and, (ii) permeable wall intrusion. The model also considers two types of source conditions: (i) infinite source solution (steady state); and, (ii) finite source solution (unsteady state). The ACTS software also provides Monte Carlo simulation for analyzing the parameter uncertainty of the model, again based on the standard modeling platform that is similar to the other pathways of the ACTS software.

We assume that a contaminant vapor source is located below the foundation of an enclosed commercial or residential dwelling constructed with a basement or with a slab-on-grade type foundation. The source of contamination considered is either the volatile contaminants originating from the contaminated soil zone or volatile emissions from a contaminant that is in solution within the groundwater below the groundwater table (Fig. 4.25).

The transport of volatile contaminants through a soil matrix can be described in terms of Eq. (4.81),

$$\frac{\partial\left(\sum_i \varepsilon_i C_i\right)}{\partial t} + \sum_i u_i \nabla C_i = \sum_i \nabla D_{i,eff} \nabla C_i + \sum_i R_i \qquad (4.81)$$

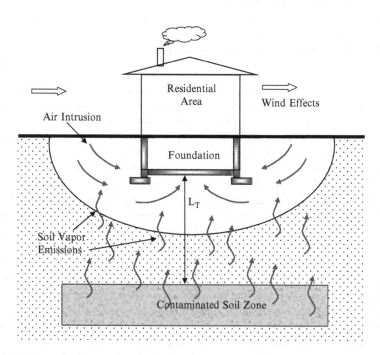

Fig. 4.25 Schematic for vapor pathway intrusion into a building due to soil contamination (Adapted from USEPA 2003)

This model is again based on the advection–diffusion equation that is described in Chapter 3, where the subscript i is used to represent the phase under consideration, t is time, ε_i is the volume fraction of the phase i, C_i is the concentration of contaminant in phase i, u_i is the Darcy velocity vector associated with phase i, $D_{i,eff}$ is the effective diffusion coefficient of the contaminant in phase i, and R_i is the reaction term in the subsurface in phase i.

Under steady state conditions the solution for the attenuation coefficient, α, for vapor intrusion through foundation cracks or openings of a building is given by (Johnson and Ettinger 1991),

$$\alpha = \frac{\left(\frac{D_T^{eff} A_B}{Q_{building} L_T}\right) \exp\left(\frac{Q_{soil} L_{crack}}{D_{crack} A_{crack}}\right)}{\exp\left(\frac{Q_{soil} L_{crack}}{D_{crack} A_{crack}}\right) + \left(\frac{D_T^{eff} A_B}{Q_{building} L_T}\right) + \left(\frac{D_T^{eff} A_B}{Q_{soil} L_T}\right)\left(\exp\left(\frac{Q_{soil} L_{crack}}{D_{crack} A_{crack}}\right) - 1\right)} \tag{4.82}$$

where α is the steady-state attenuation coefficient (dimensionless), D_T^{eff} is the total overall effective diffusion coefficient (cm^2/s), A_B is area of the enclosed space below grade (cm^2), $Q_{building}$ is the building ventilation rate (cm^3/s), L_T is the source-building separation (cm), Q_{soil} is the volumetric flow rate of soil gas into the enclosed space (cm^3/s), L_{crack} is the enclosed space foundation or slab thickness (cm), A_{crack} is the estimated total area of the crack (cm^2) and D_{crack} is the effective diffusion coefficient through the cracks (cm^2/s).

For vapor intrusion into the building through permeable below-grade walls, Eq. (4.82) given above can be extended to,

$$\alpha = \frac{\left(\frac{D_T^{eff} A_B}{Q_{building} L_T}\right) \exp\left(\frac{Q_{soil} L_F}{D_F A_B}\right)}{\exp\left(\frac{Q_{soil} L_F}{D_F A_B}\right) + \left(\frac{D_T^{eff} A_B}{Q_{building} L_T}\right) + \left(\frac{D_T^{eff} A_B}{Q_{soil} L_T}\right)\left(\exp\left(\frac{Q_{soil} L_F}{D_F A_B}\right) - 1\right)} \tag{4.83}$$

where D_F is the effective diffusion coefficient through the porous foundation floor and walls (cm^2/s) and L_F is the average foundation/wall thickness (cm).

Based on this attenuation factor, the steady state vapor concentration in the building can be estimated by,

$$C_{building} = \alpha C_{source} \tag{4.84}$$

where $C_{building}$ is the estimated concentration in the building (g/cm^3-v) and C_{source} is the source concentration in the soil (g/cm^3-v). For a finite source duration a time average solution is also provided (Johnson and Ettinger 1991). For this case the time average solution for a finite duration source attenuation coefficient, $\alpha_{\Delta t}$ is given by,

$$\alpha_{\Delta t} = \left(\frac{\rho_b C_R \Delta H_c A_B}{Q_{building} C_{source} \Delta t}\right)\left(\frac{L_T^0}{\Delta H_c}\right)\left((\beta^2 + 2\zeta \Delta t)^{1/2} - \beta\right) \tag{4.85}$$

where $\alpha_{\Delta t}$ is the time average finite source attenuation coefficient (dimensionless), ρ_b is the dry bulk soil density at the source zone of contamination (g/cm^3), C_R is the initial soil concentration (g/g), ΔH_c is the initial thickness of contamination (cm), Δt is the exposure interval (s), L_T^0 is the source-building separation distance at time zero (cm) and,

$$\beta = \left(\frac{D_T^{eff} A_B}{L_T^0 Q_{soil}}\right)\left(1 - \exp\left(-\frac{Q_{soil} L_{crack}}{D_{crack} A_{crack}}\right)\right) + 1 \tag{4.86}$$

$$\zeta = \frac{D_T^{eff} C_{source}}{\left(L_T^0\right)^2 \rho_b C_R} \tag{4.87}$$

Then the time-averaged vapor concentration in the building is estimated by,

$$C_{building} = \alpha_{\Delta t} C_{source} \tag{4.88}$$

If the time for source depletion is less than the exposure interval, then the time-averaged building vapor concentration may be estimated by,

$$C_{building} = \frac{\rho_b C_R \Delta H_c A_B}{Q_{building} \Delta t} \tag{4.89}$$

The time for source depletion is estimated by,

$$\Delta t_D = \frac{\left(\Delta H_c / L_T^0 + \beta\right)^2 - \beta^2}{2\zeta} \tag{4.90}$$

In the following sections, the computation of the parameters used in the equations above is described.

4.4.12 Vapor Concentration at the Contamination Source

In soil contamination without a residual phase, the vapor concentration at the contamination source, C_{source}, may be estimated by,

$$C_{source} = \frac{H'_{TS} C_R \rho_b}{\theta_w + K_d \rho_b + H'_{TS} \theta_a} \tag{4.91}$$

where C_{source} is the vapor concentration at the source of contamination (g/cm^3-v), H'_{TS} is Henry's law constant at the system temperature, which in this case is the soil temperature (dimensionless), C_R is the initial soil concentration (g/g), ρ_b is the dry soil bulk density (g/cm^3), θ_w is the water-filled soil porosity (cm^3/cm^3), K_d is the soil–water partition coefficient (cm^3/g). This coefficient is obtained using the organic carbon partition coefficient, i.e., $K_d = K_{oc}f_{oc}$, θ_a is the air-filled soil porosity (cm^3/cm^3), K_{oc} is the soil organic carbon partition coefficient (cm^3/g), f_{oc} is the soil organic carbon weight fraction.

For the groundwater contamination case, the vapor concentration at the source of contamination C_{source} is estimated by,

$$C_{source} = H'_{TS}C_w \tag{4.92}$$

where H'_{TS} is Henry's law constant at the system temperature, which in this case is the groundwater temperature (dimensionless), C_w is the groundwater contaminant concentration (g/cm^3-w). In both cases, Henry's law constant at the system (soil/groundwater) temperature is estimated by,

$$H'_{TS} = \frac{\exp\left(-\dfrac{\Delta H_{v,TS}}{R_c}\left(\dfrac{1}{\widehat{T}_S} - \dfrac{1}{\widehat{T}_R}\right)\right)H_R}{R\widehat{T}_S} \tag{4.93}$$

where $\Delta H_{v,TS}$ is the enthalpy of vaporization at the system temperature (cal/mol), $\widehat{T}_S$ is the system temperature (K), $\widehat{T}_R$ is the Henry's law constant reference temperature (K), H_R is the Henry's law constant at the reference temperature (atm-m^3/mol), R_c is the engineering gas constant ($R_c = 1.9872$ cal/mol - K) and R is the ideal gas constant $\left(R = 8.205\ E^{-5} atm\ \text{m}^3/\text{mol K}\right)$.

The enthalpy of vaporization at the system temperature can be calculated by,

$$\Delta H_{v,TS} = \Delta H_{v,b}\left[\frac{(1-\widehat{T}_S/\widehat{T}_C)}{(1-\widehat{T}_B/\widehat{T}_C)}\right]^n \tag{4.94}$$

where $\Delta H_{v,b}$ is the enthalpy of vaporization at the normal boiling temperature (cal/mol), $\widehat{T}_S$ is system temperature (K), $\widehat{T}_C$ is the critical temperature (K), $\widehat{T}_B$ is the normal boiling point (K) and n is a constant (dimensionless) as can be determined according to Table 4.5.

For the soil gas contamination, the vapor concentration at the source of contamination C_{source} may be directly estimated by a measurement of the soil gas concentration beneath the building floor. Soil contamination with residual phase vapor concentration at the source of contamination is more complicated as discussed in Section 4.4.16.

Table 4.5 Values of exponent n used in Eq. (4.94) given as a function of $\widehat{T}_B/\widehat{T}_C$	$\widehat{T}_B/\widehat{T}_C$	N
	< 0.57	0.30
	$0.57 - 0.71$	$0.74\ \widehat{T}_B/\widehat{T}_C - 0.116$
	>0.71	0.41

4.4.13 Diffusion Through the Capillary Zone

For groundwater contamination, a saturated capillary zone above the water table may exist whereby groundwater is held within the soil pores at less than atmospheric pressures. The effective diffusion coefficient is calculated by lumping the gas-entry and aqueous-phase together, and the water-filled soil porosity in the capillary zone is estimated by,

$$\theta_{w,cz} = \theta_r + \frac{\theta_s - \theta_r}{\left(1 + (\alpha_l h)^N\right)^M} \tag{4.95}$$

where $\theta_{w,cz}$ is the water-filled porosity in the capillary zone (cm^3/cm^3), θ_r is the residual soil water content (cm^3/cm^3), θ_s is the saturated soil water content (cm^3/cm^3), α_l is the point of inflection in the water retention curve where $(d\theta_w/dh)$ is maximum (cm^{-1}), h is the air-entry pressure head (cm), N is the van Genuchten shape parameter (dimensionless), $(M = 1 - (1/N))$. These parameters for a specified soil type can be found in Table 4.6.

The total effective diffusion coefficient across the capillary zone may then be calculated by,

$$D_{cz}^{eff} = D_a(\theta_{a,cz}^{3.33}/n_{cz}^2) + (D_w/H_{TS}')(\theta_{w,cz}^{3.33}/n_{cz}^2) \tag{4.96}$$

in which D_{cz}^{eff} is the effective diffusion coefficient across the capillary zone (cm^2/s), D_a is diffusivity in air (cm^2/s), D_w is diffusivity in water (cm^2/s), $\theta_{a,cz}$ is the soil air-filled porosity in the capillary zone (cm^3/cm^3), $\theta_{w,cz}$ is the soil water-filled porosity in the capillary zone (cm^3/cm^3) and n_{cz} is the total soil porosity in the capillary zone (cm^3/cm^3).

Table 4.6 van Genuchten soil water retention parameters

Soil type	Saturated water content θ_S	Residual water content θ_R	Van Genuchten parameters		
			α_l (cm^{-1})	N	M
Clay	0.459	0.098	0.01496	1.253	0.2019
Clay loam	0.442	0.079	0.01581	1.416	0.2938
Loam	0.399	0.061	0.01112	1.472	0.3207
Loamy sand	0.039	0.049	0.03475	1.746	0.4273
Silt	0.489	0.050	0.00658	1.679	0.4044
Silty loam	0.439	0.065	0.00506	1.663	0.3987
Silty clay	0.481	0.111	0.01622	1.321	0.2430
Silty clay loam	0.482	0.090	0.00839	1.521	0.3425
Sand	0.375	0.053	0.03524	3.177	0.6852
Sandy clay	0.385	0.117	0.03342	1.208	0.1722
Sandy clay loam	0.384	0.063	0.02109	1.330	0.2481
Sandy loam	0.387	0.039	0.02667	1.449	0.3099

Table 4.7 Centroid compositions, mean particle diameters and dry bulk density of soils

Soil texture	% Clay	% Silt	% Sand	Arithmetic mean particle diameter (cm)	Dry bulk density (g/cm^3)
Clay	64.83	16.55	18.62	0.0092	1.43
Clay loam	33.50	34.00	32.50	0.016	1.48
Loam	18.83	41.01	40.16	0.020	1.59
Loamy sand	6.25	11.25	82.50	0.040	1.62
Silt	6.00	87.00	7.00	0.0046	1.35
Silty loam	12.57	65.69	21.74	0.011	1.49
Silty clay	46.67	46.67	6.66	0.0039	1.38
Silty clay loam	33.50	56.50	10.00	0.0056	1.63
Sand	3.33	5.00	91.67	0.044	1.66
Sandy clay	41.67	6.67	51.66	0.025	1.63
Sandy clay loam	26.73	12.56	60.71	0.029	1.63
Sandy loam	10.81	27.22	61.97	0.030	1.62

The mean rise of the capillary zone may be estimated by,

$$L_{cz} = \frac{2\alpha_2 cos(\lambda)}{\rho_w g R_{int}} \tag{4.97}$$

where L_{cz} is the mean rise of the capillary zone (cm), α_2 is the surface tension of water (g/s) ($\alpha_2 = 73$), λ is the angle of the water meniscus within the capillary tube in degrees and is assumed to be $0°$, ρ_w is the density of water (g/cm^3), g is the gravitational acceleration (cm/s^2), R_{int} is the mean interparticle pore radius (cm) ($R_{int} = 0.2D$) and D is the mean particle diameter (cm).

If the groundwater temperature is between 5°C and 25°C, then the mean rise of the capillary zone may be used as,

$$L_{cz} = \frac{0.15}{R_{int}} \tag{4.98}$$

The mean particle diameter for specified soil types can be found in Table 4.7.

4.4.14 Diffusion Through the Unsaturated Zone

The effective diffusion coefficient within the layered unsaturated zone may be estimated by,

$$D_i^{eff} = D_a(\theta_{a,i}^{3.33}/n_i^2) + (D_w/H'_{TS})(\theta_{w,i}^{3.33}/n_i^2) \tag{4.99}$$

where D_i^{eff} is the effective diffusion coefficient across soil layer i (cm^2/s), D_a is the diffusivity in air (cm^2/s), D_w is the diffusivity in water (cm^2/s), $\theta_{a,i}$ is the air-filled soil porosity of layer i (cm^3/cm^3), $\theta_{w,i}$ is the water-filled soil porosity of layer i (cm^3/cm^3) and n_i is the total soil porosity of the layer i (cm^3/cm^3).

Given the definitions above, the overall effective diffusion coefficient for systems composed of n distinct soil layers between the source of contamination and the enclosed foundation can be estimated as,

$$D_T^{eff} = \frac{L_T}{\sum_{i=1}^{n} L_i / D_i^{eff}} \tag{4.100}$$

where D_T^{eff} is the overall effective diffusion coefficient (cm^2/s), L_i is the thickness of the soil layer i (cm), and L_T is the distance between the source of contamination and the bottom of the foundation (cm) (Fig. 4.25).

4.4.15 Building Ventilation Rate and Volumetric Flow Rate

The building ventilation rate may be calculated by,

$$Q_{building} = \frac{(L_B W_B H_B E_R)}{3600} \tag{4.101}$$

where L_B is the length of the building (cm), W_B is the width of the building (cm), H_B is the height of the building (cm) and E_R is the air exchange rate (h^{-1}).

The volumetric flow rate of soil gas entering the building for vapor intrusion from the foundation cracks and openings can be estimated by,

$$Q_{soil} = \frac{2\pi \Delta P k_v X_{crack}}{\mu \ln(2Z_{crack}/r_{crack})} \tag{4.102}$$

where ΔP is the pressure differential between the soil surface and the enclosed space (g/cm-s^2), k_v is the soil vapor permeability (cm^2), X_{crack} is the floor-wall seam perimeter (cm), μ is the viscosity of air (g/cm-s), Z_{crack} is the crack depth below grade (cm), and r_{crack} is the equivalent crack radius (cm) as estimated by,

$$r_{crack} = \eta(A_B/X_{crack}) \tag{4.103}$$

and $(\eta = A_{crack}/A_B)$ where $(0 \leqslant \eta \leqslant 1)$. The volumetric flow rate of soil gas entering the building for vapor intrusion from the permeable foundation and walls is specified by users.

4.4.16 Soil Contamination with a Residual Phase

A residual phase mixture occurs when the sorbed phase, aqueous phase and vapor phase of each chemical reaches saturation in the soil , which results in residual phase (or NAPL phase or solid). When a residual phase is present, the vapor concentration is independent of the soil concentration but proportional to the mole fraction of the individual component of the residual phase mixture. In this case, the equilibrium vapor concentration must be calculated numerically for a series of time steps. For each time-step, the mass of each constituent that is volatilized is calculated using Raoult's law and the appropriate mole fraction. At the end of each time-step, the total mass lost is subtracted from the initial mass and the mole fraction is recomputed for the next time-step. The computational steps for this case can be given as follows

- Initially the user-defined initial soil concentration of each component in the mixture is checked to see if a residual phase is present using Eq. (4.104),

$$\alpha_i y_i = \frac{M_i}{\left[P_i^v(\widehat{T}s)\theta_a V / R\widehat{T}s + M^{H_2O}/\alpha_i + (K_{d,i}M_{soil}/\alpha_i MW^{H_2O})\delta(M^{H_2O}) \right]} \quad (4.104)$$

in which M_i is the initial moles of component i in the soil (mol), $P_i^v\left(\widehat{T}s\right)$ is the vapor pressure of i at average soil temperature (atm), θ_a is the air-filled soil (cm³/cm³), V is the volume of contaminated soil (cm³), R is the ideal gas constant ($R = 8.205$ E - 5 atm - m³/mol − K), $\widehat{T}s$ is the average soil temperature (K), M^{H2O} is the total moles of contaminant in the soil moisture in the dissolved phase (mol), α_i is the activity coefficient of the ith soil layer in water (dimensionless), $K_{d,i}$ is the soil–water partition coefficient of i the (cm³/g), M_{soil} is the total mass of contaminated soil (g), MW^{H2O} is the molecular weight of water (18 g/mol) and,

$$\delta(M^{H_2O}) = \begin{cases} 1 & if \ M^{H_2O} > 0 \\ 0 & if \ M^{H_2O} = 0 \end{cases} \quad (4.105)$$

If $\sum_{i=1}^{n} \alpha_i y_i < 1$, the mixture does not contain a residual phase and the models are not applicable. Otherwise, the mole fraction of each component (x_i) is determined by iteratively solving the following equations:

$$\alpha_i y_i = \frac{M_i}{\left[P_i^v(\widehat{T}s)\theta_a V / R\widehat{T}s + M^{HC} + M^{H_2O}/\alpha_i + (K_{d,i}M_{soil}/\alpha_i MW^{H_2O})\delta(M^{H_2O}) \right]}$$

$$(4.106)$$

$$x_i = \frac{M_i^{HC}}{M^{HC}} \quad (4.107)$$

$$\sum x_i = 1 \quad (4.108)$$

where M_i^{HC} is the number of moles of component i in residual phase, M^{HC} is the number of moles of all components in residual phase.

At the initial time-step, the equilibrium vapor concentration at the source of emission is calculated by Raoult's law,

$$C_{source} = \frac{x_i P_i^v(\widehat{T}s) MW_i}{R\widehat{T}s} \qquad (4.109)$$

- At the beginning of each time-step, the number of moles of each chemical remaining in the soil from the previous time-step is again checked to see if a residual phase is present. When a residual phase is no longer present, the equilibrium vapor concentration at the source of emission is calculated by

$$C_{source} = \frac{\alpha_i y_i P_i^v(\widehat{T}s) MW_i}{R\widehat{T}s} \qquad (4.110)$$

- Ancillary calculations: The activity coefficient of component i in water for compounds that are liquid or solid at average temperature is estimated by,

$$\alpha_i = \frac{1}{y_i} = (55.55 \, \text{mol/L}) MW_i / S_i \qquad (4.111)$$

in which S_i is the solubility of the component i (g/L).

For gases at average room temperature, the activity coefficient can be estimated by

$$\alpha_i = \frac{1}{y_i} = (55.55 \, moles/L)(MW_i(1\,atm)/S_i P_i^v(\widehat{T}s)) \qquad (4.112)$$

The vapor pressure can be estimated by,

$$P_i^v\left(\widehat{T}s\right) = P^v\left(\widehat{T}_R\right) \exp\left(\left(\frac{\widehat{T}_B \widehat{T}_R}{\widehat{T}_B - \widehat{T}_R}\right)\left(\frac{1}{\widehat{T}_B} - \frac{1}{\widehat{T}_R}\right) \ln\left(\frac{P^v\left(\widehat{T}_R\right)}{P_B}\right)\right) \qquad (4.113)$$

in which $P^v\left(\widehat{T}s\right)$ is the vapor pressure at the desired temperature $\widehat{T}s$ (atm), $P^v\left(\widehat{T}_R\right)$ is the vapor pressure at the reference temperature $\widehat{T}_R$ (atm), $\widehat{T}_B$ is the boiling point temperature (K), $\widehat{T}_R$ is the vapor pressure reference temperature (K), $\widehat{T}s$ is the desired temperature (K), and P_B is the normal boiling point pressure at 1 atm.

- Based on these equations, the steady-state attenuation coefficient α is calculated using Eq. (4.82) or (4.83) and the building concentration for each component in the mixture can estimated by Eq. (4.84).

4.4.17   Parameter Uncertainty Analysis

The parameters used in these models have significant uncertainty which results in uncertainty of the building vapor concentration estimates obtained from the models given above. A Monte Carlo simulation may be used to analyze the effect of parameter uncertainty on the building concentration. A Monte Carlo simulation randomly generates a series of data based on a specified interval and probability distribution of the parameters and calculates the building concentration using the above model for each set of parameter values. It then completes statistics analysis to obtain the statistics for the building vapor concentration, calculating values such as its mean, variance and probability distribution. Comparing these statistics with the statistics of the parameters, it may be seen how the uncertainty of the parameters propagates to the building vapor concentration through the soil pathway. The practical ranges and default values for some of the parameters that are used in the equations above are given in Table 4.8.

Indoor Vapor Intrusion Model Menu Options: For the case without a residual phase, the indoor vapor intrusion window is shown in Fig. 4.26, which includes three folders for model parameters: (i) Contaminant Source parameter folder; (ii) Soil parameter folder; and (iii) Crack and Building parameter folder. For the soil contamination with residual phase, the indoor vapor intrusion window is shown in Fig. 4.27, which includes four model parameter folders: (i) Contaminant Source parameter folder; (ii) Soil parameter folder; (iii) Crack and Building parameter folder; and, (iv) the Results folder in which the results of the calculation are given.

Indoor Vapor Intrusion Model Assumptions and Limitations: The Johnson–Ettinger Model (JEM) was developed for use as a screening level model and is consequently based on a number of simplifying assumptions regarding contaminant distribution and occurrence, subsurface characteristics, transport mechanisms, and building construction. The assumptions of the JEM as implemented in EPA's spreadsheet version are listed in Table G-1 along with the implications of and limitations posed by these assumptions. Also provided in the table is an assessment of the likelihood that the assumptions can be verified through field evaluation.

Table 4.8 Estimations of parameters used in indoor vapor intrusion model

Parameter	Practical range of values	Default value
Soil water-filled porosity (θ_w)	0.02–0.43 cm^3/cm^3	0.3 cm^3/cm^3
Soil vapor permeability (k_v)	10^{-6}–10^{-12} cm	10^{-8} cm
Soil-building pressure difference (ΔP)	0–20 P_a	4 P_a
Floor-wall seam gap (w)	0.05–1.0 cm	0.1 cm
Soil organic carbon fraction (f_{oc})	0.001–0.006	0.002
Soil total porosity (n)	0.34–0.54 cm^3/cm^3	0.43 cm^3/cm^3
Soil dry bulk density (ρ_b)	1.25–1.75 g/ cm^3	1.5 g/cm^3
Building footprint area	80–200 + m^2	100 m^2
Building mixing height – basement scenario	2.44–4.88 m	3.66 m
Building mixing height – slab-on-grade scenario	2.13–3.15 m	2.44 m
Indoor air exchange rate (E_R)	0.18–1.26 h^{-1}	0.25 h^{-1}

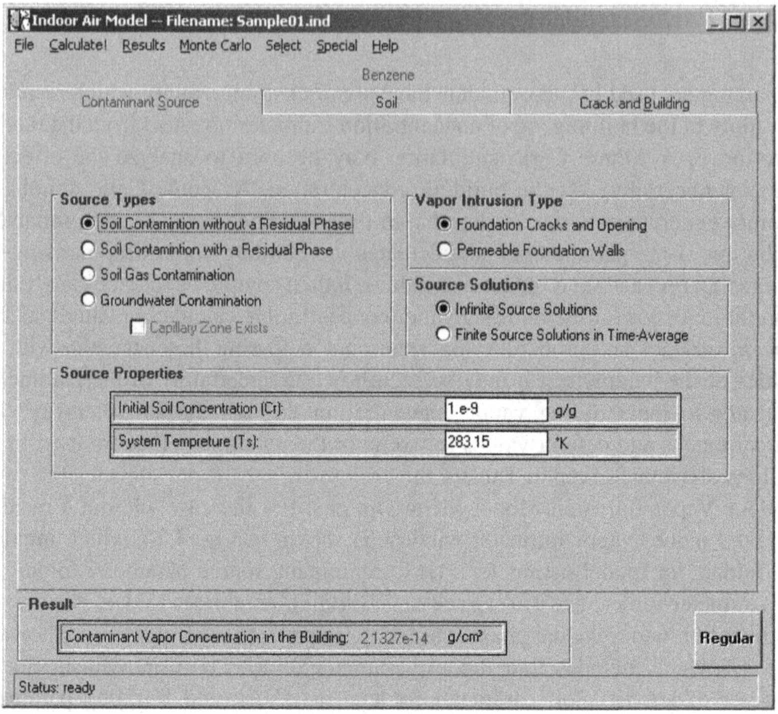

Fig. 4.26 Input window for indoor vapor intrusion model without residual phase

The JEM assumptions are typical of most simplified models of subsurface contaminant transport, with the addition of a few assumptions regarding vapor flux entry conditions into buildings.

The JEM as implemented by the US EPA assumes that the subsurface is characterized by homogeneous soil layers with isotropic properties. The first tier spreadsheet versions accommodate only one layer; the advanced spreadsheet versions accommodate up to three layers. Sources of contaminants that can be modeled include dissolved, sorbed, or vapor sources where the concentrations are below the aqueous solubility limit, the soil saturation concentration, and/or the pure component vapor concentration. The contaminants are assumed to be homogeneously distributed at the source. All but one of the spreadsheets assumes an infinite source. The exception is the advanced model for a bulk soil source, which allows for a finite source. For the groundwater and bulk soil models, the vapor concentration at the source is calculated assuming equilibrium partitioning. Vapor from the source is assumed to diffuse directly upward (one-dimensional transport) through uncontaminated soil (including an uncontaminated capillary fringe if groundwater is the vapor source) to the base of a building foundation, where convection carries the vapor through cracks and openings in the foundation into the building. Both diffusive and convective transport processes are assumed to be at steady state.

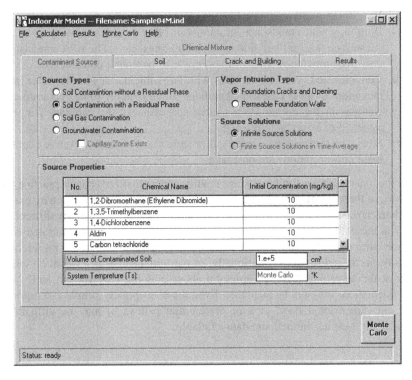

Fig. 4.27 Input window for indoor vapor intrusion model with residual phase

Neither sorption nor biodegradation is accounted for in the transport of vapor from the source to the base of the building.

The assumptions described above and the data given in Table 4.8 suggest a number of conditions that, under most scenarios, would preclude the application of the JEM as implemented by the US EPA. These include:

i. The presence or suspected presence of residual or free-product nonaqueous phase liquids (LNAPL, DNAPL, fuels, solvents, etc) in the subsurface.

ii. The presence of heterogeneous geologic materials (other than the three layers in the advanced spreadsheets) between the vapor source and building. The JEM does not apply to geologic materials that are fractured, contain macropores or other preferential pathways, or are composed of karst.

iii. Site conditions where significant lateral flow of vapors occurs. These can include geologic layers that deflect contaminants from a strictly upward motion and buried pipelines or conduits that form preferential paths. Permeability contrasts between layers greater than 1,000 times also are likely to cause lateral flow of vapors. The model assumes that the source of contaminants is directly below the potential receptors.

iv. Very shallow groundwater where the building foundation is wetted by the groundwater.

v. Very small building air exchange rates (e.g., $<0.25 \text{ h}^{-1}$)

vi. Buildings with crawlspace structures or other significant openings to the subsurface (e.g. earthen floors, stone buildings, etc.). The EPA spreadsheet only accommodates either slab on grade or basement construction.

vii. Contaminated groundwater sites with large fluctuations in the water table elevation. In these cases, the capillary fringe is likely to be contaminated, whereas in the groundwater source spreadsheets, the capillary fringe is assumed to be uncontaminated.

viii. Sites with transient flow rates and/or sites where non-conservative contaminant concentrations are observed.

In theory, the above limitations are readily conceptualized, but in practice the presence of these limiting conditions may be difficult to verify even when extensive site characterization data are available. Conditions that are particularly difficult to verify in the field include the presence of residual NAPLs in the unsaturated zone and the presence and influence of macropores, fractures and other preferential pathways in the subsurface. Additionally, in the initial stages of evaluation, especially at the screening level, information about building construction and water table fluctuations may not be available. Even the conceptually simple assumptions (e.g., one-dimensional flow, lack of preferential pathways) may be difficult to assess when there are limited site data available.

4.5 "Chemicals" Database

The air pathway models discussed in this chapter may require several chemical properties that need to be entered as input data. In some cases this input may be overwhelming, especially when the reader is not fully aware of the intermediate steps involved in adjusting some of the parameters used in the analysis to the current conditions specified in the problem such as the conditions of temperature and pressure. To minimize this task a "Chemicals" database is prepared and included in the ACTS software such that when a chemical is selected from the database all the necessary parameters are properly used in the model during the intermediate computation steps. This process which is transparent to the user simplifies the data entry effort considerably, (see Appendix 3). The "Chemicals" database is provided in triplicate form. One of these is the master database and cannot be modified by the user. The other two are editable and can be customized by the user by adding chemicals to the list. When these are added, is important for the user to enter all the properties of the chemical in proper units, as it is identified in the column headings. While editing the databases data for existing chemicals may also be revised if necessary. Another feature of the use of this database is the possibility of choosing multiple chemicals and performing the analysis for each chemical separately for the same problem, (see Appendix 3).

4.6 Applications

The environmental pathway models discussed in this chapter cover a wide range of air emission and air pathway transformation and transport models. Providing applications for each of these cases would be an almost impossible task due to the multitude of cases that can be covered using these models. In this section, several applications are selected, and their solutions are provided to demonstrate the use of important features of the ACTS software. As the reader gets familiar with the ACTS software, they will recognize that the features and procedures discussed below are standardized for all other pathway applications of the ACTS software. These procedures can be repeated in other studies that involve other environmental pathways to extend the analysis to a more sophisticated level. Thus, the purpose here is to introduce the reader to some applications in air pathway transformation and transport analysis using ACTS software and in doing so help familiarize the reader with important features of the software.

Example 1: Soil contaminated with benzene was buried underground. The area is now considered for residential development. Thus potential exposure analysis based on air emissions of benzene contamination needs to be analyzed. The surface area of the burial region is about 200 m^2 and the contaminated soil was covered with 400 m of clean soil. Background air concentration of benzene at the soil surface can be assumed to be zero, and the total benzene concentration in the soil is determined to be 100 mg/kg of soil based on field studies. Field studies also indicate that the porosity of the soil is 0.4, the soil water content is 15%, the organic carbon content of the soil is 0.1% and the soil density is 1.8 g/m^3. The ambient temperature in the region is about 20°C.

To complete an exposure and health effects study, the air emission of benzene and the concentration of benzene vapor at the soil surface need to be known. Based on these estimates, exposure risk through inhalation can be studied later on. It is also anticipated that some of the field parameter values given above, which are based on field studies, are approximate. For example, the benzene concentration in the buried soil is not known precisely but is estimated to be in the range of 10–150 mg/kg, while the soil porosity in the area may also vary in the range of 0.25–0.45. Under these conditions, what will the variability of the emissions in the soil surface be based on the variability of these two parameters? Further, if the average wind speed in the area is approximately 5 m/s what would the concentrations at the soil surface be in a nearby region? The wind speed is also given as an estimate and may vary in the range of 1 m/s to 18 m/s. Given this information and the information on uncertain parameters, provide a deterministic solution to the problem and an uncertainty based analysis of the problem using simple air pathway models and assumptions.

Solution: In this case, assuming that a simple analysis will be sufficient, the "Farmer" model will be used to estimate the benzene emission at the soil surface and the "Box" air dispersion model will be used to estimate the concentrations in the area of interest. The solution will start with the selection of the chemical

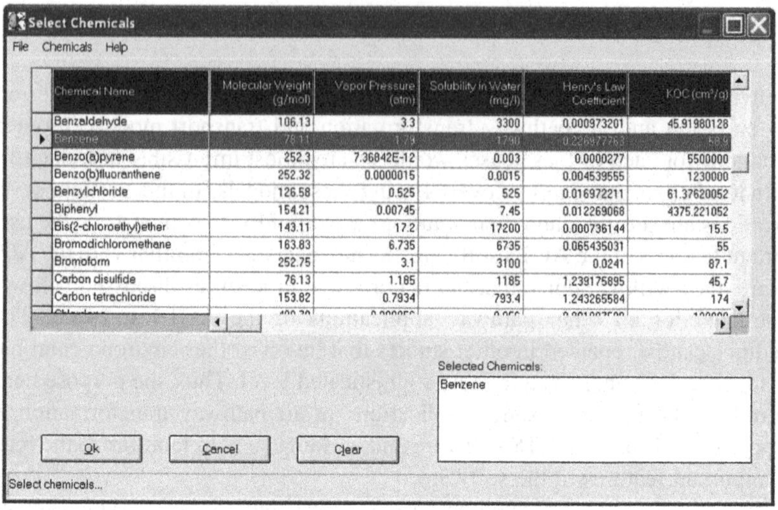

Fig. 4.28 Selection of benzene from "Chemicals" database window

benzene from the chemical database as shown above (Fig. 4.28). This step will ensure that the appropriate properties of benzene are automatically transferred to the models that will be used in this study. The user may refer to Appendix 3 to get familiar with the procedures used in making these selections.

This step is followed by the selection of the Farmer model and the entering of the appropriate data necessary to calculate the Farmer model based on emission rates. When the Farmer data entry window is opened one will notice that the air diffusion coefficient box is already populated. This data is automatically entered by the software when the chemical benzene is selected from the chemical database: At the top of this window the word Benzene appears indicating the chemical selection made. Using the data given in the problem description, the data entry boxes in this window can be populated as seen in Fig. 4.29. Based on the data entered in this window benzene emissions at the soil surface can be calculated by clicking on the calculate button. This will result in the deterministic estimate of benzene emission rates at the soil surface, which is calculated as 1.2064 kg/year.

The next step of this analysis will be the use of the emission rate calculated in a dispersion model to calculate the benzene concentrations at the soil surface. Again, if providing a quick and simple answer to the problem is the goal, then one may use the Box model to estimate the benzene concentrations at the soil surface. Before initiating this step, the emission rate calculated in the first step will be saved in a file by selecting the "Save as" option under the file menu. We have saved this solution under the file name "EXAMPLE 1_1.FRM." Notice that the extension FRM is standard and automatically selected by the ACTS software since the model used in this application is the Farmer model. The next step is the opening of the "Box" model window (Fig. 4.30).

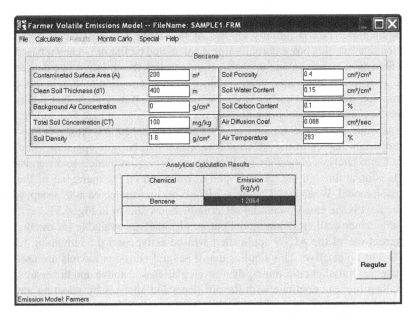

Fig. 4.29 Deterministic benzene emission rate obtained from the Farmer model

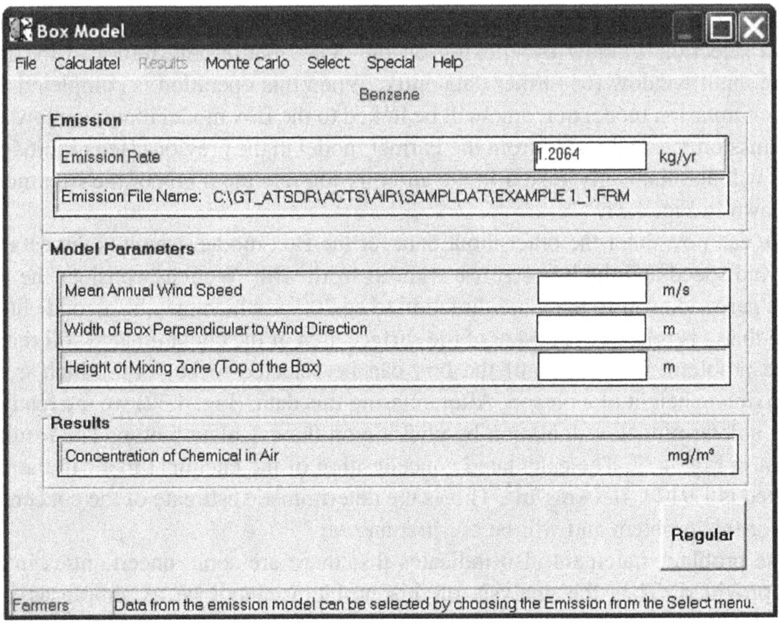

Fig. 4.30 Box model input window

When the "Box" model input window is opened one will notice that the "Emission Rate" box is initially empty. At this stage it is possible to enter an emission rate into this box externally if it is calculated earlier or is known based on some field data. However, one may also choose to link the emission rate data entry box to a previously calculated value obtained from an emission model. In this case the emission model used earlier is the Farmer model. It may seem that this is an unnecessary step, since only one number needs to be entered here and we already know its value, as it was obtained from the Farmer model earlier. However, this is a very important feature of the ACTS software as it will become apparent when the Monte Carlo analysis is performed. Linking air dispersion models to air emission models is done through the "Select" pull down menu on the menu bar of the air dispersion model. When the "Chemical/Model" option is selected at this step, what appears next is the emission model selection window shown in Fig. 4.31.

In this window all emission model output files that are available for use during the current use of the ACTS application will be active, and the remaining listed models will be inactive. This implies that if several emission models are used to calculate the emission rates during the first step discussed above and the results are saved, then one can continue with the air dispersion analysis by selecting any of these models at this stage. In this manner, the effect of different air emission rates obtained from various air emission models can be compared if such a comparison is needed. In this example we only have one file saved in our folder. That is why only the Farmer model air emission estimate option is active. We can now link this file to our "Box" model input window by clicking the radio button to the left of the Farmer model option and also by clicking on the "Use Emission from Emission Model" option at the bottom of the window (Fig. 4.31). We can now close the emission model selection window by clicking on the "OK" button and return to the Box model input window for further data entry. When this operation is completed, the Farmer emission model outcome will be linked to the Box model input window and the emission rate calculated from the Farmer model in the previous step (1.2064 kg/year) will automatically appear in the air emission rate input box of the Box model as shown in Fig. 4.31.

We can now enter the other input data for the Box model, which is the data for the wind speed and the box size we want to work with. We may consider the box width perpendicular to the wind direction to be 200 m which may be a good choice given that it is related to the size of the surface area of the contaminated soil region in the problem. The height of the box can be selected to be 2 m, which is the approximate height of a person. After entering this data, (Fig. 4.32) we are ready to make a deterministic calculation by clicking on the calculate button. The result is shown in Fig. 4.32. The calculated concentration of the chemical in air in the box selected is $1.9127 \ 10^{-5}$ mg/m^3. This is the deterministic estimate of the concentration for this problem and will be our first answer.

The problem statement also indicates that there are some uncertainties in the data provided. Thus the analysis of this problem cannot be as simple as it is represented in the calculation given above. We have to reevaluate the solution based on the uncertainties described in the problem using the Monte Carlo analysis

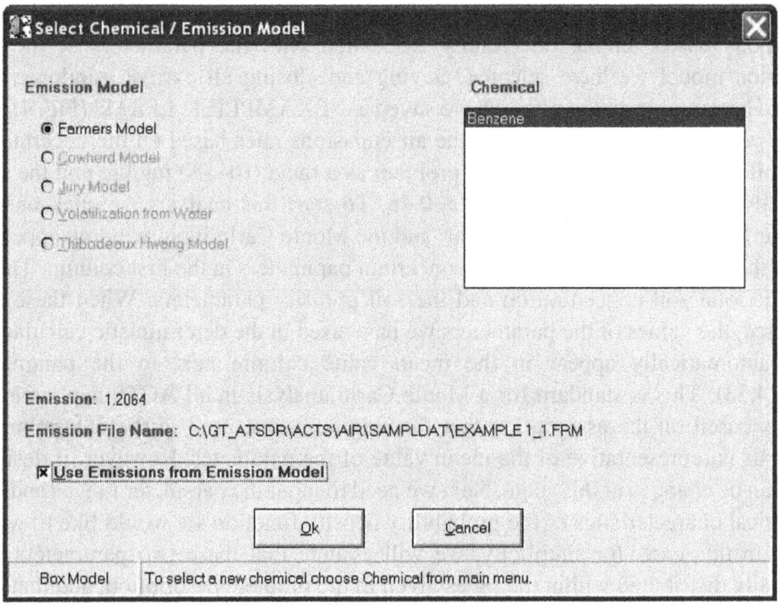

Fig. 4.31 Emission model selection window

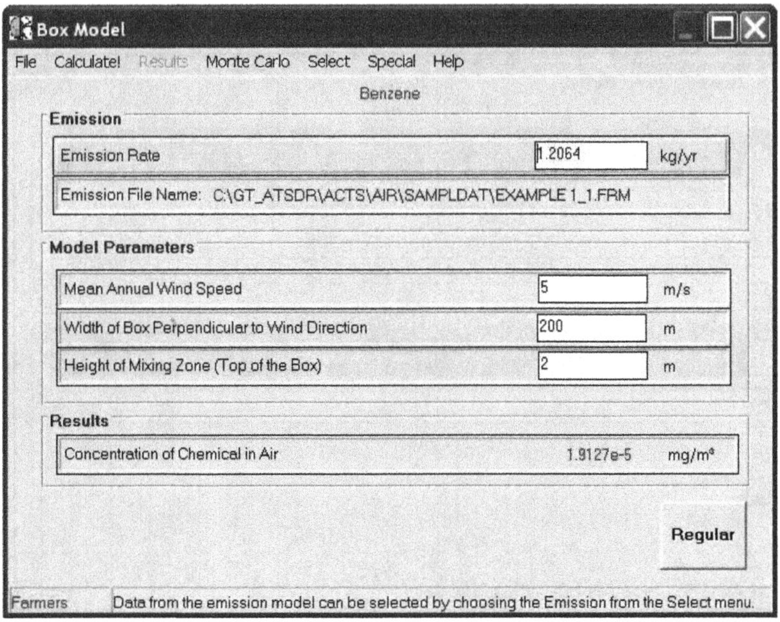

Fig. 4.32 Box model output

procedures. For this analysis, we need to start from the estimation of the air emissions based on the uncertainty associated with the parameters of the air emission model we have selected. Saving and closing all current windows, we can go back and open the file we have saved as "EXAMPLE 1_1.FRM" (Fig. 4.29). Now we would like to recalculate the air emissions rates based on the uncertainty the soil concentration, given in the problem as a range 10–150 mg/kg, and the soil porosity, which has a range of 0.25–0.45. To start this analysis we click on the Monte Carlo button on the menu bar, and the Monte Carlo input window appears.

In this window we select the two uncertain parameters in the first column. These are the total soil concentration and the soil porosity parameters. When these are selected, the values of the parameters we have used in the deterministic calculation will automatically appear in the mean value column next to the parameter (Fig. 4.33). This is standard for a Monte Carlo analysis in all ACTS applications, and is based on the assumption that the original value used in the deterministic analysis is representative of the mean value of the parameter. However, if desired this can be changed at this stage. Next we need to input the parameter ranges and the statistical characteristics of the probability density function we would like to work with. In this case, for simplicity, we will assume that these two parameters are normally distributed within the rages given in the problem description, and that the statistical characteristics are as shown in Fig. 4.34. Similar data entry procedures should also be followed for the porosity parameter. After the appropriate data is entered, we click on the generate button to generate the two normal distributions

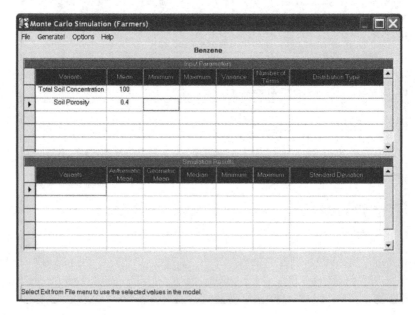

Fig. 4.33 Initial Monte Carlo input window

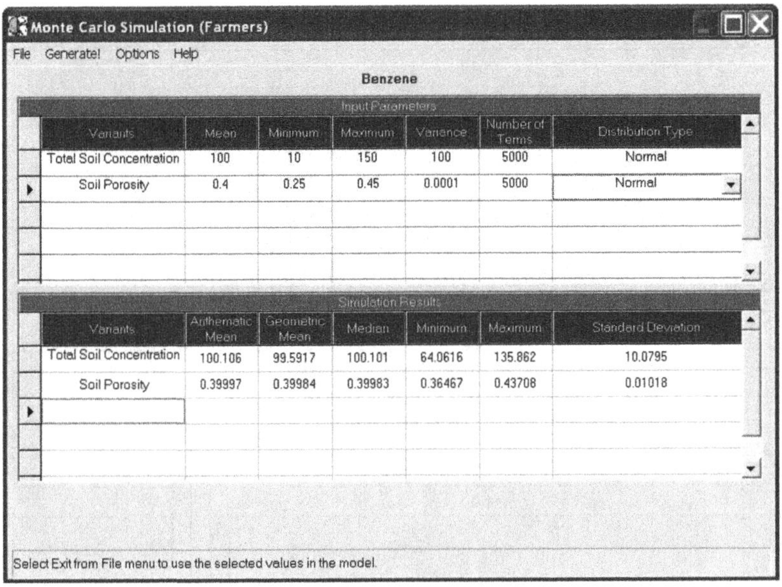

Fig. 4.34 Monte Carlo window after calculations are performed

that represent these two parameters (Fig. 4.35). Characteristics of the generated probability density function will also appear in the output box below, which are the arithmetic mean, geometric mean, median, minimum and maximum of the generated probability density functions (PDF). At this stage there are two options available to the user: (i) perform a single stage Monte Carlo analysis; or, (ii) perform a two stage Monte Carlo analysis. The difference between these two options is associated with the choice of just using the arithmetic mean, geometric mean, median, minimum or maximum of the distributions generated to represent the values of the two uncertain parameters of soil concentration and porosity used in Farmer's model (option i) or using all of the random soil concentration and porosity values generated in Farmer model (option ii). This selection will be done by double clicking the option we want to work with, in the output window of the Monte Carlo analysis. In this case, we would like to work with a two stage Monte Carlo analysis, so we will double click on the parameter name to make that selection rather than on the other representative value boxes. When this is done, the parameter name box turns red, indicating the selected option (note this selection is not shown in Fig. 4.34). For example, if one would like to work with the arithmetic mean, double clicking on the arithmetic mean box would turn the box to red, indicating the selection. If the arithmetic mean was the selection, the arithmetic means of the two PDFs would have been automatically transferred to the Farmer model window as the soil concentration and porosity input value.

At this stage we may also want to see and evaluate the distributions generated by the software for these two parameters. This evaluation is a good idea, since we

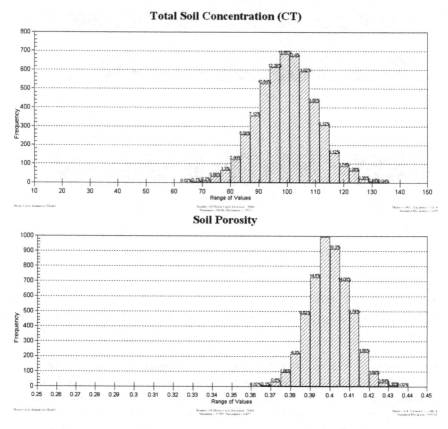

Fig. 4.35 Frequency plots of the normal distributions generated for the two parameters

need to confirm the appropriate selection of the statistical parameters used in the probability density function generation, such as 100 for the variance of the soil concentration parameter, 0.0001 for the variance of the soil porosity parameter and 5,000 for the total number of random variates used in the analysis. The graphs of the two distributions can be obtained by making the "Graph" selection under the "Option" pull down menu in the Monte Carlo window (see Appendix 3). The frequency plots of the normal distributions generated for these two parameters are shown in Fig. 4.35. There are other options to plot these distributions as well, such as PDF plots, cumulative PDF plots or complementary cumulative PDF plots. The user may try these different options as he or she becomes familiar with the software. At this stage it is also possible to change the title and other features of these plots. The reader is referred to Appendix 3 for further details of these operations.

Now that we have decided to perform a two stage Monte Carlo analysis by selecting the two parameters in the Monte Carlo window, we are ready to exit the

Monte Carlo data generation window. We will simply close the Monte Carlo window to proceed. When we exit and return to the Farmer's emission model window we notice that the boxes for the soil concentration and the soil porosity no longer contains a numerical value, but instead these two boxes are replaced with the words "Monte Carlo," indicating that the 5,000 random numbers generated for these two parameters are now ready to be used as data entries in the emission calculations for a two stage Monte Carlo analysis. Now we can click on the calculate button again and recalculate emissions based on these 5,000 random entries of the two uncertain parameters. The outcome will also be a probability density function which is indicated by the words "Monte Carlo" that will appear in the results box (Fig. 4.36). The first thing we may want to do is to see the distribution obtained for the emission rates. The graph of this distribution can be obtained by selecting the "Graph" option in the "Results" pull down menu (Fig. 4.36).

As can be seen from Fig. 4.37, rather than a single deterministic outcome as we obtained before (1.2064 kg/year), we now have a probability density function for emission rates which looks to be a normal distribution in the range of 0.7–1.82 kg/year. This distribution gives us information on the variability of the emission rate outcome based on the variability of the two parameters we have selected. Now we can save this outcome and continue to use the box dispersion model with the air emission PDF distribution we have just calculated. It is appropriate to save this file under a different file name since we may want to keep our first deterministic calculation intact for future reference. We choose to identify this new file as

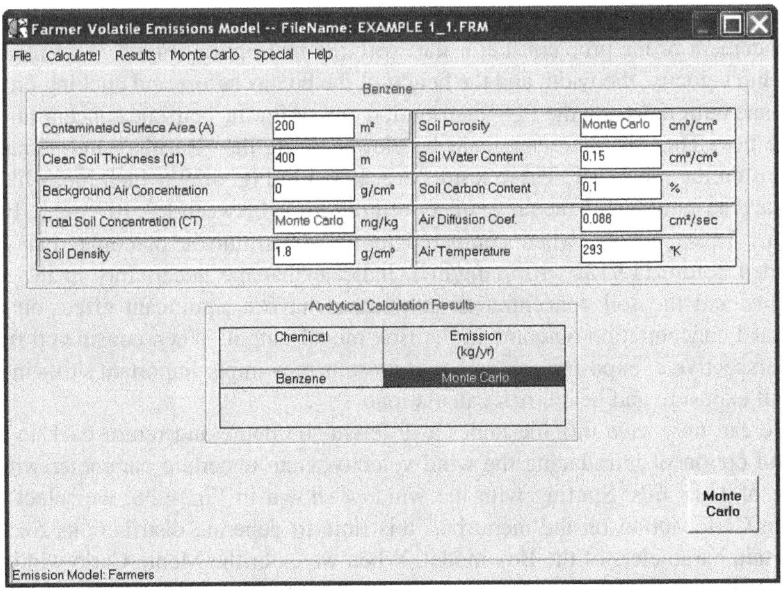

Fig. 4.36 Monte Carlo output for emission calculations

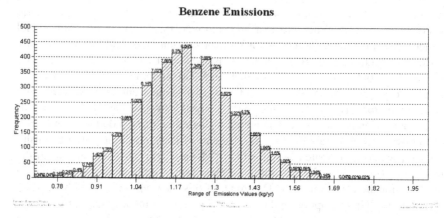

Fig. 4.37 Benzene emissions output

"EXAMPLE 1_2_MC.FRM". The emission calculation window can now be closed and the Box dispersion model window can be opened.

Opening the box dispersion model window and pulling the Monte Carlo based emission file into the Box dispersion model as input follows the procedures described earlier. Now the user should recognize the importance of linking the emission outcome to the dispersion analysis process because we need to transfer 5,000 data points to the Box model as air emission rates. We again have two options at this stage: (i) using the emission PDF distribution with a constant wind velocity, of 5 m/s; or, (ii) using the emission PDF distribution along with a PDF distribution for wind velocity, since this parameter is also given as an uncertain parameter in the statement of the problem. Let's start with the first option (Fig. 4.38). Entering the wind velocity, the width and the height of the box as before and clicking on the calculate button, we get the PDF distribution output for the benzene concentrations in the box. This outcome can again be viewed using the "Results" and "Graph" options on the menu bar. These results are shown in Fig. 4.39, which looks like a normal distribution with the range of concentrations in between $1.2 \ 10^{-5}$–$2.8 \ 10^{-5}$ mg/m^3. These results, when compared to the deterministic outcome that was obtained before ($1.9127 \ 10^{-5}$ mg/m^3), indicate that the uncertainty in the soil porosity and the soil concentration parameters have a significant effect on the expected concentration outcome in the Box model output. When considered from the perspective of exposure analysis, these results may imply important shifts in the overall exposure and health risk calculations.

We can now save this file under a different file name and return back to the second option of introducing the wind velocity as an uncertain parameter with a range of 1–18 m/s. Starting with the window shown in Fig. 4.38, we select the Monte Carlo option on the menu bar, this time to generate distributions for the uncertain parameters of the Box model. When we enter the Monte Carlo window, we will select only the wind velocity as an uncertain parameter and generate a lognormal distribution for the wind velocity parameter. The database used for this

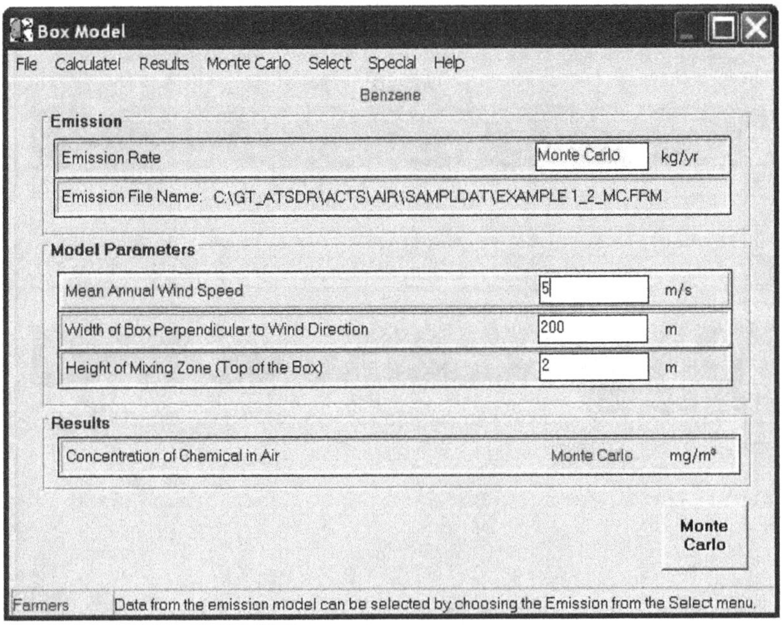

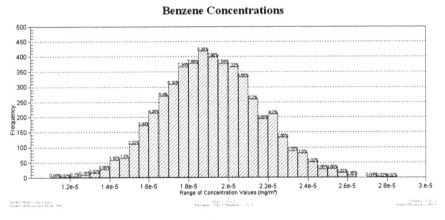

Fig. 4.38 Benzene concentration calculation with constant wind velocity and emission rates entered as a PDF distribution

Benzene Concentrations

Fig. 4.39 Benzene concentration frequency distribution with constant wind velocity and emission rates entered as a PDF distribution

generation is shown in Fig. 4.40 and the resulting lognormal distribution for the wind velocity is shown in Fig. 4.41. Notice that one has to choose 5,000 random numbers at this step since the emissions file used has 5,000 random numbers in its database. The two random number sets must match.

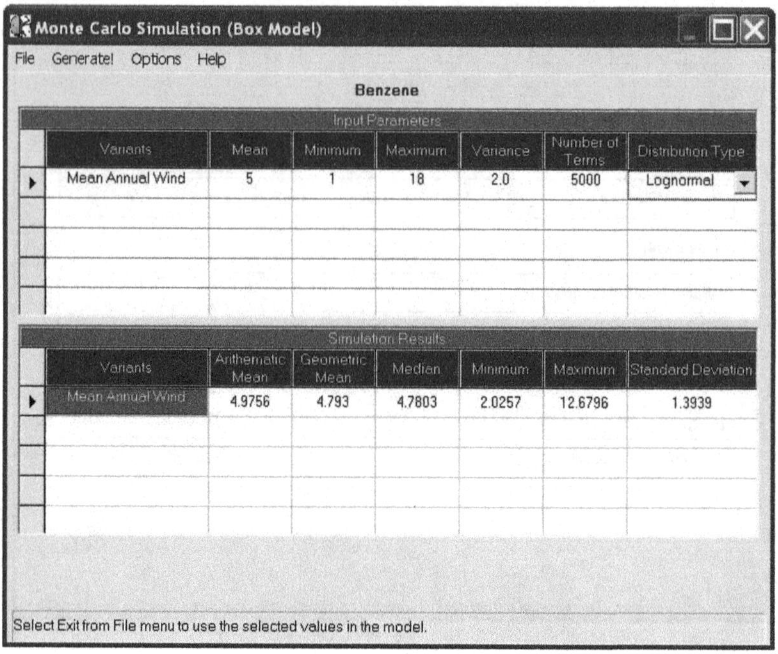

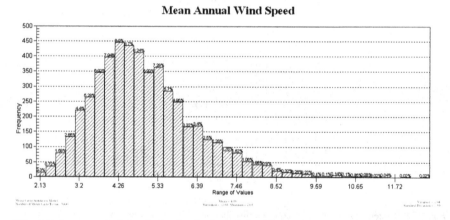

Fig. 4.40 Monte Carlo distribution generation window for wind velocity

Mean Annual Wind Speed

Fig. 4.41 Lognormal wind velocity distribution obtained from Monte Carlo window

Now we are ready to transfer the data generated for the wind velocity into the Box model. This is accomplished by selecting the wind velocity parameter name in the Monte Carlo window by double clicking on the name box and closing the Monte Carlo window. One should also remember that we could have selected the arithmetic mean or the geometric mean as discussed earlier for a one stage Monte Carlo analysis. When we exit the Monte Carlo window and return to the Box model data input window, we recognize that the wind velocity box is now replaced with the word Monte Carlo, indicating that the PDF for wind velocity is successfully transferred to the Box model window. We can now click on the calculate button to evaluate the concentration distribution in the box we have selected, which treats the wind velocity and the emission rates as uncertain parameters. The resulting PDF for the concentrations is shown in Fig. 4.42.

As can be seen from the comparison of Figs. 4.37 and 4.42, the results for benzene concentrations are significantly different. This analysis, which can be done very quickly using the ACTS software, may provide important outcomes when uncertainty based health risk studies are performed. Eventually the benzene concentration obtained from the ACTS software will be linked to the RISK software to evaluate the health risk of inhalation exposure to benzene vapor at the contaminated site we are studying. This preliminary example provides an insight into the way the problems may be analyzed using the ACTS software.

The problem solved above using the box model may also be analyzed using the Gaussian line source model. In this case we may be interested in obtaining the concentration distribution in an area or a region that assumes the source emissions calculated earlier to be a line source at the boundary of the box region we used before. Obviously this would be a different problem, since in this case the source is not an aerial source but a line source. Nevertheless, for demonstration purposes we will investigate the outcome of this case. For this example we will only conduct a deterministic analysis although a Monte Carlo analysis is also possible using the uncertainty involved with the parameters of the problem as described above. That analysis will be left as an exercise.

Opening the Gaussian line source input window and linking the emission rate file to this window, we may start entering the other data that is necessary for the line source Gaussian model.

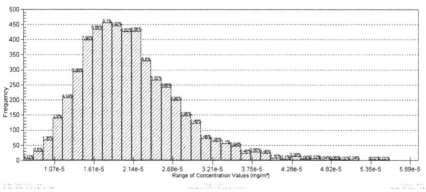

Fig. 4.42 Benzene concentrations obtained from the Box model when wind velocity and emission rates are introduced as a PDF

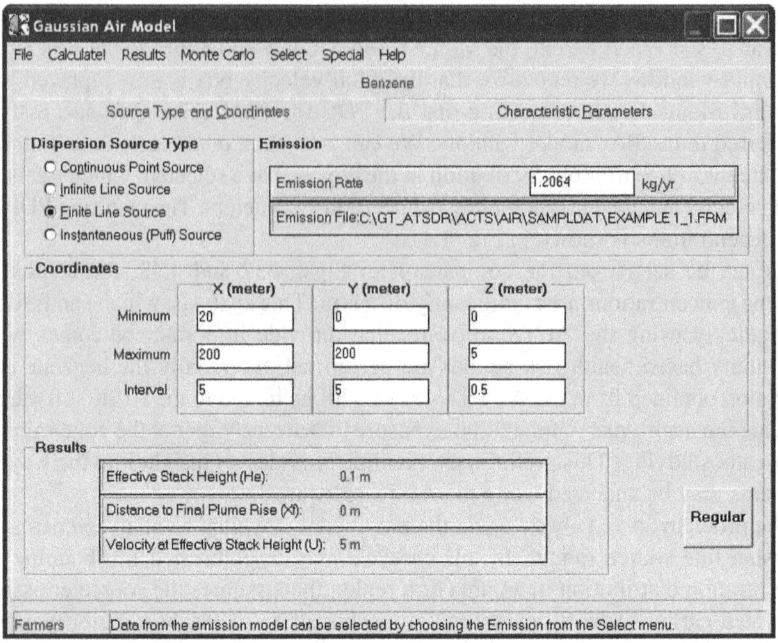

Fig. 4.43 Gaussian line source input window for coordinates and emission rate

As can be seen from Figs. 4.43 and 4.44, the data input necessary for Gaussian models is more complicated than for Box models. This is to be expected, based on the discussion provided earlier in this chapter. It is also important to recognize that although Gaussian models are described in reference to stack emission originating from industrial sources, these models can also be used in the analysis of other problems, such as the line source assumption we have made for the ground level source that may be associated with benzene emissions from a buried soil contamination. Again, we emphasize that this final step in the analysis is included here for demonstration purposes. After making the appropriate selections and entering the appropriate data into the Gaussian line source models we are now ready to make the deterministic analysis of this problem by clicking on the calculate button.

After the completion of this step, the results again can be viewed either numerically or graphically using the "Results" pull down menu on the menu bar. Since the analysis is done in a three dimensional domain described within the confines of the x-, y-, z-coordinates given in Fig. 4.43, they can be viewed in many different ways, such as the $(X–Y)$, $(X–Z)$ and $(Y–Z)$ cross-section contour plots or normal concentration plots in one coordinate direction. Two of the cases are shown below for demonstration. In Fig. 4.45, the concentration profile is shown in the x-axis direction at the centerline of the contaminant plume, as the plume originates from an emission source on the left as a line source. For this case concentration magnitudes at different elevations in the z-direction are plotted at

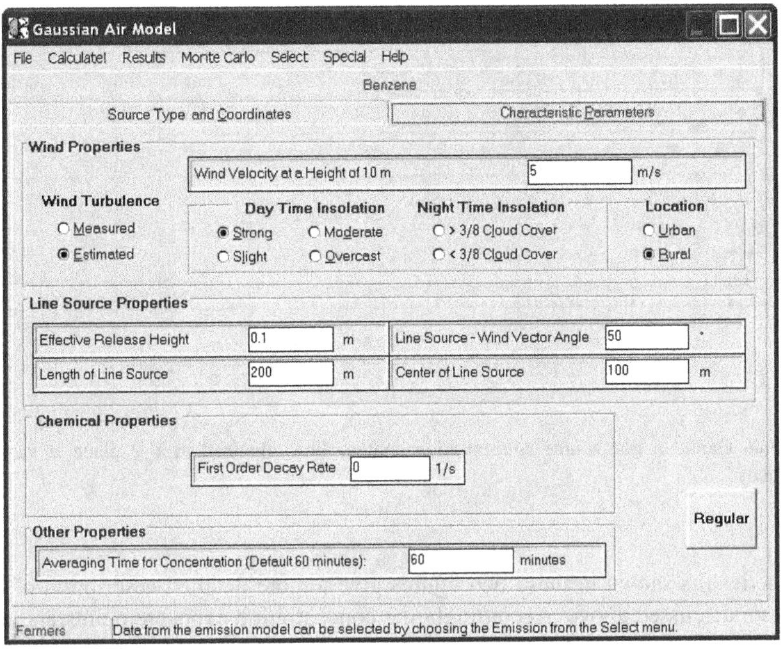

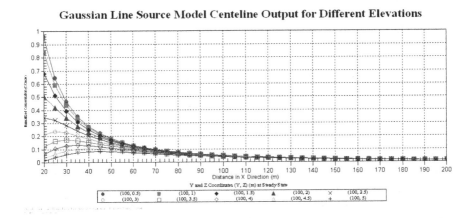

Fig. 4.44 Gaussian line source input window for atmospheric conditions and other parameters

Gaussian Line Source Model Centeline Output for Different Elevations

Fig. 4.45 Gaussian line source concentration output obtained in x-direction at various elevations

the centerline of the plume. In Fig. 4.46 concentration contours are shown again along the x-axis at the centerline of the contaminant plume in the (X–Z) plane of the solution domain.

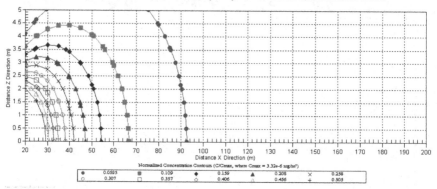

Fig. 4.46 Gaussian line source concentration contour lines obtained in $X–Z$ plane at various elevations

The results shown in these two figures give a more detailed description of the problem discussed above. Accordingly the concentration exposure points are now defined at various mesh points selected by the user. This outcome is a completely different representation of the solution of the problem discussed above. The choice depends on the user and the detailed outcome that is necessary to analyze the problem. As stated earlier, a Monte Carlo analysis could also have been conducted to solve this problem by using the Gaussian line source model. That analysis is left to the user as an exercise.

Example 2: Mercury emissions originating from a stack height of 100 m are of concern for an urban community. Estimates of mercury contamination at ground level or average human height level are necessary to perform inhalation exposure study for the community. The following information is available for the stack emission and atmospheric conditions. The emission rate is estimated to be 1.49 10^6 kg/year, the exhaust stack velocity is 13 m/s, the exhaust stack temperature is 395 K, the exhaust inner stack diameter is 3.0 m, the ambient air temperature is 291 K, the ambient air temperature gradient is 0.01 K/m, the mixing height is 3,000 m, the first order decay rate can be neglected and the day time insolation can be estimated based on rural and moderate insolation conditions.

Solution: The solution will start with the selection of the chemical mercury from the chemical database. The stack emission source can be assumed to be a continuous point source and the Gaussian model will be used to analyze this problem. After selecting this model, the data given in the problem description can be entered directly into the input data entry boxes. The outcome of this data entry process is shown in Fig. 4.47.

Once the data is input, the calculations can be done by clicking on the calculate button in the menu bar. After this step, the results at any point on the selected computational grid (Fig. 4.47) can be analyzed or plotted. In Figs. 4.48

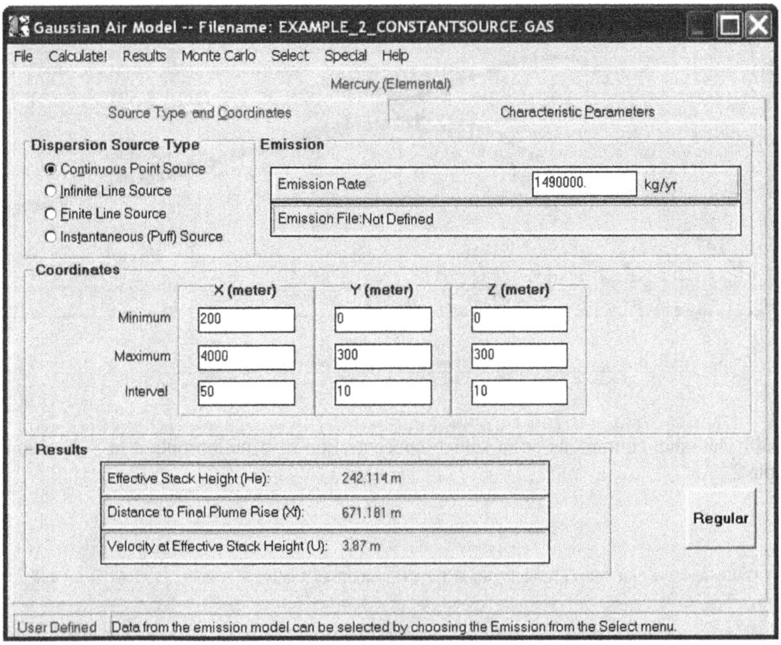

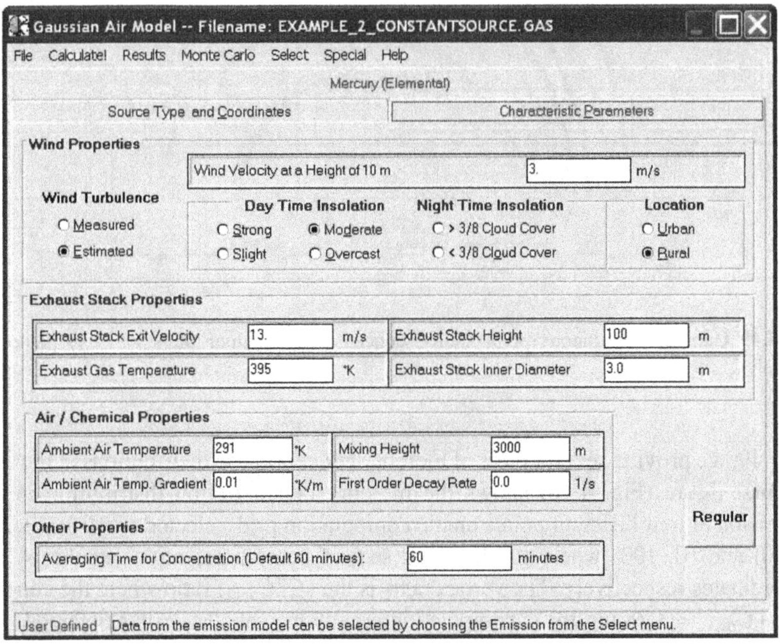

Fig. 4.47 Gaussian continuous point source input data windows for Example 2

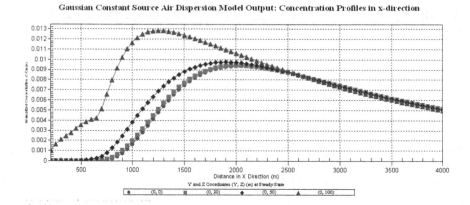

Fig. 4.48 Gaussian continuous point source concentration distribution output in *x*-direction for Example 2

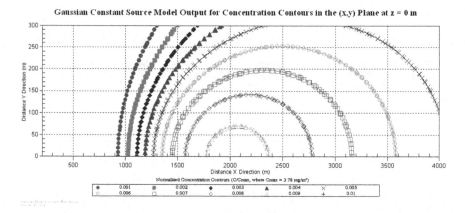

Fig. 4.49 Gaussian continuous point source concentration contour plots in (x, y) plane for Example 2

and 4.49, we provide several plots of mercury concentration distributions at the site. The first figure (Fig. 4.48) shows the mercury concentration distribution in the *x*-direction at four different points on the computation grid, selected as (0, 0), (0, 20), (0, 50) and (0, 100) where the first and second coordinates represent the *y*- and *z*-coordinates respectively. The vertical axis is the (C/C_{max}) ratio where the concentration $C_{max} = 3.78$ mg/m^3 is computed internally to scale the results obtained.

In Fig. 4.49 contour plots for the concentration distribution in the (x, y) plane of the region at ground elevation ($z = 0$ m) are shown.

A surface plot of the concentration distribution in the (x, y) plane at $z = 0$ m is shown in Fig. 4.50.

Gaussian Constant Source Air Dispersion Model a Surface Plot Representation of the Output

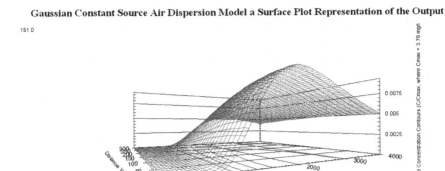

Fig. 4.50 Gaussian continuous point source concentration contour plots in (x, y) plane for Example 2

This problem may also be analyzed in a Monte Carlo analysis mode if some of the parameters of the problem are selected as uncertain parameters. That mode of analysis will be left as an exercise for the reader to explore.

Example 3: A methanol spill in a lake resulted in the contamination of the lake waters. It is estimated that the methanol level in the lake water is about $100 \ g/cm^3$ with an estimated range of $10–200 \ g/cm^3$, and the ambient air concentration level is about $1 \ g/cm^3$ with an estimated range of $0.5–4.0 \ g/cm^3$. The ambient temperature is $20°C$ and the wind speed is about $2 \ cm/s$ at the lake surface with an estimated range of $0.5–10 \ cm/s$. What are the deterministic and probabilistic results for the volatilization of methanol from the water surface for this pollution source?

Solution: The solution will start with the selection of the chemical methanol from the chemical database followed by the selection of volatilization from the water surfaces model under the emission models option. The data entry outcome of the input window for this problem is shown in Fig. 4.51. After the data entry, clicking the calculate button will yield the deterministic result for this problem, which is 4,677.7 kg/year. As indicated in the description of the problem, there are some uncertainties in the data for the concentration levels and the wind velocity. A Monte Carlo analysis can be performed to evaluate the effects of these uncertain parameters on the solution. By clicking on the Monte Carlo button in the model window we enter the Monte Carlo mode of analysis. We can generate the PDF for these three parameters (Fig. 4.52) by selecting the three parameters in the first column of the Monte Carlo window and assigning the characteristic parameters of the probability density function to be generated. We return to the volatilization model input window by clicking on the name boxes for these parameters. We again recognize that the three parameter input boxes for the two concentrations and the velocity are identified as Monte Carlo input. The probability density distributions obtained for the three parameters are shown in Figs. 4.53–4.55 for gas, liquid and velocity parameters respectively.

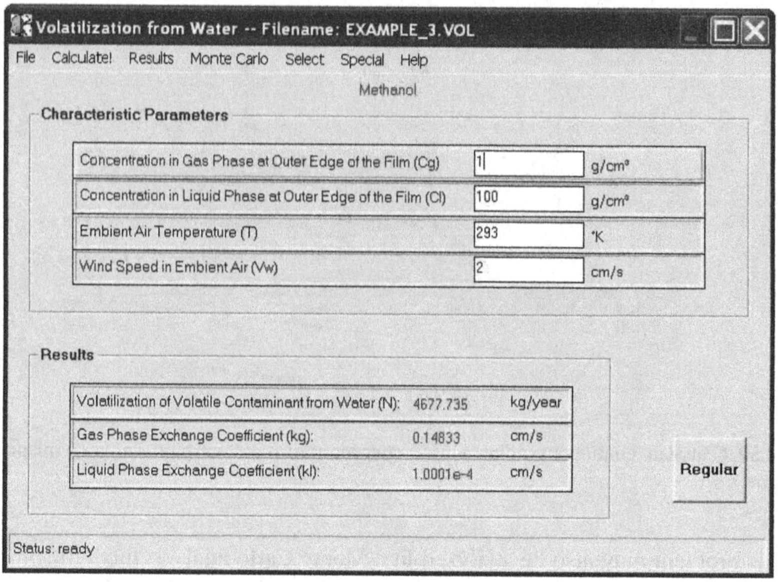

Fig. 4.51 Volatilization from water surfaces input window

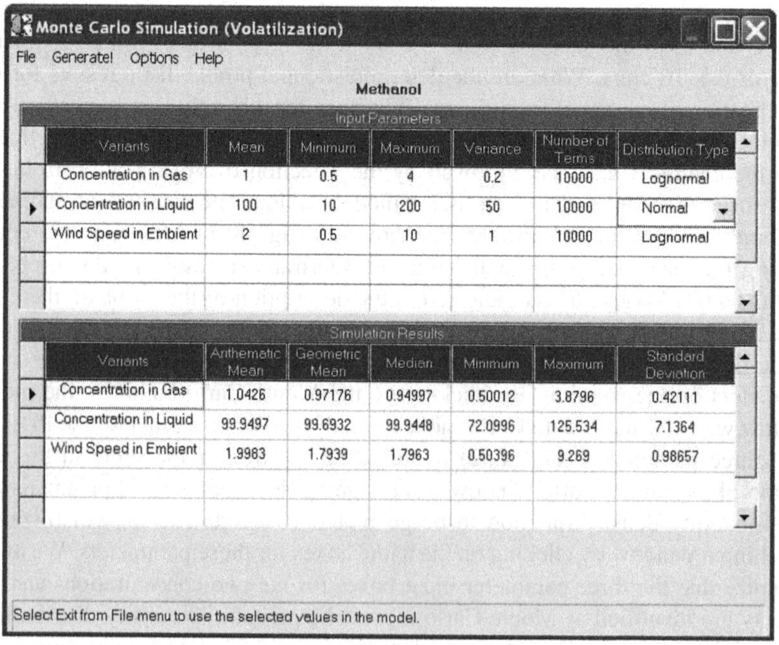

Fig. 4.52 Monte Carlo analysis input and output window

Concentration in Gas Phase at the Outer Edge of the Film (Cg)

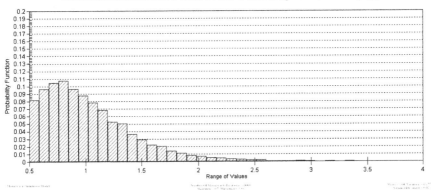

Fig. 4.53 Probability density function generated for gas phase concentration

Concentration in Liquid Phase at the Outer Edge of the Film (Cl)

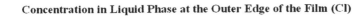

Fig. 4.54 Probability density function generated for liquid phase concentration

Wind Speed in Ambient Air (Vw)

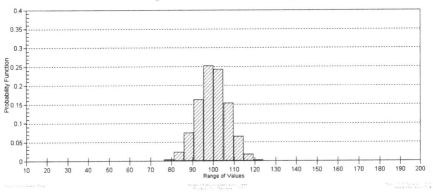

Fig. 4.55 Probability density function generated for wind speed in Ambient air

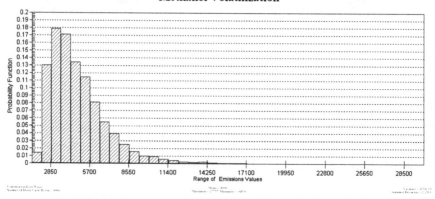

Fig. 4.56 Probability density distribution of methanol volatilization

With the random inputs for the three parameters, we can calculate the probability density function output for volatilization estimates. The results are shown in Fig. 4.56.

Other possible ways to analyze this problem using additional uncertain parameter data is left as an exercise.

References

Abdel-Magid IS, Mohammed A-WH et al (1997) Modeling methods for environmental engineers. CRC Lewis Publishers, Boca Raton, FL, 518 pp

Bird RB, Stewart WE et al (2002) Transport phenomena. Wiley, New York, p 895

Briggs GA (1975) Plume rise predictions. In: Haugen DA (ed) Lectures on air pollution and environmental impact analysis. American Meteorological Society, Boston, MA, pp 59–111

Carslaw HS, Jaeger JC (1959) Conduction of heat in solids. Oxford University Press, London, 510 pp

Clark MM (1996) Transport modeling for environmental engineers and scientists. Wiley, New York, 559 pp

Cole FW (1970) Introduction to meteorology. Wiley, New York, 388 pp

Cowherd C (1983) A new approach to estimating wind-generated emissions from coal storage piles. In: Proceedings, APCA specialty conference in fugitive dust issues in the coal use cycle, APCA Publications Department, Pittsburg, PA

Cowherd C, Muleski GE et al (1985) Rapid assessment of exposure to particulate emissions from surface contamination sites. Midwest Research Institute, Kansas City, MO

Cramer HE (1959) Engineering estimates of atmospheric dispersal capacity. Am Ind Hyg Assoc 20(3):183–189

Csanady GT (1973) Turbulent diffusion in the environment. D. Reidel, Dordrecht, The Netherlands/Boston, MA, 248 pp

Domenico PA, Schwartz FW (1990) Physical and chemical hydrogeology. Wiley, New York, 824 pp

Draxler RR (1979a) Estimating vertical diffusion from routine meteorological tower measurements. Atmos Environ 13(11):1559–1564

Draxler RR (1979b) Modeling the results of 2 recent mesoscale dispersion experiments. Atmos Environ 13(11):1523–1533

Dunnivant FM, Anders E (2006) A basic introduction to pollutant fate and transport. Wiley-Interscience, Hoboken, NJ, 480

Farmer WJ, Yang MS et al (1980) Hexachlorobenzene – its vapor-pressure and vapor-phase diffusion in soil. Soil Sci Soc Am J 44(4):676–680

Farmer WJ, Yang MS et al (1978) Land disposal of hexachlorobenzene wastes: controlling vapor movement in soils. Land disposal of hazardous wastes. In: Proceedings of the 4th annual research symposium, San Antonio, TX

Fletcher HB, Dotson WL (1971) Hermes – a digital computer code for estimating regional radiological effects from the nuclear power industry, U. S. AEC. HEDL-TME-71-1968

Gifford FA (1976) Turbulent diffusion typing schemes: a review. Nucl Saf 17(1):68–86

Günther O (1961) Environmental information systems. Springer, Berlin

Hanna SR (1989) Confidence limits for air quality model evaluations as estimated by bootstrap and Jacknife resampling methods. Atmos Environ 23:1385–1398

Hanna SR, Briggs GA et al (1982) Handbook on atmospheric dispersion. U.S. Department of Energy, Washington, D.C., DOE/TIC-11223: 102

Heinsohn RJ, Kabel RL (1999) Sources and control of air pollution. Prentice Hall, Englewood Cliffs, NJ, 696 p

Hemond HF, Fechner-Levy EJ (2000) Chemical fate and transport in the environment. Academic, San Diego, CA, 433 pp

Iman RL, Helton JC (1988) An investigation of uncertainty and sensitivity analysis techniques for computer models. Risk Anal 8:71–90

Johnson PC, Ettinger RA (1991) Heuristic model for predicting the intrusion rate of contaminant vapors into buildings. Environ Sci Technol 25(8):1445–1452

Jury WA, Russo D et al (1990) Evaluation of organic chemicals residing below the soil surface. Water Resour Res 26(1):13–20

Kaiser CC (1979) Examples of the successful application of a simple model for the atmospheric dispersion of dense, cold vapors to the accidental release of anhydrous ammonia from pressurized containers. United Kingdom Atomic Energy Authority Safety and Reliability Directorate, London

Louvar JF, Louvar BD (1998) Health and environmental risk analysis: fundamentals with applications. Prentice Hall PTR, Englewood Cliffs, NJ, 678 pp

Lyman WJ, Reehl WF et al (1990) Handbook of chemical property estimation methods. American Chemical Society, Washington, D.C

Lyons TJ, Scott WD (1990) Principles of air pollution meteorology. CRC Press, Boca Raton, FL

Masters GM (1991) Introduction to environmental engineering and science. Prentice Hall, Englewood Cliffs, NJ

Millington RJ, Quirk JP (1961) Permeability of porous solids. Trans Faraday Soc 57:1200–1207

Milton I, Stegun IA (1964) Handbook of mathematical functions with formulas, graphs, and mathematical tables. National Bureau of Standards, Washington, D.C

Nirmalakhandan N (2002) Modeling tools for environmental engineers and scientists. CRC Press, Boca Raton, FL, 312 pp

NRC (1982) Atmospheric dispersion models for potential accidental consequence assessments of nuclear power plants NRC regulatory guide 1.145. USNRC. US Nuclear Regulatory Commission, Washington, D.C

Pasquill F (1961) The estimation of the dispersion of windborne material. Meteorol Mag 90(1063):33–49

Pasquill F (1976) Atmospheric dispersion parameters in Gaussian modeling: Part II. Possible requirements for change in turner workbook values. USEPA600/4-76-030b. U.S. Environmental Protection Agency, Research Triangle Park, NC

Philp RB (1995) Environmental hazards and human health. CRC Lewis Publishers, Boca Raton, FL

Schnelle KB, Dey PR (2000) Atmospheric dispersion modeling compliance guide. McGraw Hill, New York

Schnoor JL (1996) Environmental modeling: fate and transport of pollutants in water, air, and soil. Wiley, New York, 682 pp

Slade DH (1968) Meteorology and atomic energy, Air resources Laboratories, Environmental Sciences Services Administration. U.S. Department of Commerce, U.S. Atomic Energy Commission, Washington, D.C

Stern AC (1976) Air pollution: volume I – air pollutants, their transformation and transport. Academic, New York

Stern AC, Boubel RW et al (1984) Fundamentals of air pollution. Academic, Orlando, FL

Sutton OG (1932) A theory of eddy diffusion in the atmosphere. P R Soc Lond a-Conta 135(826):143–165

Taylor GI (1953) Dispersion of soluble matter in solvent flowing slowly through a tube. Proc R Soc Lond Ser A 219:186–203

Thibodeaux DP (1982) Cotton dust particle-size distribution in oil mills. J Am Oil Chem Soc 59(6):A466–A472

Thibodeaux LJ (1979) Chemodynamics. Wiley, New York

Thibodeaux LJ, Hwang ST (1982) Landfarming of petroleum wastes – modeling the air emissions problem. Environ Prog 1(1):42–46

Thorneloe SA, Barlaz MA et al (1993) Global methane emissions from waste management. Atmospheric methane: sources, sinks, and role in global change, NATO ASI Series, vol 13. Springer, New York

Thorneloe SA, Doorn M et al (1994) Methane emissions from landfills and open dumps. In: EPA report to congress on international anthropogenic methane emissions: estimates for 1990, USEPA, Office of Policy, Planning and Evaluation. EPA-230-R-93-010

Turner DB (1994) Workbook of atmospheric dispersion estimates: an introduction to dispersion modeling. Lewis Publishers, Ann Arbor, MI

USEPA (1984) Evaluation and selection of models for estimating air emissions from hazardous waste treatment, storage and disposal facilities. Office of Air Quality Planning and Standards, Research Triangle Park, NC

USEPA (1985) Rapid assessment of exposure to particulate emissions from surface contaminated sites. Office of Health and Environmental Assessment, Washington, D.C

USEPA (1987) Guideline on air quality models. Office of Air Quality Planning and Standards, Research Triangle Park, NC, 450/2-78-027R: 21

USEPA (1988) Superfund exposure assessment manual. Office of Health and Environmental Assessment, Washington D.C

USEPA (1991) Air emissions from municipal solid waste landfills. Background information for proposed standards and guidelines. U.S. Environmental Protection Agency, Office of Air Quality Planning and Standards, Research Triangle Park, NC. EPA-450/3-90-011a (NTIS PB91-197061)

USEPA (1997) Guiding principles for Monte Carlo analysis. Office of Health and Environmental Assessment, Washington D.C. EPA/630/R-97/001

USEPA (1998) Municipal solid waste landfills, Volume 1: Summary of the requirements for the new source performance standards and emission guidelines for municipal solid waste landfills. U.S. Environmental Protection Agency, Office of Air Quality Planning and Standards, Research Triangle Park, NC. EPA-453/R-96-004

USEPA (2003) RCRA supplemental guidance for evaluating the vapor intrusion to indoor air pathway. EPA/600/SR-93/140

Vesilind PA, Peirce JJ et al (1994) Environmental engineering. Butterworth-Heinemann, Boston, MA

Viegle WJ, Head JH (1978) Derivation of the Gaussian plume model. JAPCA 28(11):1139–1141

Weber WJ, DiGiano FA (1996) Process dynamics in environmental systems. Wiley, New York, NY, 943 pp

Zheng C, Bennett GD (1995) Applied contaminant transport modeling: theory and practice. Van Nostrand Reinhold, New York, 440 pp

Chapter 5
Groundwater Pathway Analysis

Just because you do not see it everyday,
it does not mean that it is not there.

Transport of contaminants through soil is affected by several transformation and transport processes which include advection, diffusion, dispersion and chemical reactions. These processes simultaneously influence the migration pattern of contaminants in the subsurface. The physical and mathematical definitions of these transformation and transport processes are covered in detail in Chapters 2 and 3 of this book using conservation principles as they apply to air, surface water and groundwater pathways. The mathematical definitions of these models and their associated initial and boundary conditions that may be used for the closure of these models have also been covered in technical publications in the literature. In this chapter we will use the mathematical definitions of these physical, chemical and biologic processes to describe several analytical models that are frequently used for dissolved phase contaminant transport analysis in the groundwater pathway. The goal is to bring this vast amount of literature together in a cohesive manner and to discuss the limitations and applications of these models while providing a user friendly computational platform to implement these models both in deterministic and stochastic analysis mode. Thus, as is the case for all other environmental pathway models covered in this book, all groundwater pathway models that are discussed in this chapter are included in the ACTS software for use in both deterministic and stochastic (Monte Carlo) based applications.

5.1 Definitions and Governing Principles

The aquifer systems that occur in the subsurface are composed of geologic formations which are deposited in geologic time scales through geologic formation mechanisms. In groundwater hydrology terminology these multilayered unconsolidated, sedimentary

M.M. Aral, *Environmental Modeling and Health Risk Analysis (ACTS/RISK)*, 187
DOI 10.1007/978-90-481-8608-2_5, © Springer Science+Business Media B.V. 2010

and rock formations are categorized as: (i) aquifers, to identify those layers which contain and also transmit water under normal drainage conditions; (ii) aquicludes, to identify those layers which contain but do not transmit water; (iii) aquitards, to identify layers which contain water and transmit water at low rates relative to the transmission rate of aquifers; and, (iv) aquifuge layers, to identify layers that neither contain nor transmit water. Based on its layering order, an aquifer may also be identified as a confined aquifer if it is enveloped between an upper and lower aquiclude or aquifuge layers. Aquifers are identified as a semi-confined aquifer if they are bounded by an aquitard layer from above or below or both. Aquifers which are unbounded with either an impervious or a semi-pervious layer from above are identified as unconfined aquifers. In this case the upper boundary of the unconfined aquifer is the water table below which the pressure distribution is considered to be hydrostatic and positive. The pressure at the water table is commonly assumed to be zero gauge pressure or atmospheric pressure. A layered aquifer system can further be characterized as having an unsaturated and a saturated zone based on the presence of an air–water mixture in the pore space of the aquifer. The unsaturated zone most commonly occurs in the upper regions of unconfined aquifers where the pore space is partially occupied by water and partially by air above the water table. Saturated zones are observed in deeper confined or semi-confined aquifers, or below the water table in unconfined aquifers where the pore space is assumed to be completely filled by water. In terms of these definitions a layered aquifer system and the expected moisture distribution in the pore space can be characterized as shown in Fig. 5.1.

As shown in Fig. 5.1, the unsaturated zone can be further divided into three zones. The soil water zone represents upper regions of the unsaturated zone where the moisture conditions are affected by the conditions above ground surface and also by plant root uptake. The lower region of the unsaturated zone is characterized as the capillary zone where the moisture conditions are a function of the capillary rise in the soil, Table 5.1. The zone in between these two regions is characterized as

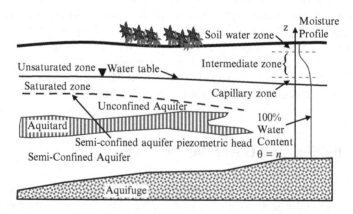

Fig. 5.1 Definition sketch for a layered aquifer system

Table 5.1 Typical capillary rise expected as a function of grain size

Material	Grain size (mm)	Capillary rise (cm)
Fine gravel	2–5	2.5
Very coarse sand	1–2	6.5
Coarse sand	0.5–1.0	13.5
Medium sand	0.2–0.5	24.6
Fine sand	0.1–0.2	42.8
Silt	0.05–0.1	105.5
Fine silt	0.02–0.05	200

the intermediate zone where the moisture gradually changes from the capillary zone level to the soil water zone level. In aquifers the pressure distribution is assumed to be hydrostatic. Pressures are greater than the atmospheric pressure below the water table and less than the atmospheric pressure above the water table due to suction created by capillary forces.

The pore space volume within the soil matrix is mainly a function of the particle size distribution of the soil and is associated with the arrangement of the granular particles within the soil matrix. Accordingly, the soil matrix in aquifers may be classified in terms of the particle size as shown in Table 5.2. The porosity of the soil matrix, n is defined as the ratio of the volume of pores within the control volume to the bulk soil volume,

$$n = \frac{\text{Volume of pores in the control volume}}{\text{Bulk soil control volume}} = \frac{\mathcal{V}_p}{\mathcal{V}_B} \tag{5.1}$$

If the pore space in the soil matrix is partially occupied by air pockets, then the effective porosity should be used, which is defined as the ratio of the pore volume filled by water to the bulk soil volume of the control volume,

$$n_e = \frac{\text{Water filled pore volume}}{\text{Bulk soil control volume}} = \frac{\mathcal{V}_{pw}}{\mathcal{V}_B} \tag{5.2}$$

In the equations given above, the symbol $\mathcal{V}$ is used for volume to distinguish this letter from the velocity symbol which may be used elsewhere in the text. Similar to the definition of porosity given above, is the definition of moisture content, θ and water saturation, S_w are given as.

$$\theta = \frac{\text{Volume of water in the control volume}}{\text{Bulk soil control volume}} = \frac{\mathcal{V}_w}{\mathcal{V}_B} \tag{5.3}$$

$$S_w = \frac{\text{Volume of water in the control volume}}{\text{Volume of pores in the control volume}} = \frac{\mathcal{V}_w}{\mathcal{V}_p} \tag{5.4}$$

Since these properties are defined in terms of volume ratios, they are dimensionless. Based on the definitions of (n, θ, S_w) the relationship between porosity, moisture content and water saturation can be given as follows,

Table 5.2 Particle size classification of soil matrix

Soil matrix	Diameter (mm)
Gravel	>2
Sand	0.05–2
Very coarse	1–2
Coarse	0.5–1
Medium	0.25–0.5
Fine	0.1–0.25
Very fine	0.05–0.1
Silt	0.002–0.05
Clay	<0.002

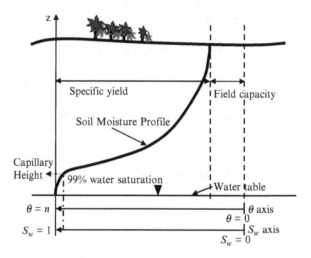

Fig. 5.2 Moisture profile in a soil column

$$\theta = \frac{\Psi_w}{\Psi_B} = \frac{\Psi_w}{\Psi_p}\frac{\Psi_p}{\Psi_B} = S_w n \tag{5.5}$$

Accordingly, the soil moisture profile or water saturation distribution between the soil surface and water table in a soil column can be characterized as shown in Fig. 5.2, in which the aquifer zone at the water table and below are characterized as 100% saturated. The soil moisture in the aquifer decreases as the point of reference moves towards the soil surface indicating a reduction in moisture.

The specific yield, as shown in Fig. 5.2, is used to define the drainable porosity of the soil column, while the field capacity refers to the amount of moisture remaining in the soil column after gravitational drainage.

In groundwater hydrology terminology, the momentum equation is defined in terms of the effective average linear velocity of water in the saturated zone, i.e. the

Darcy law. Darcy velocity, which is the outcome of the Darcy law, is defined as the ratio of the volumetric flow rate of water, Q $[L^3 T^{-1}]$ to the total (gross) cross-section area of the aquifer through which the volumetric discharge is occurring. According to the Darcy law, the Darcy velocity is proportional to the hydraulic conductivity of the soil times the gradient of the piezometric head in the aquifer in the flow direction. In the x-direction, the Darcy law can be given as,

$$q_x = \frac{Q}{A} = -K_{xx}\frac{\partial h}{\partial x} \tag{5.6}$$

where $h = (P/\gamma + z)$ $[L]$ is the piezometric head, P $[ML^{-1}T^{-2}]$ is the pressure and γ $[ML^{-2}T^{-2}]$ is the specific weight of water, K_{xx} $[LT^{-1}]$ is the hydraulic conductivity of the aquifer in the x-direction and z is the elevation in the vertical direction (gravitational direction). The effective average linear velocity, which is also referred to as the pore velocity, is defined as the ratio of the Darcy velocity to the porosity or the effective porosity depending on the definition of the water occupancy conditions of the pore volume as described earlier. In the x-direction the pore velocity can be given as,

$$v_x = \frac{Q}{nA} = -\frac{K_{xx}}{n}\frac{\partial h}{\partial x} \tag{5.7}$$

The hydraulic conductivity term used in Eqs. (5.6) and (5.7) is defined as a function of the properties of the soil matrix, such as particle packing and particle size, as well as the properties of the fluid in the medium, such as viscosity and density. In the definition of the hydraulic conductivity parameter, the relationship between the properties of these two media can be defined as shown in Eq. (5.8),

$$K_{xx} = \kappa\left(\frac{\rho g}{\mu}\right) \tag{5.8}$$

where κ $[L^2]$ is the intrinsic permeability of the soil, ρ $[ML^{-3}]$ and μ $[ML^{-1}T^{-1}]$ are the density and viscosity of the fluid in the pore space, and g $[LT^{-2}]$ is the gravitational acceleration. With this definition the intrinsic permeability becomes a soil property while the hydraulic conductivity is still a function of the soil and fluid properties. Typical hydraulic conductivity values for various soil media are given in Table 5.3. The hydraulic conductivity values given in Table 5.3 are based on the assumption that the fluid in the soil matrix is water, which can be calculated from Eq. (5.8) given intrinsic permeability.

The directional Darcy velocity definition given in Eqs. (5.6) and (5.7) indicates that this definition is more complicated when one considers the anisotropy of the geologic formations. In fact for an anisotropic medium the Darcy velocity is defined in terms of directional values of hydraulic conductivities as well as directional values of hydraulic gradients in each coordinate direction. Based on

Table 5.3 Typical hydraulic conductivity values for various subsurface media

Material	K (cm/s)	K (m/day)
Unconsolidated material		
Gravel	10^{-1}–10^{1}	10^{2}–10^{4}
Sand	10^{-4}–10^{0}	10^{-1}–10^{3}
Silt	10^{-7}–10^{-3}	10^{-4}–10^{0}
Clay and glacial till	10^{-11}–10^{-6}	10^{-8}–10^{-3}
Sedimentary rock		
Sandstone	10^{-8}–10^{-3}	10^{-5}–10^{0}
Limestone, dolomite	10^{-7}–10^{-1}	10^{-4}–10^{2}
Karst limestone	10^{-4}–10^{0}	10^{-1}–10^{3}
Shale	10^{-11}–10^{-6}	10^{-8}–10^{-3}
Crystalline rock		
Basalt	10^{-9}–10^{-5}	10^{-6}–10^{-2}
Fractured basalt	10^{-5}–10^{0}	10^{-2}–10^{3}
Dense crystalline rock	10^{-12}–10^{-8}	10^{-9}–10^{-5}
Fractured crystalline rock	10^{-6}–10^{-2}	10^{-3}–10^{1}

this concept, the three dimensional Darcy velocity in an anisotropic medium can be given as,

$$q_x = -K_{xx}\frac{\partial h}{\partial x}; \quad q_y = -K_{yy}\frac{\partial h}{\partial y}; \quad q_z = -K_{zz}\frac{\partial h}{\partial z} \tag{5.9}$$

where K_{xx}, K_{yy}, K_{zz} are the hydraulic conductivities in the x-, y-, z-coordinate directions respectively. In more complicated cases, for example in an anisotropic soil matrix with principal hydraulic conductivity directions not matching with the principal x-, y-, z-coordinate directions selected in the domain, the three dimensional Darcy equation will be defined using a matrix, Eq. (5.10).

$$\begin{Bmatrix} q_x \\ q_y \\ q_z \end{Bmatrix} = - \begin{bmatrix} K_{xx} & K_{xy} & K_{xz} \\ K_{yx} & K_{yy} & K_{yz} \\ K_{zx} & K_{zy} & K_{zz} \end{bmatrix} \begin{Bmatrix} \partial h/\partial x \\ \partial h/\partial y \\ \partial h/\partial z \end{Bmatrix} \tag{5.10}$$

In this notation, for example, the hydraulic conductivity K_{xy} represents the value of hydraulic conductivity in the x-direction relative to piezometric head gradient in the y-direction and so forth. A more detailed analysis and discussion of these concepts can be found in Bear (1972, 1979); Charbeneau (2000).

The contaminant transport models which are used to simulate dissolved phase contaminant migration in aquifers are characterized by advection–dispersion equation. Advection-dispersion is a generic term, so one should also include the transformation processes within this characterization for it to be complete as it is used in the context of groundwater pathway analysis. The transformation term is associated with the biologic or chemical reactions that may take place in the subsurface as the contaminants migrate through the pores of the soil due to advection and dispersion processes. The derivation of the advection–dispersion-reaction equation is based on the conservation of mass principles covered in Chapter 3 (Bear 1972, 1979). In three-dimensions this equation can be given as:

$$\frac{\partial C}{\partial t} + v_x \frac{\partial C}{\partial x} + v_y \frac{\partial C}{\partial y} + v_z \frac{\partial C}{\partial z} = \frac{\partial}{\partial x}\left(D_x \frac{\partial C}{\partial x}\right) + \frac{\partial}{\partial y}\left(D_y \frac{\partial C}{\partial y}\right) + \frac{\partial}{\partial z}\left(D_z \frac{\partial C}{\partial z}\right)$$
$$+ \sum_{w=1}^{N} (C_s - C)Q_w \delta(x_w, y_w, z_w) + \sum R_{reaction}$$

$$(5.11)$$

In Eq. (5.11), v_x, v_y, v_z $[LT^{-1}]$ are the pore velocities as defined earlier and D_x, D_y, D_z $[L^2T^{-1}]$ are the hydrodynamic dispersion coefficients in the x-, y-, z-directions respectively, Q_w $[L^3T^{-1}]$ is the source and sink strength in the solution domain, C_s $[ML^{-3}]$ is the source concentration, C $[ML^{-3}]$ is the dissolved phase contaminant concentration in the aquifer pore space and $R_{reaction}$ $[ML^{-3}T^{-1}]$ represent various reactions that may affect the migration of the solute in the subsurface as described in Chapter 3.

The hydrodynamic dispersion terms in Eq. (5.11) represent a combination of processes, namely the mechanical dispersion and molecular diffusion processes as given below,

$$D_x = \alpha_x v_x + D_m$$
$$D_y = \alpha_y v_x + D_m \qquad (5.12)$$
$$D_z = \alpha_z v_x + D_m$$

where D_m $[L^2T^{-1}]$ is the molecular diffusion term for the chemical in the soil water and the terms $(\alpha_x v_x,\ \alpha_y v_x,\ \alpha_z v_x)$ $[L^2T^{-1}]$ are the dispersion terms. The mixing that may occur due to the channeling property of the aquifer pore space is identified in terms of dispersivity, which is characterized as $(\alpha_x,\ \alpha_y,\ \alpha_z)$ $[L]$. As shown in Fig. 5.3, mechanical mixing due to the channeling effect of the pore space arrangement contributes to the overall mixing process. In some cases, assuming that the primary groundwater flow direction is in the x-axis direction, the dispersivity in the primary groundwater flow direction α_x is also identified as the longitudinal

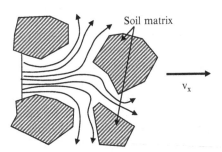

Fig. 5.3 Dispersion effects in a pore space

dispersivity term, α_L, while for dispersivities in transverse directions, i.e. the y- and z-directions, the dispersivity terms (α_y, α_z) may also be identified as the transverse dispersivity terms $(\alpha_{Ty}, \alpha_{Tz})$. The rule of thumb to estimate the dispersivity terms can be given as,

$$\alpha_L = 0.1L \quad \text{and} \quad \alpha_{Ty} = \alpha_{Tz} = 0.1\alpha_L \tag{5.13}$$

where L is the length of the solution domain or the contaminant transport distance that is analyzed in the longitudinal direction. The longitudinal dispersion term may also be estimated using the following empirical equations (Bear 1979; Charbeneau 2000),

$$\begin{aligned} \alpha_L &= 0.831 (\log L)^{2.414} \quad \text{for} \quad L > 100 \text{ m} \\ \alpha_L &= 0.01691 (\log L)^{1.53} \quad \text{for} \quad L < 100 \text{ m} \end{aligned} \tag{5.14}$$

As is the case with the hydraulic conductivity definition for an anisotropic media the longitudinal and transverse dispersion terms may need to be defined in terms of a matrix for an anisotropic media as given below,

$$\mathbf{D_H} = \begin{bmatrix} D_{xx} & D_{xy} & D_{xz} \\ D_{yx} & D_{yy} & D_{yz} \\ D_{zx} & D_{zy} & D_{zz} \end{bmatrix} \tag{5.15}$$

The mathematical definitions of the hydrodynamic dispersion terms of the matrix above take the form given below in terms of longitudinal and transverse dispersivities and the pore velocities in three dimensions (Bear 1979; Charbeneau 2000).

$$\begin{aligned} D_{xx} &= \frac{\left(\alpha_L v_x^2 + \alpha_T v_y^2 + \alpha_T v_z^2\right)}{\sqrt{\left(v_x^2 + v_y^2 + v_z^2\right)}}; \quad & D_{yy} &= \frac{\left(\alpha_T v_x^2 + \alpha_L v_y^2 + \alpha_T v_z^2\right)}{\sqrt{\left(v_x^2 + v_y^2 + v_z^2\right)}} \\ D_{zz} &= \frac{\left(\alpha_T v_x^2 + \alpha_T v_y^2 + \alpha_L v_z^2\right)}{\sqrt{\left(v_x^2 + v_y^2 + v_z^2\right)}}; \quad & D_{xy} = D_{yx} &= \frac{(\alpha_L - \alpha_T)(v_x v_y)}{\sqrt{\left(v_x^2 + v_y^2 + v_z^2\right)}} \end{aligned} \tag{5.16}$$

Equation (5.11) represents the most general form of the advection–dispersion–reaction equation. The solution of this parabolic partial differential equation, which may be used to represent problems in complex heterogeneous domains, can only be obtained using numerical procedures, excluding some special cases. In general, this mathematical model cannot be used in analytical modeling studies, since the solution of this parabolic partial differential equation is not possible when analytical methods are used, again excluding some special cases. On the other hand, the analytical solution of this three dimensional equation can be easily obtained when we make some simplifying assumptions, such as the assumption that the groundwater flow in the aquifer is unidirectional, the aquifer domain is homogeneous and that there are no sources or sinks in the aquifer. With these assumptions Eq. (5.11) can be written as,

$$\frac{\partial C}{\partial t} + v_x \frac{\partial C}{\partial x} = D_x \frac{\partial^2 C}{\partial x^2} + D_y \frac{\partial^2 C}{\partial y^2} + D_z \frac{\partial^2 C}{\partial z^2} + \sum R_{reaction} \qquad (5.17)$$

Notice that, although we are only considering the longitudinal velocity in Eq. (5.17), this equation is still a three dimensional partial differential equation with respect to dispersion effects.

Most analytical solutions of the advection–dispersion-reaction equation that are reported in the literature are based on either the full form or a reduced form of Eq. (5.17) as described in Chapter 3. Equation (5.17) is a second order parabolic partial differential equation, thus it is also necessary to describe the initial and boundary conditions for this mathematical model for closure (see Chapter 3). These initial and boundary conditions can be given as:

$$
\begin{aligned}
C &= C_0(x, y, z, 0) & \forall\, (x, y, z); & \qquad \text{Initial Condition (I. C.)} \\
C &= C_1(x_b, y_b, z_b, t) & \forall\, (x_b, y_b, z_b); & \qquad \text{Dirichlet B. C.} \\
D_{x_i} \frac{\partial C}{\partial x_i} &= C_2(x_b, y_b, z_b, t) & x_i = x, y, z; \forall\, (x_b, y_b, z_b); & \qquad \text{Neuman B. C.} \\
v_{x_i} C - D_{x_i} \frac{\partial C}{\partial x_i} &= C_3(x_b, y_b, z_b, t) & x_i = x, y, z; \forall\, (x_b, y_b, z_b); & \qquad \text{Cauchy B. C.}
\end{aligned}
$$

$$(5.18)$$

where (x_b, y_b, z_b) are the coordinates of the boundaries of the solution domain and (C_0, C_1, C_2, C_3) are known functions which define the boundary condition value for the concentration at the respective boundaries of the solution domain. When the reduced or the complete form of Eq. (5.17) is considered with various combinations of the boundary and initial conditions given in Eq. (5.18) the outcome is several mathematical models that can be used in the definition of contaminant transport problems in the groundwater pathway which are of significant practical importance in engineering applications.

In these applications, at least for certain two-dimensional cases, it may be necessary to describe the upstream boundary condition as a probability density function in the transverse direction instead of as a constant Dirichlet boundary condition at a point or within a finite length in the transverse direction, as given in Eq. (5.18). For these cases, where it is assumed that the lateral extent of the source is not known precisely, the contaminant source concentration can be defined as a Gaussian distribution in the transverse direction, and this distribution is assumed to be uniform over the vertical mixing depth or the source penetration depth $H\ [L]$ of the aquifer. Mathematically, this boundary condition can be given as:

$$C(0,y,z,t) = \begin{cases} C_0 \exp\left[-(y-y_c)^2/(2\sigma^2)\right] & -\infty \le y \le +\infty\,;\, 0 \le z \le H \\ 0 & -\infty \le y \le +\infty\,;\, H \le z \le B \end{cases} \qquad (5.19)$$

In Eq. (5.19), C_0 $[ML^{-3}]$ is the maximum dissolved phase contaminant concentration of the solute at the source and occurs at the center of the Gaussian distribution. The standard deviation σ is a measure of the width of the source in the transverse y-coordinate direction, where the concentration values are assumed to be variable. A typical physical case for which the Gaussian distribution based upstream boundary condition can be used is shown in Fig. 5.4, in which the solute concentration at the upstream boundary is known through the observations made at several monitoring wells and the data is approximately represented as a Gaussian distribution. For these cases the standard deviation of the distribution can be determined from the field data as given in Eq. (5.20) or it can be estimated based on the observation of the width of the source concentration in the aquifer,

$$\sigma = \frac{y - y_c}{\sqrt{-2\ln(C/C_0)}} \qquad (5.20)$$

where C is the concentration observed at the boundary at a distance $(y - y_c)$ away from the point of maximum concentration. This boundary condition is included in the ACTS software for two dimensional applications and turns out to be a very useful boundary condition as demonstrated in an Agency for Toxic Substances and Disease Registry (ATSDR) study (Anderson et al. 2007), in which historical contamination analysis of a pesticide contamination event at a site in Georgia was analyzed using the ACTS software.

In some other applications it may also be necessary to introduce a time dependent Dirichlet boundary condition for the dissolved contaminant concentration value, C_0 as an upstream boundary condition. This case may be modeled using the superposition method, which can be accessed through the "Boundary Condition" folder in the ACTS software for groundwater pathway models. For example,

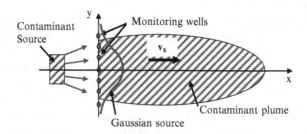

Fig. 5.4 Gaussian source upstream boundary condition

using the superposition method, the upstream boundary condition C_0 can be selected to be constant for all times (continuous release):

$$C(0, y, z, t) = C_0 \quad t > 0 \tag{5.21}$$

or it can be selected as a finite pulse upstream source which can be described as:

$$\begin{aligned} C(0, y, z, t) &= C_0 & 0 \leq t \leq T_s \\ C(0, y, z, t) &= 0 & t > T_s \end{aligned} \tag{5.22}$$

in which T_s represents the source concentration duration. This concept can also be extended to define a time dependent Dirichlet boundary condition value $C_{0,i}$ which changes over several periods, Eq. (5.23).

$$\begin{aligned} C(0, y, z, t) &= C_{0,0} & 0 \leq t \leq t_1 \\ C(0, y, z, t) &= C_{0,1} & t_1 \leq t \leq t_2 \\ C(0, y, z, t) &= C_{0,2} & t_2 \leq t \leq t_3 \\ C(0, y, z, t) &= 0 & t_3 \leq t \leq T_s \end{aligned} \tag{5.23}$$

All of the cases given above can be analyzed using the superposition method, which is an option that is included in all groundwater pathway models of the ACTS software.

The superposition method may be applied to the solution of linear differential equations, which is the case in saturated groundwater contaminant transport models that are implemented in the ACTS software. Mathematically, the superposition method can be defined as the addition of time lapse solutions of the advection–dispersion-reaction equation,

$$C(x, y, z, t) = C_{0,0} S(x, y, z, t) + \left(C_{0,1} - C_{0,0} \right) S(x, y, z, [t - t_1]) \tag{5.24}$$

in which $C_{0,0}$ is the initial solute concentration at the boundary, t_1 is the time at which solute concentration changes at the boundary, $C_{0,1}$ is the solute concentration at the boundary after $(t = t_1)$, and $S(x, y, z, t)$ is the analytical solution of the model selected in which the concentration is defined as a function of space and time. Although a one time superposition index is given in Eq. (5.24), the superposition interval can be selected to be more than one as given in Eq. (5.23). ACTS software allows the user to implement several superposition calculations as can be seen on the boundary condition input window of each module.

A simple example may provide further insight to the superposition method (Wexler 1989). Let's assume that the solute is passing through a 100 cm long soil column for a period of 10 h with flow velocity equal to 0.5 cm/h, hydrodynamic dispersion coefficient is equal to 0.05 cm^2/h, and $C_{0,0}$ equal to 100 mg/L. At the end of the 10 h period, the concentration of the influent is increased to $C_{0,1}$, which is equal to 300 mg/L. Of interest is the concentration at 10 cm from the entrance section at the end of a total elapsed time of 20 h.

The analytical solution for the transport of a conservative solute in a semi-infinite column with a Dirichlet boundary condition in which the downstream boundary effects are negligible can be given as,

$$S(x,t) = \frac{1}{2}\left(erfc\left[\frac{x - v_x t}{2\sqrt{D_x t}}\right] + \exp\left[\frac{v_x x}{D_x}\right] erfc\left[\frac{x + v_x t}{2\sqrt{D_x t}}\right]\right) \tag{5.25}$$

Here the "erfc" is the complementary error function. Based on Eq. (5.24), the superposition solution can be given as,

$$\begin{aligned}
C(10\,cm,\,20\,h) &= 100\,mg/L \times S(10\,cm,\,20\,h) \\
&\quad + (300\,mg/L - 100\,mg/L) \times S(10\,cm,\,[20 - 10]\,h)
\end{aligned} \tag{5.26}$$

The superposition approach described above can be extended to include additional time steps and contaminant concentration levels at these time steps. Using this data, which is entered by the user in the boundary condition input window of the ACTS software, the superposition calculations are automatically calculated internally in all linear analytical models that are included in the software.

5.2 Groundwater Pathway Models

Biologic and chemical transformation has various effects on contaminant concentration as the contaminants migrate through the soil media. One of these effects is the decay property of a chemical, which is discussed in Chapter 3. The other property is the possibility of the formation of daughter by-products. Accordingly, the groundwater pathway models can be classified as multispecies or single species models depending on whether one includes or excludes the formation of daughter by-products of the chemicals studied in the analysis. Multispecies analysis occurs when the daughter by-product formation is considered and single species analysis occurs when the daughter by-product generation is ignored. In the ACTS software there is the option of modeling both cases for all models that are included in the software.

From the opening window of the ACTS software, when the user selects the groundwater pathway icon on the menu bar to access the saturated groundwater pathway models, or if the "Groundwater Path" option is selected under the "Pathways" pull down menu, the groundwater pathway module starts with the window shown in Fig. 5.5. In this window there are three options that are available to the user. The "File" menu option allows the user to create a "New" application data file, "Open" an old application data file, "Edit" an existing data file that is open, "Close" an open application data file or "Exit" the groundwater pathway module. When the

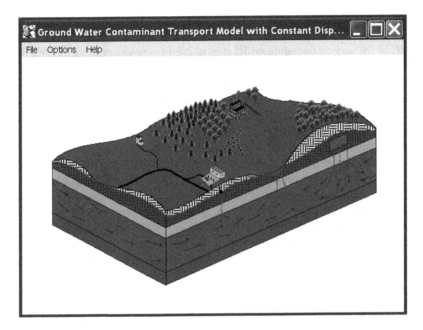

Fig. 5.5 Groundwater pathway opening window

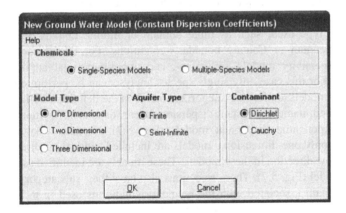

Fig. 5.6 Constant dispersion groundwater pathway model selection window

new file option is selected, the user is given further options to enter into the specific model types as shown in Fig. 5.6. As can be seen on this window (Fig. 5.6) in the groundwater pathway models there is an option of working with single species or multi-species models. This option is not similar to the multiple chemicals option we have used in the air pathway models in Chapter 4. In air pathway models, the selection of multiple chemicals from a chemical database implied that the selected air pathway model would be executed for all chemicals selected using their chemical properties, which are directly obtained from the chemical database. In the groundwater pathway module the "multi-species" option refers to the analysis of applications in which degradation by-products are of concern. For example if PCE is the source contaminant its degradation by-products are expected to be TCE, DCE, VC and ethane. The formation of these by-products and the analysis of the migration of these by-products can also be performed by the selection of the multi-species option in the groundwater pathway models. In that case the next data entry window will have an additional folder when compared to the single species case. In this data entry folder, the user will enter the degradation rates of the chemical sequences that will be considered in the analysis. This is an important and very powerful option by which all single species analytical solutions are extended to the analysis of the degradation by-product cases. This procedure is available for all one-, two- and three-dimensional models. The mathematical procedures used in these calculations are described in Section 5.4 since these processes are generic to all of the saturated constant dispersion models which will be discussed first.

Models that are included in the groundwater pathway of the ACTS software are based on a specific set of assumptions, boundary conditions and parameters which form the basis of the input database required to execute the selected model. For example, the groundwater pathway module consists of saturated constant dispersivity models, saturated variable dispersivity models and unsaturated constant dispersivity models. These classifications are further subdivided into sub-category models, identified as one-, two- and three-dimensional models which can be run for single or multi-species applications (Fig. 5.7). Under each group, there are still further subdivisions that categorize the models based on the boundary conditions and aquifer type used, such as finite domain or infinite domain aquifers with Dirichlet, Neuman or Cauchy boundary conditions. For the case of saturated variable dispersivity models, only one-, and two-dimensional applications are considered. For this case only single species analysis can be performed. Variable dispersivity models are further categorized based on the definition of the dispersivity model used, namely, constant dispersion, linear dispersion, asymptotic dispersion and exponential dispersion models (Fig. 5.7). For the unsaturated zone simulations only one-dimensional models are included, since the z-direction is the dominant flow direction in these models. These models are identified as the Marino and Jury model (Fig. 5.7). These selections also have their subcategory groupings which are again associated with the boundary conditions used in them. Overall, these combinations yield a significant collection of groundwater pathway transformation and transport models that can be used in the analysis of various applications (Fig. 5.7).

Monte Carlo simulations can also be used on a number of parameters for each of these models. The ACTS graphics package included in the software provides a user friendly interface to review the results of the computations. Alternatively, numerical results can be accessed and viewed through the use of a text editor available to the user on his or her computer. A number of sample data files are provided to enhance the interpretation of different functions of the groundwater module. All groundwater pathway models used in the ACTS software are generic models which are reported in the literature. Thus, their application in site-specific cases requires

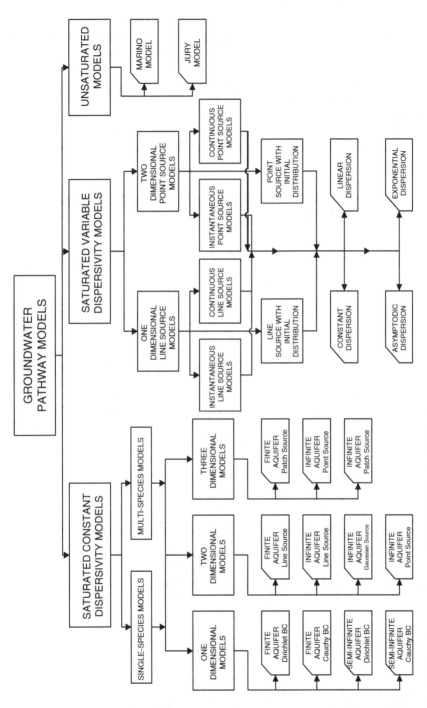

Fig. 5.7 Groundwater pathway models included in the ACTS software

knowledge of the assumptions and limitations that are inherent in these models. In this chapter, a description of the models included in the groundwater pathway module is given in sufficient detail to help users to understand the limitations of these models and the procedures that are necessary to follow to implement them in specific applications. The user may supplement the information provided here with the discussion provided in other groundwater modeling reference books or technical publications such as Anderson (1984), Bear (1972, 1979), Charbeneau (2000), Dagan (1986), Fetter (1999), Freeze and Cherry (1979), Huyakorn et al. (1987), Maslia and Aral (2004), Schnoor (1996) and Wexler (1989).

5.3 Saturated Constant Dispersion Coefficient Contaminant Transport Models

Saturated constant dispersion models refer to a category of models in which the dispersion coefficient is assumed to be a constant throughout the solution domain of the problem. As it is reported in the literature, in groundwater pathway analysis this assumption usually does not hold (Bear 1972; Dagan 1986). For this purpose another class of models are developed and included in the ACTS software , through which the dispersion coefficient can be chosen to be a variable within the solution domain. A review of these models is given in Section 5.5, but we will first start with a review of the constant dispersion models. The use and application of these models are more common in the groundwater pathway analysis literature. The saturated constant dispersion models are further categorized into one-, two- and three-dimensional models as discussed below (Figs. 5.6 and 5.7). Also, as described earlier degradation byproduct analysis can be performed for all these models if the multi-species option is selected in the window shown in Fig. 5.6.

5.3.1 One-Dimensional Contaminant Transport Models with Constant Dispersion Coefficient

One-dimensional contaminant transport models may be used in cases where the aquifer is relatively shallow such that the contaminants in the aquifer can be assumed to be uniformly mixed in the vertical direction. In this case the transverse concentration gradients are also considered to be negligible. For this case two types of aquifers can be analyzed, namely a finite length aquifer and a semi-infinite aquifer. Either selection allows the user to implement two types of boundary conditions, a Cauchy boundary condition and a Dirichlet boundary condition Eq. (5.18), as shown in Fig. 5.8. In these applications and throughout the rest of this chapter the following definitions are used.

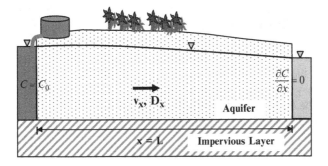

Fig. 5.8 Definition sketch for a one-dimensional groundwater pathway model

Finite Length Aquifer: In a finite length aquifer system, the downstream boundary is assumed to be close to the upstream contaminant source boundary, such that the conditions specified on the downstream boundary will have an effect on the magnitude and distribution of concentrations in the solution domain of interest within the solution time period. If this effect is not desired, then a semi-infinite domain aquifer option can be used for the solution. Accordingly, another interpretation of this condition can be associated with the solution time period used. If the solution period is selected to be large, then the contaminant front may migrate close to the downstream boundary within this period and the downstream boundary effect may again become important. If this is not a desired condition, then a semi-infinite domain aquifer should be selected to analyze the problem.

Semi-infinite Aquifer: In a semi-infinite aquifer system, the downstream boundary is assumed to be far away from the upstream contaminant source, such that the downstream boundary condition will have a negligible effect on the concentration distribution in the solution domain within the solution time period. To give another frame of reference, a semi-infinite aquifer system can be used when the number of displaced pore volumes in the aquifer is less than 0.25.

Infinite Aquifer: An infinite aquifer refers to the case where both the upstream and downstream boundaries are considered to be far away from the source concentration, such that the effect of the boundary conditions at these boundaries does not affect the solution in the solution domain. Here it is assumed that the boundary condition is placed somewhere in the middle of the solution domain.

Depending on the aquifer type and boundary conditions selected, the user may analyze one of the following four problems, which utilize the analytical solutions given in Sections 5.3.1.1 through 5.3.1.4. All four analytical solutions discussed in these sections are based on the reduced form of the three-dimensional differential equation (5.17) given below,

$$R \frac{\partial C}{\partial t} + v_x \frac{\partial C}{\partial x} - D_x \frac{\partial^2 C}{\partial x^2} + R \lambda\, C = 0 \qquad (5.27)$$

in which v_x and D_x are the pore velocity and the hydrodynamic dispersion coefficients in the longitudinal direction and C is the solute concentration. The groundwater pore velocity in the longitudinal direction is calculated internally in the ACTS software given the Darcy velocity in the longitudinal direction and the porosity, R is the retardation coefficient which is defined in Eq. (5.28) and λ is the first order decay coefficient, which is defined in Eq. (5.29). In this case the solution domain may be characterized as shown in Fig. 5.8.

The retardation coefficient is defined as (see also Chapter 3),

$$R = 1 + \frac{\rho_b K_d}{n} \tag{5.28}$$

and the effective decay coefficient is defined as,

$$\lambda = \frac{\lambda_1 n + \lambda_2 \rho_b K_d}{n + \rho_b K_d} + \lambda_b \tag{5.29}$$

In Eqs. (5.28) and (5.29) $\rho_b \, [ML^{-3}]$ is the bulk density of the porous media and $K_d \, [L^3 M^{-1}]$ is the distribution coefficient of the contaminant in the solid and liquid phases. In Eq. (5.29), $\lambda_1 \, [T^{-1}]$ is the first-order decay constant for the dissolved phase; $\lambda_2 \, [T^{-1}]$ is the first-order decay constant for the sorbed phase, $\lambda_b \, [T^{-1}]$ is the first-order lumped biodegradation rate in the saturated zone, and n is the aquifer porosity as defined earlier. The reader should also recognize that in Fig. 5.8, an unconfined aquifer is shown. In this case, since the water table is characterized as a parabolic surface, the Darcy velocity cannot be constant in the aquifer. Thus, the constant longitudinal velocity used for these applications is an approximation. This is an assumption which may hold for most field applications in which the aquifer is shallow. For a constant thickness confined aquifer case this assumption and approximation is not necessary since the longitudinal velocity will be constant.

Given Eq. (5.27) and the solution domain shown in Fig. 5.8, the following boundary conditions can be used to describe a one-dimensional application.

Dirichlet Boundary Condition: This type of boundary condition is commonly used at the inflow boundaries, where the concentration value is known. Mathematically this condition can be given as,

$$C = C_0 \qquad at \quad x = 0; \quad t \geq 0 \tag{5.30}$$

This boundary condition is sometimes referred to as the "first-type" boundary condition as well, Eq. (5.18).

Cauchy Boundary Condition: This boundary condition is also identified as the "mixed" or "third-type" boundary condition. For example, this boundary condition can be used in applications in which an aquifer is in contact with the contaminant source over a relatively thin semi-pervious confining layer through which seepage is occurring. An example of this could be the case of a contaminated river or a lake, where the bottom sediments or deposits act as the semi-pervious layer and the leakage through the semi-pervious region controls the influx of contaminants into the aquifer. Mathematically this condition can be described as,

$$v_x C - D_x \frac{\partial C}{\partial x} = C_0 \qquad at \quad x = 0, L \quad or \quad x \Rightarrow +\infty; \quad t \geq 0 \qquad (5.31)$$

Neuman Boundary Condition: The Cauchy boundary condition given above may be simplified to a special case referred to as a Neuman boundary condition, which is most commonly used to define a downstream boundary condition,

$$\frac{\partial C}{\partial x} = 0 \qquad at \quad x = L \quad or \quad x \Rightarrow +\infty; \quad t \geq 0 \qquad (5.32)$$

This boundary condition is also identified as the "second-type" boundary condition in the groundwater contaminant transport analysis literature. This boundary condition implies zero contaminant flux through the boundary on which it is defined.

Based on these boundary conditions and the finite and semi-infinite aquifer domain options that can be selected, the user may select one of the four applications described below. Further, in these applications, as described earlier, the superposition method can also be used to change the concentration magnitudes at the source boundary for single or multiple chemical options over several periods when a Dirichlet boundary condition is used.

The analytical solutions of these applications are included in the ACTS software as modules, which can be accessed through the menus of the ACTS software. The use of these models will provide the user with numerical results of a deterministic solution to the problem in spatial and temporal dimensions, i.e., all parameters of the problem are defined as deterministic values in the input window of each model. Each of these cases may also be analyzed in a Monte Carlo sense, which implies that a certain set of parameters or all parameters of the problem can be represented in terms of probability distributions. For these cases, the user is given the option to select from six probability distributions imbedded in the ACTS software. The user may generate as many random variables as desired, based on the specified mean, variance and the range of an input parameter. This option will provide the user with the ability to perform sensitivity and uncertainty analysis on certain variants of the selected model. A detailed description of the theory behind the Monte Carlo analysis can be found in Chapter 7 of this book. Numerical results obtained for each of these cases can either be viewed using a text editor available in the WINDOWS[TM] environment, or these results may be viewed through the graphics module of the ACTS software in a graphical format. In the case of deterministic solutions, the user may select to view contour plots or breakthrough curves in the spatial and time domains. In the case of Monte Carlo analysis, numerical results may be viewed in terms of probability distributions at a selected point in space and time. A more detailed description of the use of the graphics package can be found in Appendix 3.

Assumptions and Limitations: The more important assumptions and limitations of the one-dimensional saturated zone transformation and transport models are as follows:

i. The flow field within the saturated zone is at steady state and unidirectional.
ii. The seepage velocity and other model parameters (e.g., the diffusion coefficient, partition coefficient) are uniform and constant (i.e., the aquifer is homogeneous and isotropic).
iii. Transport is assumed to be strictly one-dimensional. Lateral and transverse advection and diffusion or dispersion is neglected.
iv. Decay of the solute may be described by a first-order decay constant. The daughter products of chemicals are neglected.
v. Heterogeneous reactions considered are first order and are represented by the retardation coefficient.
vi. The solution domain considered is either of finite length or of semi-infinite length.

5.3.1.1 Finite Aquifer with a Dirichlet Boundary Condition

The governing equation and the boundary conditions of this problem are given as:
Governing equation,

$$\frac{\partial C}{\partial t} + v\frac{\partial C}{\partial x} - D\frac{\partial^2 C}{\partial x^2} + \lambda C = 0 \tag{5.33}$$

Boundary conditions,

$$
\begin{aligned}
C &= C_0 & at\ x = 0; & \quad t \geq 0 \\
\frac{\partial C}{\partial x} &= 0 & at\ x = L; & \quad t \geq 0
\end{aligned}
\tag{5.34}
$$

and the initial condition,

$$C = 0 \quad at\ 0 < x < L; \quad t = 0 \tag{5.35}$$

Given the one-dimensional model in Eq. (5.27), one should notice that velocity v $[LT^{-1}]$ and hydrodynamic dispersion coefficient D $[L^2T^{-1}]$ in Eq. (5.33) are defined as,

$$v = \frac{v_x}{R}; \quad D = \frac{D_x}{R} \tag{5.36}$$

in which v_x is the pore velocity in the x-direction, which is calculated internally using the porosity and the Darcy velocity values entered Eq. (5.7), D_x is the hydrodynamic dispersion coefficient in the x-direction and R is the retardation coefficient. The analytical solution of this problem is given as (Bear 1972; Wexler 1989),

$$
C(x,t) = C_0 \left\{ \frac{\exp\left[\frac{(v-U)x}{2D}\right] + \frac{(U-v)}{(U+v)}\exp\left[\frac{v+U}{2D}x - \frac{UL}{D}\right]}{\left[1 + \frac{U-v}{U+v}\exp\left[\frac{-UL}{D}\right]\right]} \right.
$$

$$
\left. - 2\exp\left[\frac{vx}{2D} - \lambda t - \frac{v^2 t}{4D}\right] \sum_{i=1}^{\infty} \frac{\beta_i \sin\left(\frac{\beta_i x}{L}\right)\left[\beta_i^2 + \left(\frac{vL}{2D}\right)^2\right]\exp\left[-\frac{\beta_i^2 Dt}{L^2}\right]}{\left[\beta_i^2 + \left(\frac{vL}{2D}\right)^2 + \left(\frac{vL}{2D}\right)\right]\left[\beta_i^2 + \left(\frac{vL}{2D}\right)^2 + \left(\frac{\lambda L^2}{D}\right)\right]} \right\}
$$

in which $U = \sqrt{v^2 + 4\lambda D}$ and β_i are the roots of the equation $\left(\beta \cot \beta + \frac{vL}{2D} = 0 \right)$

$$(5.37)$$

One should also notice that this solution can be used to model a conservative contaminant by assuming a zero decay coefficient. In that case $U = v$ in Eq. (5.37) and the solution given above is still valid. One should also notice that in the ACTS input the half-life of the contaminant is entered to calculate the decay rate, which is an internal calculation in ACTS. Thus, a conservative contaminant is characterized as a contaminant which has a very large half-life. The same argument is also valid for the definition of the retardation coefficient. A retardation coefficient of $R = 1$ implies an application without a heterogeneous reaction, and $R > 1$ will imply the case in which a heterogeneous reaction is considered. Both of these cases can be solved using the Eq. (5.37) since only the definitions of the advection, dispersion and reaction terms of the above solution are changing, Eq. (5.36). This will always be the case for all models considered in this chapter, and the reader should keep these variations in mind when different applications are considered.

5.3.1.2 Finite Aquifer with a Cauchy Boundary Condition

The governing equation and the boundary conditions of this problem are given as:
Governing equation,

$$
\frac{\partial C}{\partial t} + v\frac{\partial C}{\partial x} - D\frac{\partial^2 C}{\partial x^2} + \lambda C = 0 \tag{5.38}
$$

Boundary conditions,

$$
vC - D\frac{\partial C}{\partial x} = vC_0 \quad at\ x = 0; \quad t \geq 0
$$
$$
\frac{\partial C}{\partial x} = 0 \quad at\ x = L; \quad t \geq 0 \tag{5.39}
$$

Initial condition,

$$C = 0 \quad at\ 0 < x < L; \quad t = 0 \tag{5.40}$$

Given the one-dimensional equation (5.27), one should again notice that the longitudinal pore velocity v and the longitudinal dispersion coefficient D in Eq. (5.38) are defined as shown in Eq. (5.36). The analytical solution of this problem is given as (Bear 1979; Cherry et al. 1984; Wexler 1989),

$$C(x,t) = C_0 \left\{ \frac{\exp\left[\frac{(v-U)x}{2D}\right] + \frac{(U-v)}{(U+v)}\exp\left[\frac{v+U}{2D}x - \frac{UL}{D}\right]}{\left[\frac{U+v}{2v} + \frac{(U-v)^2}{2v(U+v)}\exp\left[\frac{-UL}{D}\right]\right]} \right.$$

$$\left. - 2\frac{vL}{D}\exp\left[\frac{vx}{2D} - \lambda t - \frac{v^2 t}{4D}\right] \sum_{i=1}^{\infty} \frac{\beta_i \left[\beta_i \cos\left(\frac{\beta_i x}{L}\right) + \left(\frac{vL}{2D}\right)\sin\left(\frac{\beta_i x}{L}\right)\right]\exp\left[-\frac{\beta_i^2 Dt}{L^2}\right]}{\left[\beta_i^2 + \left(\frac{vL}{2D}\right)^2 + \left(\frac{vL}{2D}\right)\right]\left[\beta_i^2 + \left(\frac{vL}{2D}\right)^2 + \left(\frac{\lambda L^2}{D}\right)\right]} \right\}$$

in which $U = \sqrt{v^2 + 4\lambda D}$ and β_i are the roots of the equation $\left(\beta \cot \beta - \frac{\beta^2 D}{vL} + \frac{vL}{2D} = 0\right)$

$$\tag{5.41}$$

The application and the use of the decay coefficient or the retardation coefficient are the same as the previous model given in Section 5.3.1.1.

5.3.1.3   Semi-infinite Aquifer with a Dirichlet Boundary Condition

The governing equation and the boundary conditions of this problem are given as:
Governing equation,

$$\frac{\partial C}{\partial t} + v\frac{\partial C}{\partial x} - D\frac{\partial^2 C}{\partial x^2} + \lambda C = 0 \tag{5.42}$$

Boundary conditions,

$$C = C_0 \qquad at \quad x = 0; \ t \geq 0$$
$$\frac{\partial C}{\partial x} = 0 \qquad at \quad x \Rightarrow \infty; \ t \geq 0 \tag{5.43}$$

Initial condition,

$$C = 0; \qquad 0 < x < \infty; \quad t = 0 \tag{5.44}$$

Again, the parameters v and D in Eq. (5.42) are defined as shown in Eq. (5.36). The analytical solution of this problem is given in Bear (1979), Wexler (1989) as,

$$C(x,t) = \frac{C_0}{2}\left\{\exp\left[\frac{x(v-U)}{2D}\right]erfc\left[\frac{x-Ut}{2\sqrt{Dt}}\right] + \exp\left[\frac{x(v+U)}{2D}\right]erfc\left[\frac{x+Ut}{2\sqrt{Dt}}\right]\right\}$$

where $U = \sqrt{v^2 + 4\lambda D}$.   (5.45)

The definition of the decay coefficient or the retardation coefficient in equation above is the same as in the previous models given in Section 5.3.1.1 or 5.3.1.2.

5.3.1.4   Semi-infinite Aquifer with a Cauchy Boundary Condition

The governing equation and the boundary conditions of this problem are given as:
The governing equation,

$$\frac{\partial C}{\partial t} + v\frac{\partial C}{\partial x} - D\frac{\partial^2 C}{\partial x^2} + \lambda C = 0 \qquad (5.46)$$

Boundary conditions,

$$vC + D\frac{\partial C}{\partial x} = vC_0 \quad at\ x = 0; \quad t \geq 0$$
$$\frac{\partial C}{\partial x} = 0 \qquad at\ x \Rightarrow \infty; \quad t \geq 0 \qquad (5.47)$$

The initial condition,

$$C = 0; \quad 0 < x < \infty; \quad t = 0 \qquad (5.48)$$

Given the form of the one-dimensional equation (5.27), one should again notice that v and D in Eq. (5.46) are defined as shown in Eq. (5.36). The analytical solution of this problem is given as (Bear 1979; Wexler 1989),

$$C(x,t) = \frac{C_0 v^2}{4\lambda D}\left\{2\exp\left[\frac{xv}{D} - \lambda t\right]erfc\left[\frac{x+Ut}{2\sqrt{Dt}}\right] + \left(\frac{U}{v} - 1\right)\exp\left[\frac{x(v+U)}{2D}\right]\right.$$

$$\left. \times erfc\left[\frac{x-Ut}{2\sqrt{Dt}}\right] - \left(\frac{U}{v} + 1\right)\exp\left[\frac{x(v+U)}{2D}\right]erfc\left[\frac{x+Ut}{2\sqrt{Dt}}\right]\right\}$$

in which $U = \sqrt{v^2 + 4\lambda D}$.   (5.49)

The definition of the decay coefficient or the retardation coefficient is similar to those in the previous models discussed above. These definitions will not be repeated.

5.3.2   Two-Dimensional Contaminant Transport Models with Constant Dispersion Coefficient

As is the case for one-dimensional models, several analytical solutions can be defined for the two-dimensional advection–diffusion-reaction equation. Two-dimensional contaminant transport models may be used in cases where the aquifer is relatively shallow, the effect of vertical dispersion of contaminants is minimal and the solute is well mixed within the shallow aquifer in the vertical direction. In these applications the transverse dispersion effect in the y-direction is not ignored. Similar to one-dimensional models, the analytical models used in the ACTS module are given for finite and infinite aquifer domains. In these cases, boundary conditions such as continuous point source, finite line source and Gaussian source cases can be analyzed. The differences between the one-dimensional and two-dimensional cases are that in the two-dimensional case, the boundary conditions are defined on a two-dimensional domain (x, y) and are assumed to be vertically uniform in the aquifer. Further, the dispersive expansion of the contaminant plume is not ignored in the transverse y-axis direction. Since the definitions of the general boundary conditions used in these models were given earlier, they will not be repeated here. For the two-dimensional case the parabolic partial differential equation governing the transformation and transport of a dissolved phase contaminant can be given as,

$$R\frac{\partial C}{\partial t} + v_x\frac{\partial C}{\partial x} = D_{xx}\frac{\partial^2 C}{\partial x^2} + D_{yy}\frac{\partial^2 C}{\partial y^2} - R\lambda C - R\frac{qC}{Bn} \tag{5.50}$$

As can be seen from the equation above, the advective transport of the contaminant is represented using the longitudinal velocity component in the x-axis direction and the dispersion terms are defined in the x- and y-axis directions. In this case, an areal extent of the aquifer is considered in the plan view, and the dilution effects may also be considered if there is infiltration into the aquifer. This is represented by the last term in Eq. (5.50), in which q is the vertical infiltration rate, n is the aquifer porosity and B is the effective aquifer thickness. Also, as described below, for each model considered, the first order reaction term will be replaced by an effective degradation term, which will combine the effect of dilution terms due to infiltration and the first order decay term defined earlier. The calculation of the effective degradation term will be handled internally in the ACTS software based on the input data provided by the user for the parameters λ, q, B and n. In this case the source width, W_s in the transverse direction is also one of the input parameters for the models considered (Fig. 5.9).

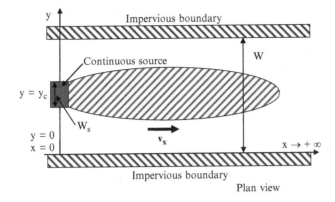

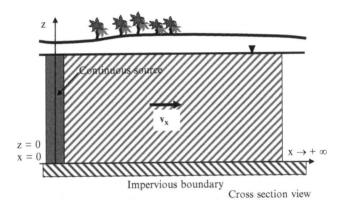

Fig. 5.9 Definition sketch for two-dimensional semi-infinite aquifer of finite width with a finite line source

Based on this general two-dimensional definition, the ACTS software has special modules for the specific scenarios given below. The use of these models will provide the user with a deterministic solution of the problem selected in spatial and temporal dimensions. In all cases the user may also choose to perform a Monte Carlo analysis on all or a selected set of model parameters. This option will provide the user with the ability to perform a sensitivity analysis on the parameters of the model. Numerical results obtained for either case can be viewed using a text editor that may be available in the WINDOWS^{TM} environment, or these results can be viewed through the graphics module that is compatible with all modules included in the ACTS software. In the case of deterministic solutions, the user may select to view contour and breakthrough plots in the spatial or time domain. In the case of Monte Carlo analysis, the numerical results may be viewed in terms of probability density function plots at a point in space and time. A more detailed description of the use of the graphics package can be found in Appendix 3.

Assumptions and Limitations: The more important assumptions and limitations of the two-dimensional saturated zone transformation and transport models are as follows:

i. The flow field within the saturated zone is at steady state and unidirectional.
ii. The seepage velocity and other model parameters (e.g., the diffusion coefficient, partition coefficient) are uniform and constant (i.e., the aquifer is homogeneous and isotropic).
iii. Advective transport is assumed to be strictly one-dimensional in the longitudinal direction. The transverse diffusion or dispersion in the y-direction is not neglected.
iv. Solute in the aquifer is well mixed in the vertical direction, thus the solution is *two*-dimensional in the (x, y) domain.
v. Decay of the solute may be described by a first-order decay constant. The daughter products of chemicals are neglected.
vi. Heterogeneous reactions considered are first order and are represented by the retardation coefficient.
vii. The solution domain considered is either finite length or infinite length.

5.3.2.1 Finite Width Source in Finite and Infinite Aquifers

In a finite aquifer system the boundaries parallel to the flow direction are considered to be close to the contaminant source so that they have an effect on the magnitude and distribution of contaminant concentrations within the aquifer. For an infinite aquifer the boundary effects are considered to be negligible in the transverse direction. For this type of aquifer we will consider a finite patch source with a finite width.

First we will discuss the finite width aquifer case. In this case the parabolic partial differential equation governing the two-dimensional fate and transport process can be given as:

$$\frac{\partial C}{\partial t} + v\frac{\partial C}{\partial x} = D_x\frac{\partial^2 C}{\partial x^2} + D_y\frac{\partial^2 C}{\partial y^2} - \bar{\lambda}C \tag{5.51}$$

Boundary conditions for this model can be given as,

$$C = C_0; \quad x = 0 \quad \text{and } y_c - \frac{W_s}{2} < y < y_c + \frac{W_s}{2}$$

$$\frac{\partial C}{\partial x} = 0; \quad x \Rightarrow \infty$$

$$\frac{\partial C}{\partial y} = 0; \quad y = 0 \tag{5.52}$$

$$\frac{\partial C}{\partial y} = 0; \quad y = W$$

The initial condition is defined as,

$$C = 0; \quad 0 < x < \infty; \quad 0 < y < W; \quad t = 0 \tag{5.53}$$

Given the two-dimensional equation (5.50), one should notice that v, D_x, D_y and $\bar{\lambda}$ given in Eq. (5.51) above are defined as,

$$v = \frac{v_x}{R}; \quad D_x = \frac{D_{xx}}{R}; \quad D_y = \frac{D_{yy}}{R}; \quad \bar{\lambda} = \lambda + \frac{q}{B\theta} \tag{5.54}$$

The analytical solution of this problem is given by Wexler (1989). It should be recognized that the analytical solution given in Eq. (5.55) for the mathematical model given in Eq. (5.51) may also be used with either D_y or $\bar{\lambda}$ equal to zero, or the retardation coefficient equal to one, which would imply negligible diffusion effects in the y-axis direction, a conservative solute with no dilution due to infiltration effects and no adsorption processes, respectively. Using this solution, the temporal variations in source concentration may also be evaluated using the method of superposition described earlier. Based on these variations, the mathematical model described above can be used to solve several different cases representing different applications in a two-dimensional domain. In Fig. 5.9 a definition sketch of the solution domain is shown.

$$C(x, y, t) = C_0 \sum_{n=0}^{\infty} L_n P_n \cos(\eta y)$$

$$\bullet \left\{ \exp\left[\frac{x(v - \beta)}{2D_x}\right] erfc\left[\frac{x - \beta t}{2\sqrt{D_x t}}\right] + \exp\left[\frac{x(v + \beta)}{2D_x}\right] erfc\left[\frac{x + \beta t}{2\sqrt{D_x t}}\right] \right\}$$

in which

$$L_n = \begin{cases} \dfrac{1}{2}, & n = 0 \\ 1, & n > 0 \end{cases}$$

$$P_n = \begin{cases} \dfrac{y_2 - y_1}{W}, & n = 0 \\ \dfrac{[\sin(\eta y_2) - \sin(\eta y_1)]}{n\pi}, & n > 0 \end{cases}$$

$$y_1 = y_c - \frac{W_s}{2}$$

$$y_2 = y_c + \frac{W_s}{2}$$

$$\eta = \frac{n\pi}{W}, \quad n = 0, 1, 2, 3, \dots$$

$$\beta = \sqrt{v^2 + 4D_x(\eta^2 D_y + \bar{\lambda})} \tag{5.55}$$

An aquifer system, which is infinite in the y-axis direction can also be analyzed. In this case, the inflow and outflow boundaries are considered to be far away from the source, so they will not have an effect on the magnitude and distribution of contaminant concentrations in the solution domain within the time period selected for the solution. Based on this assumption, one may describe the mathematical model using the partial differential equation as given in Eq. (5.51),

$$\frac{\partial C}{\partial t} + v\frac{\partial C}{\partial x} = D_x\frac{\partial^2 C}{\partial x^2} + D_y\frac{\partial^2 C}{\partial y^2} - \bar{\lambda}C \tag{5.56}$$

The boundary conditions for this case are defined as,

$$C = C_0; \quad x = 0 \quad \text{and } y_c - \frac{W_s}{2} < y < y_c + \frac{W_s}{2}$$

$$\frac{\partial C}{\partial x} = 0; \quad x \Rightarrow \infty \tag{5.57}$$

$$\frac{\partial C}{\partial y} = 0; \quad y \Rightarrow \pm\infty$$

The initial condition is defined as,

$$C = 0; \quad 0 < x < \infty; \quad -\infty < y < \infty; \quad t = 0 \tag{5.58}$$

Similar to the case discussed above, one should notice that v, D_x, D_y and $\bar{\lambda}$ given in Eq. (5.56) are defined as shown in Eq. (5.54). The analytical solution of this problem is given as (Wexler 1989).

$$
\begin{aligned}
C(x,y,t) = {} & \frac{C_0 x}{4\sqrt{\pi D_x}} \exp\left[\frac{vx}{2D_x}\right] \\
& \bullet \int_{\tau=0}^{\tau=t} \tau^{-\frac{3}{2}} \exp\left[-\left(\frac{v^2}{4D_x} + \bar{\lambda}\right)\tau - \frac{x^2}{4D_x\tau}\right] \\
& \times \left\{ erfc\left[\frac{y_1 - y}{2\sqrt{D_y\tau}}\right] - erfc\left[\frac{y_2 + y}{2\sqrt{D_y\tau}}\right] \right\} d\tau
\end{aligned}
\tag{5.59}
$$

$$\text{where} \quad y_1 = y_c - \frac{W_s}{2} \quad \text{and} \quad y_2 = y_c + \frac{W_s}{2}$$

The definition sketch for this problem is shown in Fig. 5.10.

5.3.2.2 Infinite Aquifer with a Gaussian Boundary Condition

For a contaminant source that exhibits a Gaussian concentration distribution along the inflow boundary, the Gaussian source boundary condition option should be selected. In this case the governing equation for the problem is defined as follows,

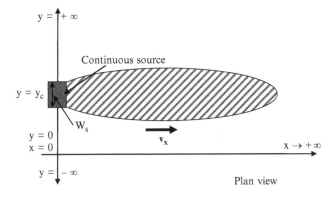

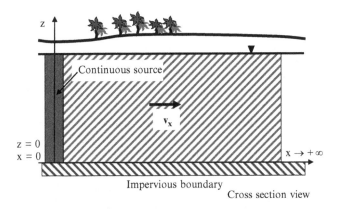

Fig. 5.10 Definition sketch for a two-dimensional infinite aquifer and a finite line source

$$\frac{\partial C}{\partial t} + v \frac{\partial C}{\partial x} = D_x \frac{\partial^2 C}{\partial x^2} + D_y \frac{\partial^2 C}{\partial y^2} - \bar{\lambda} C \qquad (5.60)$$

which is the same partial differential equation given in Eq. (5.51). The boundary conditions for this problem are different and are given below,

$$C = C_0 \exp\left[\frac{-(y - y_c)^2}{2\sigma^2}\right]; \quad x = 0$$

$$\frac{\partial C}{\partial x} = 0; \quad x \Rightarrow \infty \qquad (5.61)$$

$$\frac{\partial C}{\partial y} = 0; \quad y \Rightarrow \pm\infty$$

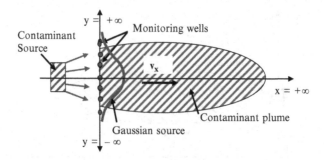

Fig. 5.11 Definition sketch for a two-dimensional infinite aquifer with a Gaussian contaminant source

Initial conditions,

$$C = 0 \quad at \; 0 < x < \infty; \quad -\infty < y < \infty; \quad t = 0 \tag{5.62}$$

in which the maximum concentration at the center of Gaussian plume source is C_0, Y_c is the y-coordinate of the center of the solute source ($x_c = 0$), and σ is the standard deviation of the Gaussian distribution which can be determined using Eq. (5.20). Given the two-dimensional equation (5.50), one should again notice that v, D_x, D_y and $\bar{\lambda}$ given in Eq. (5.60) are defined as shown in Eq. (5.54). The analytical solution of this problem is given in Eq. (5.63) (Wexler 1989). The definition sketch of this model is shown in Fig. 5.11.

$$C(x,y,t) = \frac{C_m x \sigma}{\sqrt{8\pi D_x}} \exp\left[\frac{vx}{2D_x}\right] \int_{\tau=0}^{\tau=t} \frac{\exp\left[-\beta\tau - \frac{x^2}{4D_x\tau} - \frac{(y-y_c)^2}{4(D_y\tau+\frac{\sigma^2}{2})}\right] d\tau}{\tau^{\frac{3}{2}}\sqrt{D_y\tau + \frac{\sigma^2}{2}}} \tag{5.63}$$

$$\text{in which} \quad \beta = \frac{v^2}{4D_x} + \bar{\lambda}$$

5.3.2.3 Infinite Aquifer with a Point Source Boundary Condition

A contaminant which originates from a point source may be modeled using the "Continuous Point" source model included in the ACTS software. A definition sketch for a continuous point source problem domain is shown in Fig. 5.12. In this case the problem is defined in terms of the mathematical model,

$$\frac{\partial C}{\partial t} + v\frac{\partial C}{\partial x} =$$
$$D_x\frac{\partial^2 C}{\partial x^2} + D_y\frac{\partial^2 C}{\partial y^2} - \bar{\lambda}C + Q C_o \delta(x - x_c)\delta(y - y_c)\delta(t - t')dt \tag{5.64}$$

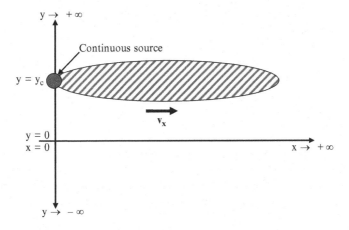

Fig. 5.12 Definition sketch for a two-dimensional infinite aquifer with a continuous point source

in which the boundary conditions are given as,

$$\frac{\partial C}{\partial x} = 0; \quad x = \pm\infty$$
$$\frac{\partial C}{\partial y} = 0; \quad y = \pm\infty$$

(5.65)

The initial condition is defined as,

$$C = 0; \quad -\infty < x < \infty; \quad -\infty < y < \infty; \quad t = 0 \qquad (5.66)$$

where x_c and y_c are the x- and y-coordinates of the point source, Q is the fluid injection rate per unit thickness of the aquifer, dt is the time interval of the release, $\delta(\bullet)$ is the Dirac delta function (Gunduz and Aral 2005), and t' is the time at which the point source boundary condition is initiated, which may be assumed to be equal to zero in most cases. A definition sketch for this problem is given in Fig. 5.12. The reader should notice that the contaminant source is not introduced to the model as a boundary condition but rather is defined as an injection rate in the aquifer domain. This type of application may be used in leaking underground storage tank problems, which are common in aquifer pollution.

The analytical solution for a continuous point source case can be derived by first solving the solute transport equation for an instantaneous point source and then by integrating the solution over time. The following equation, modified from Bear (1979), represents the analytical solution for an instantaneous point source integrated with respect to time (Wexler 1989).

$$C(x, y, t) = \frac{C_o Q}{4\pi \sqrt{D_x D_y}} \exp \left[\frac{v(x - x_c)}{2D_x} \right]$$

$$\bullet \int_{\tau=0}^{\tau=t} \frac{1}{\tau} \exp \left[-\left(\frac{v^2}{4D_x} + \bar{\lambda} \right) \tau - \frac{(x - x_c)^2}{4D_x \tau} - \frac{(y - y_c)^2}{4D_y \tau} \right] d\tau \qquad (5.67)$$

Given the two-dimensional Eq. (5.64), one should again notice that v, D_x, D_y and $\bar{\lambda}$ given in Eq. (5.67) are defined as shown in Eq. (5.54), and that the retardation coefficient definition may be used in a similar manner. The principle of superposition can also be used in this model to simulate variable boundary concentration values.

5.3.3 Three-Dimensional Contaminant Transport Models with Constant Dispersion Coefficient

Relatively few analytical solutions are available for the three-dimensional form of the solute transport equation given in Eq. (5.11). A three-dimensional contaminant transport model may be used in cases where the aquifer is relatively deep and the transverse diffusion effects in both the y- and z-directions cannot be ignored. For the three-dimensional models two types of aquifers are considered, i.e. a finite aquifer and an infinite aquifer. For a finite aquifer a line source is the only available option for the boundary condition, which gives a "finite/line source" model. For an infinite aquifer, two types of boundary conditions can be considered, i.e. a continuous point source and a finite/line source boundary condition. For these cases the general fate and transport equation can be given as,

$$R \frac{\partial C}{\partial t} + v_x \frac{\partial C}{\partial x} = D_{xx} \frac{\partial^2 C}{\partial x^2} + D_{yy} \frac{\partial^2 C}{\partial y^2} + D_{zz} \frac{\partial^2 C}{\partial z^2} - \lambda RC$$

$$+ Q RC_o \, \delta(x - x_c) \delta(y - y_c) \delta(z - z_c) \delta(t - t') \, dt \qquad (5.68)$$

In the first solution discussed below, the aquifer is assumed to be of infinite extent along all three coordinate directions. Fluid is injected into the aquifer through a point source at a constant rate C_0. In the second model, the aquifer is assumed to be semi-infinite with the solute source located along the inflow upstream boundary. The physical domain of a semi-infinite aquifer can be either finite in both width and depth, extending from $y = 0$ to $y = W$ and from $z = 0$ to $z = H$, or it can be considered to be of infinite width and depth. A definition sketch for the idealized three-dimensional aquifer domain for semi-infinite and finite width and height cases is shown in Fig. 5.13.

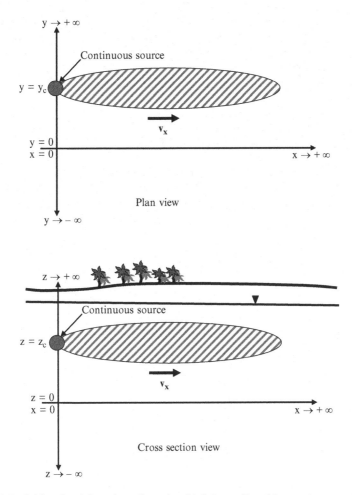

Fig. 5.13 Definition sketch for a three-dimensional infinite aquifer with a continuous point source

Assumptions and Limitations: The more important assumptions and limitations of the three-dimensional saturated zone transformation and transport models are as follows:

i. The flow field within the saturated zone is at steady state and unidirectional.
ii. The seepage velocity and other model parameters (e.g., the diffusion coefficient, partition coefficient) are uniform and constant (i.e., the aquifer is homogeneous and isotropic).
iii. Advective transport is assumed to be strictly one-dimensional in the longitudinal direction. The transverse diffusion or dispersion in the y- and z-directions is not neglected.
iv. The solution is three-dimensional in the (x, y, z) domain.

v. Decay of the solute may be described by a first-order decay constant. The daughter products of chemicals are neglected.

vi. Heterogeneous reactions considered are first order and are represented by the retardation coefficient.

vii. The solution domain considered is either finite length or infinite length.

5.3.3.1 Infinite Aquifer with a Point Source Boundary Condition

A contaminant which originates from a point source, may be modeled using the continuous point source model. The definition sketch of a continuous point source model is shown in Fig. 5.13. In this case the problem is defined as given in Eq. (5.69),

$$
\frac{\partial C}{\partial t} + v\frac{\partial C}{\partial x} = D_x\frac{\partial^2 C}{\partial x^2} + D_y\frac{\partial^2 C}{\partial y^2} + D_z\frac{\partial^2 C}{\partial z^2} - \bar{\lambda}C
$$
$$
+ Q\,C_o\,\delta(x - x_c)\delta(y - y_c)\delta(z - z_c)\delta(t - t')\,dt
$$

(5.69)

The boundary conditions of this model are defined as,

$$
\frac{\partial C}{\partial x} = 0; \quad x \Rightarrow \pm\infty
$$
$$
\frac{\partial C}{\partial y} = 0; \quad y \Rightarrow \pm\infty
$$
$$
\frac{\partial C}{\partial z} = 0; \quad z \Rightarrow \pm\infty
$$

(5.70)

The initial condition is given as,

$$
C = 0; \quad -\infty < x < \infty; \quad -\infty < y < \infty; \quad t = 0
$$

(5.71)

in which x_c, y_c and z_c are the x-, y- and z-coordinates of the point source, Q is the volumetric injection rate, dt is the time interval of the source release, $\delta(\bullet)$ is the Dirac delta (impulse) function (Gunduz and Aral 2005), and t' is the time at which the point source boundary condition is initiated, which can be assumed to be equal to zero. In Fig. 5.13 a definition sketch of the solution domain is shown. One should also notice that v, D_x, D_y, D_z and $\bar{\lambda}$ given in Eq. (5.69) are defined similar to the previous definitions as shown in Eq. (5.54). The principle of superposition may also be used in this model to simulate variable boundary concentration values.

The analytical solution of the continuous point source problem is given in Eq. (5.72) (Wexler 1989).

$$C(x, y, z, t) = \frac{C_o Q}{8\pi \gamma \sqrt{D_y D_z}} \exp\left[\frac{v(x - x_c)}{2D_x}\right]$$

$$\bullet \left\{ \exp\left[\frac{\bar{\lambda}\beta}{2D_x}\right] erfc\left[\frac{(\bar{\lambda} + \beta t)}{2\sqrt{D_x t}}\right] + \exp\left[-\frac{\bar{\lambda}\beta}{2D_x}\right] erfc\left[\frac{(\bar{\lambda} - \beta t)}{2\sqrt{D_x t}}\right] \right\}$$

in which $\quad \gamma = \left[(x - x_c)^2 + \frac{D_x(y - y_c)^2}{D_y} + \frac{D_x(z - z_c)^2}{D_z} \right]^{\frac{1}{2}}$

$$\beta = \left[v^2 + 4D_x\bar{\lambda} \right]^{\frac{1}{2}}$$

$$(5.72)$$

5.3.3.2 Finite Aquifer with a Finite Patch Contaminant Source Boundary Condition

For a contaminant source of a finite width and height, the Finite/Patch model can be used. An example of a finite/patch source is an effluent leaking from a buried source as shown in Fig. 5.14. In this case the governing differential equation of the problem is given as shown in Eq. (5.73),

$$\frac{\partial C}{\partial t} + v\frac{\partial C}{\partial x} = D_x \frac{\partial^2 C}{\partial x^2} + D_y \frac{\partial^2 C}{\partial y^2} + D_z \frac{\partial^2 C}{\partial z^2} - \bar{\lambda}C \qquad (5.73)$$

The boundary conditions of the problem are defined as,

$$C = C_o; \quad x = 0; \quad y_c - \frac{W_s}{2} < y < y_c + \frac{W_s}{2}; \quad z_c - \frac{H_s}{2} < z < z_c + \frac{H_s}{2}$$

$$\frac{\partial C}{\partial x} = 0; \quad x \Rightarrow \infty$$

$$\frac{\partial C}{\partial y} = 0; \quad y = 0; \quad y = W \qquad (5.74)$$

$$\frac{\partial C}{\partial z} = 0; \quad z = 0; \quad z = H$$

The initial condition is given as,

$$C = 0; \quad 0 < x < \infty; \quad 0 < y < W; \quad 0 < z < H; \quad t = 0 \qquad (5.75)$$

in which y_c and z_c are the coordinates of the center of the source at $x_c = 0$, H_s is the height of the source, W_s is the width of the source, H is the depth of the aquifer and W is the width of the aquifer. In Fig. 5.14 a definition sketch of the solution domain is provided. The solution to Eq. (5.73) can be given as shown in Eq. (5.76) (Wexler 1989).

$$C(x,y,z,t) = C_o \sum_{m=0}^{\infty} \sum_{n=0}^{\infty} L_{mn} \, O_m \, P_n \, \cos(\varsigma \, z) \cos(\eta \, y)$$

$$\bullet \left\{ \exp\left[\frac{x(v - \beta)}{2D_x}\right] erfc\left[\frac{(x - \beta \, t)}{2\sqrt{D_x t}}\right] \right.$$

$$\left. + \exp\left[-\frac{x(v + \beta)}{2D_x}\right] erfc\left[\frac{(x + \beta \, t)}{2\sqrt{D_x t}}\right] \right\}$$

(5.76)

in which

$$L_{mn} = \begin{cases} \dfrac{1}{2} & m = 0 \,;\, n = 0 \\[2mm] 1 & m = 0 \,;\, n > 0 \\[2mm] 1 & m > 0 \,;\, n = 0 \\[2mm] 2 & m > 0 \,;\, n > 0 \end{cases}$$

$$O_m = \begin{cases} \dfrac{z_2 - z_1}{H} \,;\, & m = 0 \\[3mm] \dfrac{[\sin(\varsigma \, z_2) - \sin(\varsigma \, z_1)]}{m \, \pi} \,;\, & m > 0 \end{cases}$$

$$P_n = \begin{cases} \dfrac{y_2 - y_1}{W} \,;\, & n = 0 \\[3mm] \dfrac{[\sin(\eta \, y_2) - \sin(\eta \, y_1)]}{n \, \pi} \,;\, & n > 0 \end{cases}$$

$$z_1 = z_c - H_s/2 \,;\quad z_2 = z_c + H_s/2 \,;\quad y_1 = y_c - W_s/2 \,;\quad y_2 = y_c + W_s/2$$

$$\varsigma = m \, \pi \, /H \quad;\quad m = 0, 1, 2, 3, \ldots$$

$$\eta = n \, \pi \, /W \quad;\quad n = 0, 1, 2, 3, \ldots$$

$$\beta = \sqrt{v^2 + 4D_x(\eta^2 D_y + \varsigma^2 \, D_z + \overline{\lambda})}$$

(5.77)

This solution may also be used to simulate cases where D_y, D_z or $\overline{\lambda}$ are equal to zero, which implies that transverse dispersion and decay effects are ignored. The principle of superposition may also be used in this model to simulate variable boundary concentration values.

5.3.3.3 Infinite Aquifer with a Finite Patch Contaminant Source Boundary Condition

For a contaminant of a finite width and height in an infinite aquifer, the finite/patch and infinite aquifer model should be used. An example of a finite/patch source is effluent flow from a leaking landfill as shown in Fig. 5.15. In this case the governing differential equation of the problem is defined as shown in Eq. (5.78),

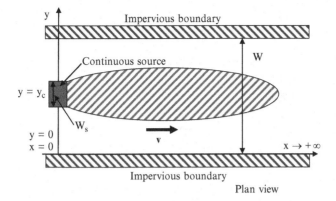

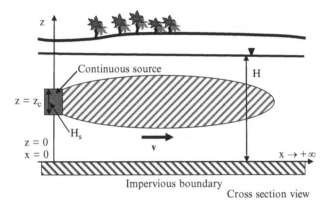

Fig. 5.14 Definition sketch for a three-dimensional finite aquifer with a finite area continuous source

$$\frac{\partial C}{\partial t} + v\frac{\partial C}{\partial x} = D_x\frac{\partial^2 C}{\partial x^2} + D_y\frac{\partial^2 C}{\partial y^2} + D_z\frac{\partial^2 C}{\partial z^2} - \bar{\lambda}C \qquad (5.78)$$

The boundary conditions for this problem are given as,

$$C = C_o; \quad x = 0; \quad y_c - \frac{W_s}{2} < y < y_c + \frac{W_s}{2}; \quad z_c - \frac{H_s}{2} < z < z_c + \frac{H_s}{2}$$

$$\frac{\partial C}{\partial x} = 0; \quad x \Rightarrow \infty$$

$$\frac{\partial C}{\partial y} = 0; \quad y \Rightarrow \pm\infty \qquad (5.79)$$

$$\frac{\partial C}{\partial z} = 0; \quad z \Rightarrow \pm\infty$$

The initial condition of this problem is given as,

$$C = 0; \quad 0 < x < \infty; \quad -\infty < y < +\infty; \quad -\infty < z < +\infty; \quad t = 0 \qquad (5.80)$$

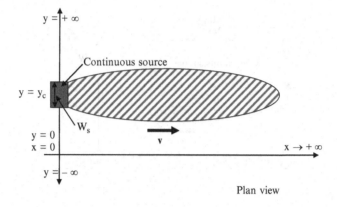

Plan view

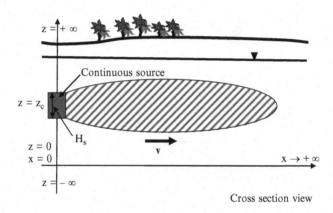

Cross section view

Fig. 5.15 Definition sketch for a three-dimensional infinite aquifer with a finite area continuous source

where y_c and z_c are the center of the solute source at $x_c = 0$, H_s is the height of the solute source and W_s is the width of the solute source. The solution of the analytical model given in Eq. (5.80) is as shown in Eq. (5.81) (Wexler 1989).

$$C(x, y, z, t) = \frac{C_o x \, \exp\left[\frac{vx}{2D_x}\right]}{8\sqrt{\pi D_x}} \int_0^t \tau^{-\frac{3}{2}} \exp\left[-\left(\frac{v^2}{4D_x} + \bar{\lambda}\right)\tau - \frac{x^2}{4D_x\tau}\right]$$

$$\bullet \left\{ erfc\left[\frac{(y_1 + y)}{2\sqrt{D_y\tau}}\right] - erfc\left[\frac{(y_2 - y)}{2\sqrt{D_y\tau}}\right] \right\}$$

$$\times \left\{ erfc\left[\frac{(z_1 + z)}{2\sqrt{D_z\tau}}\right] - erfc\left[\frac{(z_2 - z)}{2\sqrt{D_z\tau}}\right] \right\} d\tau$$

(5.81)

in which $\quad y_1 = y_c - \dfrac{W_s}{2}$

$$y_2 = y_c + \frac{W_s}{2}$$

$$z_1 = z_c - \frac{H_s}{2}$$

$$z_2 = z_c + \frac{H_s}{2}$$

One should also notice that v, D_x, D_y, D_z and $\bar{\lambda}$ given in Eq. (5.78) are defined similar to the previous definitions as shown in Eq. (5.54). The principle of superposition may also be used in this model to simulate time dependent boundary concentration values.

5.4 Multi-species Biodegradation By-Product Models

In Section 5.3 we have discussed one-, two- and three-dimensional groundwater advection–dispersion-reaction models for contaminants that may undergo degradation (decay) or adsorption as the contaminant plume evolves in the subsurface. It is also important to recognize that biodegradation processes that may occur in the subsurface may trigger other reactions that would yield daughter chemical byproducts of the parent chemical which may need to be traced as a separate contaminant plume in addition to the parent contaminant plume. In these cases the daughter by-products may not be in the system as an original contaminant source, but they may appear due to a biodegradation process. An example of this parent–daughter by-product sequence can be seen for tetrachloroethylene (PCE), where the biodegradation process may yield the PCE (tetrachloroethylene) → TCE (trichloroethylene) → DCE (dichloroethylene) → VC (vinyl chloride) → ethane sequence (Fig. 5.16).

More recently, various parent daughter byproduct models and their simultaneous analytical and numerical solutions have been proposed in the literature for the analysis of multi-species plumes (Clement 2001; Sun et al. 2007). Using some of the procedures described in these studies it is possible to develop restricted analytical solutions for all constant dispersion models that are discussed in the previous sections of this chapter. These multi-species analytical models are identified as the "Multi-species Models" in the ACTS software and can be accessed from the first input window of the constant dispersion models (Fig. 5.6). In this section we review the analytical solution procedures that are used for these multispecies models as they are implemented in the ACTS software. Applications for the multispecies problems are given in Section 5.7.

Multi-species reactive transport equations in porous media, in analogy to Eq. (5.17), can be given as,

$$R\frac{\partial C_i}{\partial t} + v_x\frac{\partial C_i}{\partial x} = D_x\frac{\partial^2 C_i}{\partial x^2} + D_y\frac{\partial^2 C_i}{\partial y^2} + D_z\frac{\partial^2 C_i}{\partial z^2} + Rf_i(C_i); \; i$$
$$= 1, 2, 3, ..., N \tag{5.82}$$

in which C_i $[ML^{-3}]$ is the concentration of ith chemical species; t is time $[T]$; v_x $[LT^{-1}]$ is the groundwater pore velocity in the longitudinal direction; $(D_x; D_y; D_z)$ $[L^2T^{-1}]$ are the dispersion coefficients in the x-, y-, z- coordinate directions; $f_i(C_i)$ is the gain or loss of the ith species due to reactions; R is the retardation factor which is species dependent but must be considered to be the same for all species due to

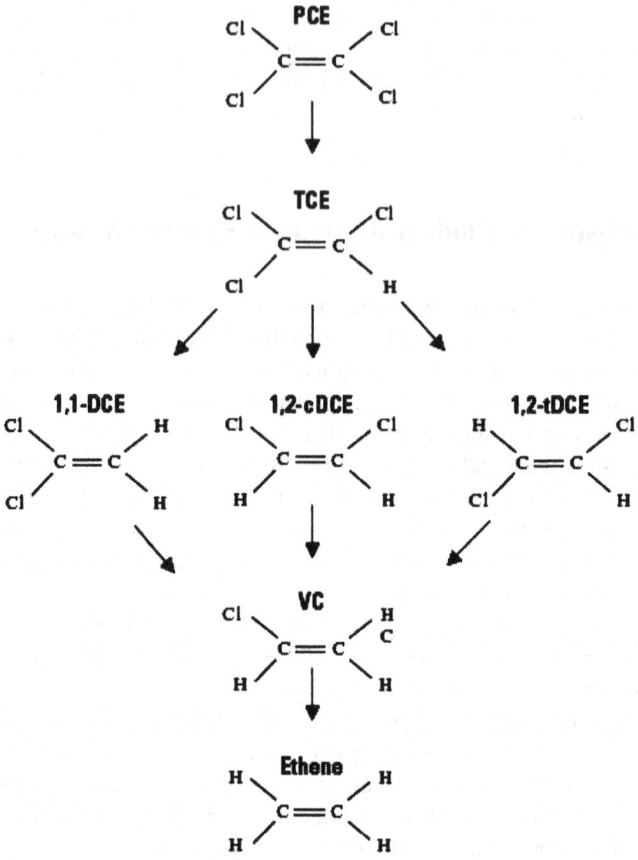

Fig. 5.16 Reductive dehalogenation of tetrachloroethylene

restrictions of the analytical solution process used, or one may assume that the adsorption desorption process can be ignored by selecting $(R = 1)$; and N is the total number of species in the system. The reaction terms of Eq. (5.82) can be written as (Clement 2001),

$$f_i(C_i) = \sum_{j=1}^{i-1} Y_{i/j} K_j C_j - K_i C_i + \sum_{j=i+1}^{N} Y_{i/j} K_j C_j; \quad i = 1, 2, 3, ..., N \tag{5.83}$$

in which $Y_{i/j}$ $[MM^{-1}]$ is the effective yield factor which describes the mass of species i produced from mass of species j, and K_j $[T^{-1}]$ is the first order destruction rate constant of species j. For a sequential degradation case, as described for the chemical PCE above, Eq. (5.82) can be given as,

$$R\frac{\partial C_i}{\partial t} + v_x\frac{\partial C_i}{\partial x} = D_x\frac{\partial^2 C_i}{\partial x^2} + D_y\frac{\partial^2 C_i}{\partial y^2} + D_z\frac{\partial^2 C_i}{\partial z^2} + R Y_i K_{i-1} C_{i-1}$$
$$- RK_i C_i \; ; \; i = 1, 2, 3, ..., N$$

(5.84)

where Y_i is the amount of species i produced from the parent species $(i-1)$ for which the values of K_o will be assumed to be equal to zero, which implies no production term for the first species in the decay chain. Eq. (5.82) or (5.84) can be written in matrix notation as follows,

$$\mathbf{R}\left\{\frac{\partial C}{\partial t}\right\} + v_x\left\{\frac{\partial C}{\partial x}\right\} - D_x\left\{\frac{\partial^2 C}{\partial x^2}\right\} - D_y\left\{\frac{\partial^2 C}{\partial y^2}\right\} - D_z\left\{\frac{\partial^2 C}{\partial z^2}\right\} = \mathbf{K}\{C\}$$

(5.85)

in which $\mathbf{R}$ and $\mathbf{K}$ are the retardation and reaction matrices as described in Eqs. (5.82)–(5.84) above. For example, for a sequential by-product generation of PCE case Eq. (5.85) can be written as,

$$\begin{bmatrix} R & 0 & 0 & 0 \\ 0 & R & 0 & 0 \\ 0 & 0 & R & 0 \\ 0 & 0 & 0 & R \end{bmatrix}\begin{Bmatrix} \partial C_1/\partial t \\ \partial C_2/\partial t \\ \partial C_3/\partial t \\ \partial C_4/\partial t \end{Bmatrix} + v_x\begin{Bmatrix} \partial C_1/\partial x \\ \partial C_2/\partial x \\ \partial C_3/\partial x \\ \partial C_4/\partial x \end{Bmatrix} = D_x\begin{Bmatrix} \partial^2 C_1/\partial x^2 \\ \partial^2 C_2/\partial x^2 \\ \partial^2 C_3/\partial x^2 \\ \partial^2 C_4/\partial x^2 \end{Bmatrix}$$

$$+ D_y\begin{Bmatrix} \partial^2 C_1/\partial y^2 \\ \partial^2 C_2/\partial y^2 \\ \partial^2 C_3/\partial y^2 \\ \partial^2 C_4/\partial y^2 \end{Bmatrix} + D_z\begin{Bmatrix} \partial^2 C_1/\partial z^2 \\ \partial^2 C_2/\partial z^2 \\ \partial^2 C_3/\partial z^2 \\ \partial^2 C_4/\partial z^2 \end{Bmatrix} + R\begin{bmatrix} -K_1 & 0 & 0 & 0 \\ Y_2 K_1 & -K_2 & 0 & 0 \\ 0 & Y_3 K_2 & -K_3 & 0 \\ 0 & 0 & Y_4 K_3 & -K_4 \end{bmatrix}\begin{Bmatrix} C_1 \\ C_2 \\ C_3 \\ C_4 \end{Bmatrix}$$

(5.86)

in which (C_1, C_2, C_3, C_4) refer to (PCE, TCE, DCE, VC) concentrations in the aquifer. Equation (5.86) is a coupled simultaneous partial differential equation which can be solved for four components or in the case of Eq. (5.85), for N components. Using matrix transformation techniques, the simultaneous equations (5.85) or (5.86) can be uncoupled and solved for individual species concentration in the transformed domain. These solutions can then be transformed back to the physical domain, and the solution can be obtained as described below.

In linear algebra, two n-by-n matrices $\mathbf{A}$ and $\mathbf{B}$ are called similar if,

$$\mathbf{B} = \mathbf{P}^{-1}\mathbf{A}\mathbf{P}$$

(5.87)

for some invertible (non-singular) n-by-n matrix $\mathbf{P}$. Similar matrices represent the same linear transformation under two different bases, with $\mathbf{P}$ defined as the change of basis matrix. The operation defined in Eq. (5.87) that involves the definition of the matrix $\mathbf{P}$ is called a similarity transformation. Further, if the similarity transformation yields a diagonal matrix, i.e. if the matrix $\mathbf{B}$ is a diagonal matrix,

$$\mathbf{B} = \mathbf{P}^{-1}\mathbf{A}\mathbf{P} = \begin{bmatrix} \lambda_1 & 0 & \dots & 0 \\ 0 & \lambda_2 & \dots & 0 \\ \vdots & \vdots & \ddots & \vdots \\ 0 & 0 & \dots & \lambda_n \end{bmatrix} \tag{5.88}$$

This implies that $\mathbf{P}$ must be a matrix whose columns constitute n linearly independent eigenvectors of $\mathbf{A}$, and $\mathbf{B}$ must be a diagonal matrix whose diagonal entries are the corresponding eigenvalues. It is straightforward to reverse the above argument i.e., if there exists a linearly independent set of n eigenvectors that are used as columns to build a nonsingular matrix $\mathbf{P}$, and if $\mathbf{B}$ is the diagonal matrix whose diagonal entries are the corresponding eigenvalues, then $\mathbf{P}\mathbf{B}\mathbf{P}^{-1} = \mathbf{A}$.

Some properties of similar matrices are the following. The determinant of the similarity transformation of a matrix is equal to the determinant of the original matrix.

$$|\mathbf{B}| = \left|\mathbf{P}^{-1}\mathbf{A}\mathbf{P}\right| = \left|\mathbf{P}^{-1}\right||\mathbf{A}||\mathbf{P}| = |\mathbf{A}| \tag{5.89}$$

The determinant of a similarity transformation minus a multiple of the unit matrix is given by:

$$\begin{aligned} |\mathbf{B} - \lambda\mathbf{I}| &= \left|\mathbf{P}^{-1}\mathbf{A}\mathbf{P} - \lambda\mathbf{I}\right| = \left|\mathbf{P}^{-1}\mathbf{A}\mathbf{P} - \mathbf{P}^{-1}\lambda\mathbf{I}\mathbf{P}\right| \\ &= \left|\mathbf{P}^{-1}(\mathbf{A} - \lambda\mathbf{I})\mathbf{P}\right| = \left|\mathbf{P}^{-1}\right|\left|(\mathbf{A} - \lambda\mathbf{I})\right||\mathbf{P}| = |(\mathbf{A} - \lambda\mathbf{I})| \end{aligned} \tag{5.90}$$

If $\mathbf{A}$ is an antisymmetric matrix $\left(a_{ij} = -a_{ji}\right)$ and $\mathbf{P}$ is an orthogonal matrix $\left(p_{ij}^{-1} = p_{ji}\right)$, then the matrix for the similarity transformation $\mathbf{B}$ is antisymmetric, i.e. $\mathbf{B} = -\mathbf{B}^{\mathrm{T}}$.

Using the matrix operation principles given above, the simultaneous partial differential equation (5.85) or the special case given in Eq. (5.86) can be decoupled. Let's assume that there is an arbitrary matrix $\mathbf{P}$, and we use its inverse in Eq. (5.85) in the transformation $\left\{\widehat{C}\right\} = \mathbf{P}^{-1}\{C\}$ or $\mathbf{P}\left\{\widehat{C}\right\} = \{C\}$. Similarly the following transformations can also be performed,

$$\left\{\frac{\partial \widehat{C}}{\partial t}\right\} = \mathbf{P}^{-1}\left\{\frac{\partial C}{\partial t}\right\}$$

$$\left\{\frac{\partial \widehat{C}}{\partial x}\right\} = \mathbf{P}^{-1}\left\{\frac{\partial C}{\partial x}\right\}$$

$$\left\{\frac{\partial^2 \widehat{C}}{\partial x^2}\right\} = \mathbf{P}^{-1}\left\{\frac{\partial^2 C}{\partial x^2}\right\}; \quad \left\{\frac{\partial^2 \widehat{C}}{\partial y^2}\right\} = \mathbf{P}^{-1}\left\{\frac{\partial^2 C}{\partial y^2}\right\}; \quad \left\{\frac{\partial^2 \widehat{C}}{\partial z^2}\right\} = \mathbf{P}^{-1}\left\{\frac{\partial^2 C}{\partial z^2}\right\}$$

(5.91)

Substituting the inverse of these transformations into Eq. (5.85) we obtain the following,

$$\mathbf{RP}\left\{\frac{\partial \widehat{C}}{\partial t}\right\} + v_x\mathbf{P}\left\{\frac{\partial \widehat{C}}{\partial x}\right\} - D_x\mathbf{P}\left\{\frac{\partial^2 \widehat{C}}{\partial x^2}\right\} - D_y\mathbf{P}\left\{\frac{\partial^2 \widehat{C}}{\partial y^2}\right\}$$

$$- D_z\mathbf{P}\left\{\frac{\partial^2 \widehat{C}}{\partial z^2}\right\} = \mathbf{KP}\left\{\widehat{C}\right\}$$

(5.92)

Multiplying Eq. (5.92) by the inverse of $\mathbf{P}$ we obtain,

$$\mathbf{P}^{-1}\mathbf{RP}\left\{\frac{\partial \widehat{C}}{\partial t}\right\} + v_x\left\{\frac{\partial \widehat{C}}{\partial x}\right\} - D_x\left\{\frac{\partial^2 \widehat{C}}{\partial x^2}\right\} - D_y\left\{\frac{\partial^2 \widehat{C}}{\partial y^2}\right\} - D_z\left\{\frac{\partial^2 \widehat{C}}{\partial z^2}\right\}$$

$$= \mathbf{P}^{-1}\mathbf{KP}\left\{\widehat{C}\right\}$$

(5.93)

If we select the matrix $\mathbf{P}$ as composed of the eigenvectors of the matrix $\mathbf{K}$, in accordance to the definitions given in Eqs. (5.87) and (5.88) (i.e. $\mathbf{K} = \mathbf{A}$), then the resultant matrix $\mathbf{P}^{-1}\mathbf{KP}$ in Eq. (5.93) will be diagonal, which will yield the decoupled form of Eq. (5.85). For the decoupling process of the simultaneous system given in Eq. (5.85) to be complete we need to check the conditions that are necessary for $\mathbf{P}^{-1}\mathbf{RP}$ to remain as the original matrix $\mathbf{R}$. For this condition to hold, the following must be true,

$$\mathbf{P}^{-1}\mathbf{R} = \mathbf{RP}^{-1}$$

(5.94)

such that,

$$\mathbf{P}^{-1}\mathbf{RP} = \mathbf{RP}^{-1}\mathbf{P} = \mathbf{R}$$

(5.95)

The condition given in Eq. (5.94) requires that the matrix $\mathbf{R}$ be a diagonal matrix with all entries on the diagonal being equal. This is an important restriction for the decoupling of the simultaneous Eqs. (5.85). Nevertheless the procedure described above yields an effective analytical process to analyze multi-species contaminant transport problems for cases where the adsorption-desorption processes are not going to be important. In these cases, the matrix **R** is an identity matrix and satisfies the conditions described in Eqs. (5.94) and (5.95).

After the orthogonalization of the matrix **K**, each differential equation in Eq. (5.93) is independent of other partial differential equations of the transformation and transport equation. Thus, temporal analytical solutions for the advection–diffusion-reaction equations can be obtained in the transformed domain (Clement 2001; Sun et al. 2007). After these solutions are obtained in the transformed domain the results can be transformed back to the physical domain using the transformation given by $\mathbf{P}\{\widehat{C}\} = \{C\}$. Considering reaction sequences that are sequential and first order, the multi-species transport equations can now be solved analytically in spatial and temporal dimensions by extending the analytical methods discussed in the previous sections of this chapter to multi-species analysis.

The analytical procedures discussed above are included for all one-, two- and three-dimensional constant dispersivity models discussed in the previous sections of this chapter. The multi-species option for these models can be directly accessed by choosing the "Multi-species Models" option in the data entry window as shown in Fig. 5.6. The difference between these applications and the previous applications is that the user will now have to enter the necessary by-product reaction constants for the sequence considered in a specific application. The other parameters of the problem, the boundary conditions used, and the required inputs for the coordinate system and its discretization will remain the same.

Assumptions and Limitations: The more important assumptions and limitations of the *two*-dimensional saturated zone transformation and transport models are as follows:

  i. The flow field within the saturated zone is at steady state and unidirectional.
 ii. The seepage velocity and other model parameters (e.g., the diffusion coefficient, partition coefficient) are uniform and constant (i.e., the aquifer is homogeneous and isotropic).
iii. Advective transport is assumed to be strictly one-dimensional in the longitudinal direction.
 iv. All analytical solutions discussed in the previous section can be analyzed using the multi-species analysis approach discussed in this section. Thus their limitations apply to this case as well.
  v. Decay of the solute may be described by a first-order decay constant. The daughter products of chemicals are not neglected.
 vi. Heterogeneous reactions considered are first order and are represented by the retardation coefficient. However, all species generated must use the same retardation coefficient. Otherwise the retardation effects must be neglected.
vii. The solution domain considered are similar to the cases discussed in the previous section.

5.5   Saturated Variable Dispersion Coefficient Groundwater Pathway Models

Analytical solutions to the advection–diffusion-reaction equation are of interest since they represent benchmark solutions to various problems in geohydrology, chemical engineering and also fluid mechanics. As described in this chapter, the migration of dissolved phase contaminants in the subsurface is also modeled by the advection– diffusion-reaction equation. As we have seen in the previous chapters, when analytical methods are used in the solution of these models some simplifying assumptions are usually made. For example, in these solutions the hydrodynamic dispersion coefficients are usually assumed to be constant with respect to space and time, and only the longitudinal velocity component in the flow field is used to represent the dominant advection component in the aquifer. However, field and laboratory experiments indicate that the hydrodynamic dispersion coefficient is not a constant, but rather is a field parameter which may change as a function distance from the contaminant source, both in the longitudinal and transverse directions. The apparent spatial variability of the hydrodynamic dispersion coefficient is identified in the literature as the scale effect, (Fried 1975). Stochastic analyses have shown that variable hydrodynamic dispersion coefficients may also be represented as a function of travel time in association with the longitudinal velocity and that they may increase until they reach an asymptotic value, (Gelhar et al. 1979). In this approach, since the velocity is a function of time and space, the representation of the hydrodynamic dispersion coefficient in terms of travel time would provide a characteristic distribution for the spatial variation of the dispersion coefficients as well. These applications are analyzed in the literature for some restricted cases where the analytical solutions of the advection– dispersion-reaction equation with time-dependent hydrodynamic dispersion coefficient were solved. In these applications the temporal variations in the dispersion coefficient are tied to spatial variations based on the constant velocity patterns used in the analysis (Barry and Sposito 1989; Pickens and Grisak 1981, 1987; Yates 1992).

More recently, analytical solution for one-dimensional contaminant transport equation with time dependent dispersion coefficients in an infinite domain aquifer has been given (Basha and Elhabel 1993). In their study the authors describe one-dimensional analytical solutions for the advection–diffusion-reaction equation using four different time dependent functions that describe the hydrodynamic dispersion coefficient. These four functions yield different spatial distributions for the hydrodynamic dispersion coefficient in an aquifer for a constant velocity field in the longitudinal direction. In this section, parallel to the analysis given in that study (Basha and Elhabel 1993), solutions for a general two-dimensional advec-tion– diffusion-reaction model are given. It is also shown that the two-dimensional solutions discussed below yield the one-dimensional solutions given by Basha and Elhabel as special cases of the general two-dimensional solution.

The analytical solutions discussed below can be used to model the transforma-tion and transport of contaminants that are characterized by hydrodynamic dispersion coefficients that may vary as a function of travel time from the contaminant source. As stated earlier, for a constant longitudinal velocity, these time dependent functions also represent a spatial variation for the dispersion coefficient in a constant velocity field. In the models studied here, as discussed in the literature (Basha and Elhabel 1993), analytical solutions of the transport equations for instantaneous and continuous point and line source boundary conditions are considered. In this analysis dispersion coefficients are defined using four standard functions, i.e. constant, linear, asymptotic and exponential functions. The flow field is assumed to be steady and uniform. Using this approach, the analytical solutions referenced in the earlier sections of this chapter are extended to variable dispersion coefficient cases for these models. Further, by using these models, particular analytical solutions may also be developed. This is true in cases where the injection rate of the contaminant in an aquifer is zero with the initial concentration distribution in the aquifer domain different than zero; in cases concerning the discharge of a contaminant in an aquifer; and in cases with an initially contaminated aquifer condition. Using these solutions, superposition principles may also be employed to arrive at the analytical solutions of more complex cases as discussed before. Using these procedures, the analytical solutions included in the ACTS software may be used as practical tools in evaluating contaminant transport problems with scale dependent dispersion coefficients. The analytical solutions discussed in this section are included in the ACTS software under the variable dispersion model category.

Assumptions and Limitations: The more important assumptions and limitations of the saturated zone transformation and transport models with variable hydrodynamic diffusion coefficients are as follows:

i. The flow field within the saturated zone is at steady state and can be two-dimensional.
ii. The seepage velocity is constant but the diffusion coefficients are variable. This variation is represented in terms of four different functions which are functions of time.
iii. Decay of the solute may be described by a first-order decay constant. The daughter products of chemicals are neglected.
iv. Heterogeneous reactions considered are first order and are represented by the retardation coefficient.
v. The solution domain considered is of infinite length.

5.5.1 Mathematical Models for Variable Dispersion Coefficients

The advection–dispersion-reaction equation analyzed in this case takes the form given in Eq. (5.96) for a steady state two-dimensional velocity field with a first order decay coefficient and time dependent dispersion coefficients. For a two-dimensional infinite aquifer, this equation was also discussed earlier in Chapter 3:

$$R\frac{\partial \overline{C}}{\partial \overline{t}} + \overline{v}_x \frac{\partial \overline{C}}{\partial \overline{x}} + \overline{v}_y \frac{\partial \overline{C}}{\partial \overline{y}} = \overline{D}_x(t)\frac{\partial^2 \overline{C}}{\partial \overline{x}^2} + \overline{D}_y(t)\frac{\partial^2 \overline{C}}{\partial \overline{y}^2} - \overline{\lambda}R\overline{C} + \dot{\overline{q}}\,(\overline{x},\overline{y},\overline{t})$$

$$C(\overline{x},\overline{y},0) = f(\overline{x},\overline{y}) \qquad -\infty < (\overline{x},\overline{y}) < +\infty \qquad (5.96)$$

in which R is the retardation coefficient (dimensionless), $(\overline{v}_x, \overline{v}_y)$ are the components of the steady state velocity vector in the $(\overline{x},\overline{y})$ coordinate direction respectively, $(\overline{D}_x, \overline{D}_y)$ are the longitudinal and transverse dispersion coefficients which are functions of time $\overline{t}$, $\overline{C}$ is the solute concentration, $\overline{q}\,(\overline{x},\overline{y},\overline{t})$ is the mass injection rate in the aquifer, that is $\dot{\overline{q}} = \frac{d\overline{q}}{dt}$ with units $[ML^{-3}T^{-1}]$, $\overline{\lambda}$ is the first order decay coefficient and $f(\overline{x},\overline{y})$ is a function representing the initial concentration distribution in the infinite domain aquifer. In order to simplify the algebra involved we can work with the non-dimensional form of Eq. (5.96) using the transformations given below:

$$x = \frac{\overline{x}}{L};\; y = \frac{\overline{y}}{L};\; t = \overline{t}\frac{D_r}{L^2};\; C = \frac{\overline{C}}{C_r};\; D_x(t) = \frac{\overline{D}_x(t)}{D_r};\; D_y(t) = \frac{\overline{D}_y(t)}{D_r};\; v_x = \overline{v}_x\frac{L}{D_r};$$

$$v_y = \overline{v}_y\frac{L}{D_r};\; \lambda = \overline{\lambda}\frac{L^2}{D_r};\; \dot{q}\,(x,y,t) = \dot{\overline{q}}\,(\overline{x},\overline{y},\overline{t})\frac{L^2}{D_rC_r};\; f(x,y) = \frac{\overline{f}(\overline{x},\overline{y})}{C_r}$$

$$(5.97)$$

in which L is a reference distance, C_r is a reference concentration and D_r is a reference hydrodynamic dispersion coefficient. Using the non-dimensional system defined above, Eq. (5.96) takes the form,

$$R\frac{\partial C}{\partial t} + v_x \frac{\partial C}{\partial x} + v_y \frac{\partial C}{\partial y} = D_x(t)\frac{\partial^2 C}{\partial x^2} + D_y(t)\frac{\partial^2 C}{\partial y^2} - \lambda RC + \dot{q}\,(x,y,t)$$

$$(5.98)$$

$$C(x,y,0) = f(x,y) \qquad -\infty < (x,y) < +\infty$$

If we assume that there is a constant of proportionality between the longitudinal and transverse hydrodynamic dispersion coefficient as given below,

$$\overline{D}_x(t) = a^2\overline{D}_y(t) \qquad (5.99)$$

in which a is a constant of proportionality. Substituting Eq. (5.99) into Eq. (5.98) along with the substitutions $(x' = x)$; $(y' = ay)$ and $(v'_y = av_y)$, Eq. (5.98) can be given as,

$$R\frac{\partial C}{\partial t} + v_x\frac{\partial C}{\partial x'} + v'_y\frac{\partial C}{\partial y'} = D_x(t)\left(\frac{\partial^2 C}{\partial x'^2} + \frac{\partial^2 C}{\partial y'^2}\right) - \lambda RC + \overset{\bullet}{q}\left(x', \frac{y'}{a}, t\right)$$

$$\hspace{10cm}(5.100)$$

$$C\left(x', \frac{y'}{a}, 0\right) = f\left(x', \frac{y'}{a}\right) \qquad -\infty < \left(x', \frac{y'}{a}\right) < +\infty$$

Dividing both sides of Eq. (5.100) by the retardation coefficient we can write,

$$\frac{\partial C}{\partial t} + U\frac{\partial C}{\partial x'} + V\frac{\partial C}{\partial y'} = D(t)\left(\frac{\partial^2 C}{\partial x'^2} + \frac{\partial^2 C}{\partial y'^2}\right) - \lambda C + \frac{\overset{\bullet}{q}\left(x', \frac{y'}{a}, t\right)}{R}$$

$$\hspace{10cm}(5.101)$$

$$C\left(x', \frac{y'}{a}, 0\right) = f\left(x', \frac{y'}{a}\right) \qquad -\infty < \left(x', \frac{y'}{a}\right) < +\infty$$

in which,

$$D(t) = D_x(t)/R; \quad U = v_x/R; \quad V = v'_y/R = av_y/R \hspace{2cm}(5.102)$$

Equation (5.101) may now be simplified utilizing a series of transformations. If we let,

$$C = \widehat{C}\exp(-\lambda t); \quad X = x' - Ut; \quad Y = y' - Vt \hspace{2cm}(5.103)$$

we obtain,

$$\frac{\partial \widehat{C}}{\partial t} = D(t)\left(\frac{\partial^2 \widehat{C}}{\partial X^2} + \frac{\partial^2 \widehat{C}}{\partial Y^2}\right) + \frac{\overset{\bullet}{q}\left(X + Ut, \frac{Y+Ut}{a}, t\right)\exp(\lambda t)}{R} \quad C\left(X + Ut, \frac{Y+Vt}{a}, 0\right)$$

$$= f\left(X + Ut, \frac{Y+Vt}{a}\right) \qquad -\infty < \left(X + Ut, \frac{Y+Vt}{a}\right) < +\infty$$

$$\hspace{10cm}(5.104)$$

Utilizing the following transformation for the time variable,

$$T = \alpha(t) = \int_0^t D(t')dt' \hspace{4cm}(5.105)$$

Equation (5.104) can be reduced to the non-homogeneous equation given below.

$$\frac{\partial \widehat{C}}{\partial T} = \left(\frac{\partial^2 \widehat{C}}{\partial X^2} + \frac{\partial^2 \widehat{C}}{\partial Y^2}\right) + \dot{Q} \tag{5.106}$$

where

$$\dot{Q} = \frac{q\left((X + U\alpha^{-1}(T)), \left(\frac{Y + V\alpha^{-1}(T)}{a}\right), (\alpha^{-1}(T))\right) \exp(\lambda\alpha^{-1}(T))}{RD(\alpha^{-1}(T))} \tag{5.107}$$

The analytical solution of Eq. (5.106) can be obtained utilizing the superposition principle (Haberman 1987).

$$\widehat{C} = \int_0^T \int_{-\infty}^{\infty} \int_{-\infty}^{\infty} \frac{\dot{Q}(\xi, \eta, \tau)}{4\pi(T - \tau)} \exp\left(-\frac{(X - \xi)^2 + (Y - \eta)^2}{4(T - \tau)}\right) d\xi d\eta d\tau$$

$$+ \frac{1}{4\pi T} \int_{-\infty}^{\infty} \int_{-\infty}^{\infty} f\left(\xi, \frac{\eta}{a}\right) \exp\left(-\frac{(X - \xi)^2 + (Y - \eta)^2}{4T}\right) d\xi d\eta \tag{5.108}$$

Substituting the definition of $\dot{Q}$ given in Eq. (5.107) we obtain,

$$\widehat{C} = \int_0^T \int_{-\infty}^{\infty} \int_{-\infty}^{\infty} \frac{\dot{q}\left(\xi + U\alpha^{-1}(\tau), \frac{\eta + V\alpha^{-1}(\tau)}{a}, \alpha^{-1}(\tau)\right) \exp(\lambda\alpha^{-1}(\tau))}{4\pi R(T - \tau)D(\alpha^{-1}(\tau))}$$

$$\times \exp\left(-\frac{(X - \xi)^2 + (Y - \eta)^2}{4(T - \tau)}\right) d\xi d\eta d\tau \tag{5.109}$$

$$+ \frac{1}{4\pi T} \int_{-\infty}^{\infty} \int_{-\infty}^{\infty} f\left(\xi, \frac{\eta}{a}\right) \exp\left(-\frac{(X - \xi)^2 + (Y - \eta)^2}{4T}\right) d\xi d\eta$$

or

$$C = \int_0^T \int_{-\infty}^{\infty} \int_{-\infty}^{\infty} \frac{\dot{q}\left(\xi + U\alpha^{-1}(\tau), \frac{\eta + V\alpha^{-1}(\tau)}{a}, \alpha^{-1}(\tau)\right) \exp(-\lambda\alpha^{-1}(\tau))}{4\pi R(T - \tau)D(\alpha^{-1}(\tau))}$$

$$\times \exp\left(-\frac{(X - \xi)^2 + (Y - \eta)^2}{4(T - \tau)}\right) d\xi d\eta d\tau$$

$$+ \frac{\exp(-\lambda\alpha^{-1}(\tau))}{4\pi T} \int_{-\infty}^{\infty} \int_{-\infty}^{\infty} f\left(\xi, \frac{\eta}{a}\right) \exp\left(-\frac{(X - \xi)^2 + (Y - \eta)^2}{4T}\right) d\xi d\eta$$

$$\tag{5.110}$$

The analytical solution given in Eq. (5.109) or (5.110) is the generalized solution of the two-dimensional advection diffusion equation for which the one-dimensional Basha and Elhabel solutions are the special cases (Basha and Elhabel 1993). Based on these analytical solutions particular solutions to the advection–dispersion-reaction equation can be given.

5.5.2 Solution for Instantaneous Point Injection of a Contaminant into an Initially Uncontaminated Aquifer

In this case $C(x, y, 0) = f(x, y) = 0$, and the instantaneous non-dimensional injection of a contaminant is given by,

$$\dot{q}(x, y, t) = \frac{M}{n} \delta(x_o, y_o, t_o) \tag{5.111}$$

where M is the non-dimensional mass injected, n is the porosity, and $\delta(\bullet)$ is the Dirac delta function (Gunduz and Aral 2005). Utilizing the initial condition and Eq. (5.110), the analytical solution can be given as,

$$C(x, y, t) = \frac{aM}{4\pi n R \alpha(t)} \exp\left(-\lambda t - \frac{(x - Ut)^2 + (ay - Vt)^2}{4\alpha(t)}\right) \tag{5.112}$$

where $C(t)$ is a function of $D(t)$ through the definition of $\alpha(t)$, which is given in Eq. (5.105). Equation (5.112) is an analytical expression which can now be used to describe solutions using several dispersion coefficient functions for a particular time dependent dispersion coefficient variation. The four special cases of the hydrodynamic dispersion coefficient considered here are given below.

Constant Dispersion Coefficient: A constant non-dimensional dispersion coefficient can be defined as,

$$D(t) = D_o + D_m \tag{5.113}$$

Given Eq. (5.113), $\alpha(t)$ can be obtained from Eq. (5.105) as,

$$\alpha(t) = \frac{D_o + D_m}{R} t \tag{5.114}$$

For this case $C(x, y, t)$ can be given as follows,

$$C(x, y, t) = \frac{aM}{4\pi n (D_o + D_m)t} \exp\left(-\lambda t - R \frac{\left(x - \frac{v_x}{R} t\right)^2 + a^2\left(y - \frac{v_y}{R} t\right)^2}{(4(D_o + D_m)t)}\right) \tag{5.115}$$

Linear Dispersion Coefficient: The non-dimensional dispersion coefficient which varies linearly with respect to time can be defined as,

$$D(t) = D_o\left(\frac{t}{k}\right) + D_m \tag{5.116}$$

in which k is an arbitrary constant that is different from zero. Given Eq. (5.116), $\alpha(t)$ can be obtained from Eq. (5.105) as,

$$\alpha(t) = \frac{D_o}{2Rk}t^2 + \frac{D_m}{R}t \tag{5.117}$$

Utilizing Eq. (5.117), Eq. (5.112) can be given as,

$$C(x, y, t) = \frac{aM}{4\pi n\left(\frac{D_o}{2k}t^2 + D_m t\right)} \exp\left(-\lambda t - R\frac{\left(x - \frac{v_x}{R}t\right)^2 + a^2\left(y - \frac{v_y}{R}t\right)^2}{4\left(\frac{D_o}{2k}t^2 + D_m t\right)}\right) \tag{5.118}$$

Asymptotic Dispersion Coefficient: An asymptotically varying non-dimensional hydrodynamic dispersion coefficient can be defined as,

$$D(t) = D_o\frac{t}{k+t} + D_m \tag{5.119}$$

Given Eq. (5.119) $\alpha(t)$ can be obtained from Eq. (5.105) as,

$$\alpha(t) = \left(\frac{D_o + D_m}{R}\right)t - \frac{D_o k}{R}\ln\left(1 + \frac{t}{k}\right) \tag{5.120}$$

Utilizing Eq. (5.120), Eq. (5.112) can be given as,

$$C(x, y, t) = \frac{aM}{4\pi n((D_o + D_m)t - D_o k\ln(1 + t/k))}$$
$$\times \exp\left(-\lambda t - R\frac{\left(x - \frac{v_x}{R}t\right)^2 + a^2\left(y - \frac{v_y}{R}t\right)^2}{4((D_o + D_m)t - D_o k\ln(1 + t/k))}\right) \tag{5.121}$$

Exponential Dispersion Coefficient: The non-dimensional hydrodynamic dispersion coefficient which varies exponentially as a function of time may be represented as,

$$D(t) = D_o\left(1 - \exp\left(-\frac{t}{k}\right)\right) + D_m \tag{5.122}$$

Given Eq. (5.122) $\alpha(t)$ can be obtained as,

$$\alpha(t) = \left(\frac{D_o + D_m}{R}\right)t + \frac{D_o k}{R}\left(\exp\left(-\frac{t}{k}\right) - 1\right) \tag{5.123}$$

For this case, the solution for $C(x, y, t)$ can then be given as,

$$C(x, y, t) = \frac{aM}{4\pi n((D_o + D_m)t + D_o k(\exp(-t/k) - 1))}$$

$$\times \exp\left(-\lambda t - R\frac{\left(x - \frac{v_x}{R}t\right)^2 + a^2\left(y - \frac{v_y}{R}t\right)^2}{4((D_o + D_m)t - D_o k(\exp(-t/k) - 1))}\right) \tag{5.124}$$

In this manner four different analytical solutions can be defined for four different representations of the time dependent dispersion coefficient of an instantaneous point injection of a contaminant source into an aquifer that is initially considered to be clean. These solutions are included to the ACTS software as variable dispersion coefficient solutions for this case of boundary and initial conditions.

5.5.3 Solution for a Continuous Point Source in an Initially Uncontaminated Aquifer

A continuous non-dimensional point source at $(x = 0; \ y = 0)$ can be represented as,

$$\dot{q}(x, y, t) = C_o \delta(x_o, y_o) \tag{5.125}$$

Substituting $f(x, y) = 0$ and Eq. (5.125) in Eq. (5.110), the analytical solution of this problem can be given as,

$$C(x, y, t) = \int_0^t \frac{aC_o}{4\pi R(\alpha(t) - \alpha(\tau))}$$

$$\times \exp\left(-\lambda(t - \tau) - \frac{(x - U(t - \tau))^2 + (ay - V(t - \tau))^2}{4(\alpha(t) - \alpha(\tau))}\right)d\tau$$

$$\tag{5.126}$$

The equation given above describes the general solution to this problem, in which $\alpha(t)$ is again defined by Eq. (5.105). Particular cases of this solution for the four dispersion coefficient functions defined earlier are given below.

Constant Dispersion Coefficient: For a constant non-dimensional dispersion coefficient, $\alpha(t)$ is given by Eq. (5.114). For this case the solution of $C(x, y, t)$ can be given as,

$$C(x, y, t) = \int_0^t \frac{aC_o}{4\pi(D_o - D_m)(t - \tau)}$$
$$\times \exp\left(-\lambda(t - \tau) - R\frac{\left(x - \frac{v_x}{R}(t - \tau)\right)^2 + a^2\left(y - \frac{v_y}{R}(t - \tau)\right)^2}{4(D_o + D_m)(t - \tau)}\right) d\tau$$

$$(5.127)$$

or,

$$C(x, y, t) = \int_0^t \frac{aC_o}{4\pi(D_o - D_m)\tau} \exp\left(-\lambda\tau - R\frac{\left(x - \frac{v_x}{R}\tau\right)^2 + a^2\left(y - \frac{v_y}{R}\tau\right)^2}{4(D_o + D_m)\tau}\right) d\tau$$

$$(5.128)$$

Linear Dispersion Coefficient: For a linear non-dimensional dispersion coefficient, $\alpha(t)$ is given by Eq. (5.117). Substituting Eq. (5.117) in Eq. (5.126) the solution for $C(x, y, t)$ can be written as,

$$C(x, y, t) = \int_0^t \frac{aC_o}{4\pi\left(\frac{D_o}{2k}(t^2 - \tau^2) + D_m(t - \tau)\right)}$$
$$\times \exp\left(-\lambda(t - \tau) - R\frac{\left(x - \frac{v_x}{R}(t - \tau)\right)^2 + a^2\left(y - \frac{v_y}{R}(t - \tau)\right)^2}{4\left(\frac{D_o}{2k}(t^2 - \tau^2) + D_m(t - \tau)\right)}\right) d\tau$$

$$(5.129)$$

or it may be written as,

$$C(x, y, t) = \int_0^t \frac{aC_o}{4\pi\left(\frac{D_o}{2k}(2t - \tau) + D_m\right)\tau} \exp\left(-\lambda\tau - R\frac{\left(x - \frac{v_x}{R}\tau\right)^2 + a^2\left(y - \frac{v_y}{R}\tau\right)^2}{4\left(\frac{D_o}{2k}(2t - \tau) + D_m\right)\tau}\right) d\tau$$

$$(5.130)$$

Asymptotic Dispersion Coefficient: For the asymptotic non-dimensional hydrodynamic dispersion coefficient, $\alpha(t)$ is given by Eq. (5.120). Substituting Eq. (5.120) in Eq. (5.126), $C(x, y, t)$ can be defined as,

$$C(x, y, t) = \int_0^t \frac{aC_o}{4\pi\left((D_o + D_m)(t - \tau) - D_o k \ln\left(\frac{k+t}{k+\tau}\right)\right)}$$
$$\times \exp\left(-\lambda(t - \tau) - R\frac{\left(x - \frac{v_x}{R}(t - \tau)\right)^2 + a^2\left(y - \frac{v_y}{R}(t - \tau)\right)^2}{4\left((D_o + D_m)(t - \tau) - D_o k \ln\left(\frac{k+t}{k+\tau}\right)\right)}\right) d\tau$$

$$(5.131)$$

Using the integral variable transformation, Eq. (5.131) can be given as shown below.

$$C(x,y,t) = \int_0^t \frac{aC_o}{4\pi\left((D_o + D_m)\tau - D_o k \ln\left(\frac{k+t}{k+t-\tau}\right)\right)}$$
$$\times \exp\left(-\lambda\tau - R\frac{\left(x - \frac{v_x}{R}\tau\right)^2 + a^2\left(y - \frac{v_y}{R}\tau\right)^2}{4\left((D_o + D_m)\tau - D_o k \ln\left(\frac{k+t}{k+t-\tau}\right)\right)}\right) d\tau$$

(5.132)

Exponential Dispersion Coefficient: For this case the function $\alpha(t)$ takes the form given in Eq. (5.123). Substituting Eq. (5.123) into Eq. (5.126) one may obtain,

$$C(x,y,t) = \int_0^t \frac{aC_o}{4\pi\left((D_o + D_m)(t-\tau) + D_o k\left(\exp\left(-\frac{t}{k}\right) - \exp\left(-\frac{\tau}{k}\right)\right)\right)}$$
$$\times \exp\left(-\lambda(t-\tau) - R\frac{\left(x - \frac{v_x}{R}(t-\tau)\right)^2 + a^2\left(y - \frac{v_y}{R}(t-\tau)\right)^2}{4\left((D_o + D_m)(t-\tau) + D_o k\left(\exp\left(-\frac{t}{k}\right) - \exp\left(-\frac{\tau}{k}\right)\right)\right)}\right) d\tau$$

(5.133)

Equation (5.133) may also be written as Eq. (5.134) by using the method of integral variable transformation.

$$C(x,y,t) = \int_0^t \frac{aC_o}{4\pi\left((D_o + D_m)\tau + D_o k\left(\exp\left(-\frac{t}{k}\right) - \exp\left(-\frac{t-\tau}{k}\right)\right)\right)}$$
$$\times \exp\left(-\lambda\tau - R\frac{\left(x - \frac{v_x}{R}\tau\right)^2 + a^2\left(y - \frac{v_y}{R}\tau\right)^2}{4\left((D_o + D_m)\tau + D_o k\left(\exp\left(-\frac{t}{k}\right) - \exp\left(-\frac{t-\tau}{k}\right)\right)\right)}\right) d\tau$$

(5.134)

5.5.4 Initial Point Concentration Distribution in an Aquifer Without Injection

In this case, the injection rate is assumed to be zero $\left(\dot{q}(x,y,t) = 0\right)$. It is further assumed that the initial non-dimensional concentration distribution in the aquifer is zero except at the point $(x = 0; y = 0)$. This condition may be represented as,

$$f(x,y) = \frac{C_o}{R}\delta(x,y)$$

(5.135)

Substituting Eq. (5.135) and $\overset{\bullet}{q}(x,y,t) = 0$ in Eq. (5.109) one may obtain,

$$C(x,y,t) = \frac{aC_o}{4\pi R\alpha(t)} \exp\left(-\lambda t - \frac{(x-Ut)^2 + (ay-Vt)^2}{4\alpha(t)}\right) \tag{5.136}$$

This solution is the same as Eq. (5.111) derived earlier, only with (M/n) replaced by C_o. Thus, solutions for point initial distribution with different dispersion coefficient functions can be derived as described in Section 5.5.2 which will not be repeated here.

5.5.5 Line Initial Concentration Distribution Without Injection

In this case, the initial non-dimensional distribution of the solute concentration in the aquifer is given as,

$$f(x,y) = \frac{C_o}{R}\delta(x) \tag{5.137}$$

This distribution implies that the initial non-dimensional concentration is different than zero only along the y-axis. Substituting Eq. (5.136) and $\overset{\bullet}{q}(x,y,t) = 0$ into Eq. (5.110), one may obtain,

$$C(x,y,t) = \frac{C_o\exp(-\lambda t)}{4\pi R\alpha(t)} \int_{-\infty}^{\infty} \exp\left(-\frac{(x-Ut)^2 - (ay-Vt-\eta)^2}{4\alpha(t)}\right)d\eta \tag{5.138}$$

This solution can be given as,

$$C(x,y,t) = \frac{C_o\exp\left(-\lambda t - \frac{(x-Ut)^2}{4\alpha(t)}\right)}{4\pi R\alpha(t)} \int_{-\infty}^{\infty} \exp\left(-\frac{(ay-Vt-\eta)^2}{4\alpha(t)}\right)d\eta \tag{5.139}$$

Let,

$$\Psi = \frac{(ay-Vt-\eta)}{\sqrt{4\alpha(t)}} \tag{5.140}$$

then Eq. (5.139) can be written as,

$$C(x,y,t) = \frac{C_o\sqrt{4\alpha(t)}}{4\pi R\alpha(t)} \exp\left(-\lambda t - \frac{(x-Ut)^2}{4\alpha(t)}\right) \int_{-\infty}^{\infty} \exp(\Psi^2)d\Psi \tag{5.141}$$

Since,

$$\int_{-\infty}^{\infty} \exp\left(\Psi^2\right) d\Psi = \sqrt{\pi} \tag{5.142}$$

Equation (5.141) can be written as,

$$C(x, y, t) = \frac{C_o}{R\sqrt{4\pi\alpha(t)}} \exp\left(-\lambda t - \frac{(x - Ut)^2}{4\alpha(t)}\right) \tag{5.143}$$

It is clear from Eq. (5.143) that this solution is independent of the y coordinate. Eq. (5.143) is similar to the solution given in the literature (Basha and Elhabel 1993) for a *one*-dimensional, time-dependent advection dispersion equation with an initial point pulse at $(x = 0)$ without injection of a contaminant source into the aquifer. The following particular solutions can now be given for the four time dependent dispersion coefficient functions which were defined earlier.

Constant Dispersion Coefficient: Utilizing Eq. (5.114) one may obtain,

$$C(x, y, t) = \frac{C_o}{\sqrt{4\pi(D_o + D_m)tR}} \exp\left(-\lambda t - R\frac{\left(x - \frac{v_x}{R}t\right)^2}{4(D_o + D_m)t}\right) \tag{5.144}$$

Linear Dispersion Coefficient: Utilizing Eq. (5.117) one may obtain,

$$C(x, y, t) = \frac{C_o}{\sqrt{4\pi\left(\frac{D_o}{2k}t + D_m\right)tR}} \exp\left(-\lambda t - R\frac{\left(x - \frac{v_x}{R}t\right)^2}{4\left(\frac{D_o}{2k}t + D_m\right)t}\right) \tag{5.145}$$

Asymptotic Dispersion Coefficient: Utilizing Eq. (5.120) one may obtain,

$$C(x, y, t) = \frac{C_o}{\sqrt{4\pi R\left((D_o + D_m)t + D_o k \ln\left(1 + \frac{t}{k}\right)\right)}}$$
$$\times \exp\left(-\lambda t - R\frac{\left(x - \frac{v_x}{R}t\right)^2}{4\left((D_o + D_m)t + D_o k \ln\left(1 + \frac{t}{k}\right)\right)}\right) \tag{5.146}$$

Exponential Dispersion Coefficient: Utilizing Eq. (5.123) one may obtain,

$$C(x, y, t) = \frac{C_o}{\sqrt{4\pi R\left((D_o + D_m)t + D_o k\left(\exp\left(-\frac{t}{k}\right) - 1\right)\right)}}$$
$$\times \exp\left(-\lambda t - R\frac{\left(x - \frac{v_x}{R}t\right)^2}{4\left((D_o + D_m)t + D_o k\left(\exp\left(-\frac{t}{k}\right) - 1\right)\right)}\right) \tag{5.147}$$

5.5.6 Analytical Solutions for an Instantaneous Line Injection into an Initially Uncontaminated Aquifer

In this case we assume that $C(x, y, 0) = f(x, y) = 0$. We further assume that the injection of a contaminant into the aquifer is along a line in the y-axis direction. Then the instantaneous non-dimensional injection of a contaminant can be defined by,

$$\dot{q}(x, y, t) = \frac{M}{n} \delta(x_o, t_o) \tag{5.148}$$

where M is the non-dimensional mass injected, n is the porosity, and $\delta(\bullet)$ is the Dirac delta function. Substituting $f(x, y) = 0$ and Eq. (5.148) into Eq. (5.110), $C(x, y, t)$ can be given as,

$$C(x, y, t) = \int_{-\infty}^{\infty} \frac{M \exp(-\lambda t)}{4\pi n R \alpha(t)} \exp\left(-\frac{(x - Ut)^2 + (ay - Vt - \eta)^2}{4\alpha(t)}\right) d\eta \tag{5.149}$$

or Eq. (5.149) can be written as,

$$C(x, y, t) = \frac{M \exp\left(-\lambda t + \frac{(x - Ut)^2}{4\alpha(t)}\right)}{4\pi n R \alpha(t)} \int_{-\infty}^{\infty} \exp\left(-\frac{(ay - Vt - \eta)^2}{4\alpha(t)}\right) d\eta \tag{5.150}$$

Let,

$$\Psi = \frac{(ay - Vt - \eta)}{\sqrt{4\alpha(t)}} \tag{5.151}$$

then Eq. (5.149) takes the form,

$$C(x, y, t) = \frac{M \sqrt{4\alpha(t)}}{4\pi n R \alpha(t)} \exp\left(-\lambda t - \frac{(x - Ut)^2}{4\alpha(t)}\right) \int_{-\infty}^{\infty} \exp(-\Psi^2) d\Psi \tag{5.152}$$

Utilizing Eq. (5.142), Eq. (5.152) can be written as,

$$C(x, y, t) = \frac{M}{n R \sqrt{4\pi \alpha(t)}} \exp\left(-\lambda t - \frac{(x - Ut)^2}{4\alpha(t)}\right) \tag{5.153}$$

This equation is similar to Eq. (5.143) if C_o is replaced by (M/n). From this equation, we can see that $C(x, y, t)$ is also independent of the y coordinate. Actually, this equation is also the solution for a one-dimensional time-dependent advection–dispersion equation with a point pulse injection at $x = 0$ concentration distribution in an initially clean aquifer (Basha and Elhabel 1993). For this case, for the four time dependent dispersion coefficient functions defined earlier, the concentration distribution of the solute in the aquifer can be described as given in Section 5.5.5 which will not be repeated here.

5.5.7 Analytical Solutions for a Continuous Line Source for an Initially Uncontaminated Aquifer

In this case we assume that the continuous non-dimensional line source is located on the y-axis. This condition can be represented as,

$$\overset{\bullet}{q}(x,y,t) = C_o\delta(x_o) \tag{5.154}$$

Substituting $f(x,y) = 0$ and the equation above into Eq. (5.109), we get,

$$
\begin{aligned}
C(x,y,t) = \int_0^t \int_{-\infty}^{\infty} & \frac{C_o\exp(-\lambda(t-\tau))}{4\pi R(\alpha(t)-\alpha(\tau))} \\
& \times \exp\left(-\frac{(x-U(t-\tau))^2 + (ay-Vt-\eta)^2}{4(\alpha(t)-\alpha(\tau))}\right) d\eta d\tau
\end{aligned}
\tag{5.155}
$$

This solution can be written as,

$$
\begin{aligned}
C(x,y,t) = \int_0^t & \frac{C_o\exp\left(-\lambda(t-\tau) - \frac{(x-U(t-\tau))^2}{4(\alpha(t)-\alpha(\tau))}\right)}{4\pi R(\alpha(t)-\alpha(\tau))} \\
& \times \int_{-\infty}^{\infty}\exp\left(-\frac{(ay-Vt-\eta)^2}{4(\alpha(t)-\alpha(\tau))}\right) d\eta d\tau
\end{aligned}
\tag{5.156}
$$

Let,

$$\Psi = \frac{(ay-Vt-\eta)}{\sqrt{4\alpha(t)-\alpha(\tau)}} \tag{5.157}$$

then Eq. (5.156) can be written as,

$$
C(x,y,t) = \int_0^t \frac{C_o\exp\left(-\lambda(t-\tau) - \frac{(x-U(t-\tau))^2}{4(\alpha(t)-\alpha(\tau))}\right)}{\pi R\sqrt{4(\alpha(t)-\alpha(\tau))}} \int_{-\infty}^{\infty}\exp(-\Psi^2) d\Psi d\tau \tag{5.158}
$$

Substituting Eq. (5.142) into Eq. (5.158), $C(x,y,t)$ can be given as,

$$C(x,y,t) = \int_0^t \frac{C_o}{R\sqrt{4\pi(\alpha(t) - \alpha(\tau))}} \exp\left(-\lambda(t-\tau) - \frac{(x - U(t-\tau))^2}{4(\alpha(t) - \alpha(\tau))}\right) d\tau$$

(5.159)

In this case $C(x,y,t)$ is also independent of the y coordinate. The equation given above is also the solution for a one-dimensional problem with a continuous injection of concentration C_o at point $x = 0$. Special cases of this solution can be given as shown below.

Constant Dispersion Coefficient: Utilizing Eq. (5.114) one may obtain,

$$C(x,y,t) = \int_0^t \frac{C_o}{R\sqrt{4\pi(D_o + D_m)(\tau/R)}} \exp\left(-\lambda\tau - \frac{R(x - U\tau)^2}{4(D_o + D_m)\tau}\right) d\tau \quad (5.160)$$

Linear Dispersion Coefficient: Utilizing Eq. (5.117) one may obtain,

$$C(x,y,t) = \int_0^t \frac{C_o}{R\sqrt{4\pi\left(\frac{D_o}{2k}(t^2 - \tau^2) + D_m(t-\tau)\right)/R}}$$
$$\times \exp\left(-\lambda(t-\tau) - \frac{R(x - U(t-\tau))^2}{4\left(\frac{D_o}{2k}(t^2 - \tau^2) + D_m(t-\tau)\right)}\right) d\tau$$

(5.161)

Asymptotic Dispersion Coefficient: Utilizing Eq. (5.120) one may obtain,

$$C(x,y,t) = \int_0^t \frac{C_o}{R\sqrt{4\pi\left((D_o + D_m)(t-\tau) - D_o k \ln\left(\frac{k+t}{k+\tau}\right)\right)/R}}$$
$$\times \exp\left(-\lambda(t-\tau) - \frac{R(x - U(t-\tau))^2}{4\left((D_o + D_m)(t-\tau) - D_o k \ln\left(\frac{k+t}{k+\tau}\right)\right)}\right) d\tau$$

(5.162)

Exponential Dispersion Coefficient: Utilizing Eq. (5.123) one may obtain,

$$C(x,y,t) = \int_0^t \frac{C_o}{R\sqrt{4\pi\left((D_o + D_m)(t-\tau) - D_o k\left(\exp\left(-\frac{t}{k}\right) - \exp\left(-\frac{\tau}{k}\right)\right)\right)/R}}$$
$$\times \exp\left(-\lambda(t-\tau) - \frac{R(x - U(t-\tau))^2}{4\left((D_o + D_m)(t-\tau) - D_o k\left(\exp\left(-\frac{t}{k}\right) - \exp\left(-\frac{\tau}{k}\right)\right)\right)}\right) d\tau$$

(5.163)

5.5.8 Numerical Examples for Variable Dispersion Coefficient Models

In the discussion above, a catalog of analytical solutions to a large class of contaminant transports problems was described. Due to space limitations numerical examples for all these cases cannot be discussed here. Thus, in order to evaluate contaminant migration patterns for the asymptotically varying dispersion coefficient case, a selected set of analytical solutions for the instantaneous point source problem will be discussed. In this example, the parameters $(R; \lambda; D_m; D_o; M; n; v_x; v_y)$ are defined as follows (1;0;0;1;0.25;0.25;0.25;0) respectively. Numerical results shown in Figs. 5.17–5.19 correspond to the case of an instantaneous point source in an initially uncontaminated aquifer for $(a^2 = 1)$, Eq. (5.121). In Fig. 5.17, numerical results obtained for $y = 0$ are summarized. In this solution $k = 0$ corresponds to a constant non-dimensional dispersion coefficient. Similarly, results obtained for $(y = 2)$ and $(y = 5)$ are given in Figs. 5.18 and 5.19 respectively. For the parameters considered above, analytical results obtained for $(a^2 = 1)$ indicate that for small values of time $(t = 10)$, as "k" increases from 0 to 20, peak concentration magnitudes increase approximately fivefold in the longitudinal direction $(y = 0)$ and about sixfold in the transverse direction $(y = 2)$. Again for small times $(t = 10)$, for $k = 50$ this increase may reach up to a level of tenfold in the longitudinal direction $(y = 0)$ and fivefold in the transverse

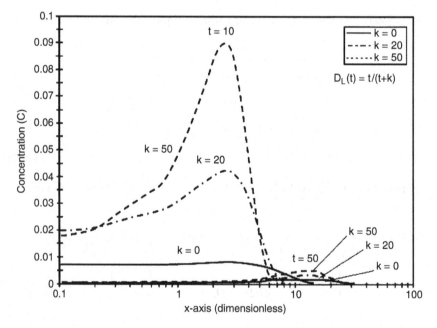

Fig. 5.17 Dimensionless concentration profiles as a function of time and x-coordinate at y = 0 for an instantaneous point source (a² = 1)

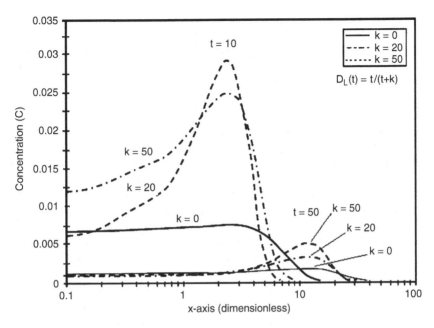

Fig. 5.18 Dimensionless concentration profiles as a function of time and x-coordinate at $y = 2$ for an instantaneous point source ($a^2 = 1$)

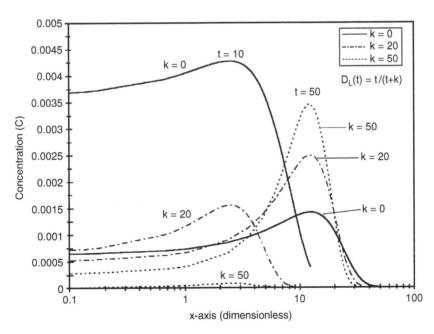

Fig. 5.19 Dimensionless concentration profiles as a function of time and x-coordinate at $y = 5$ for an instantaneous point source ($a^2 = 1$)

direction $(y = 2)$. For $(y = 2)$, the peak concentration for the case of $k = 50$ is less than the peak concentration for $k = 20$. The results summarized in Fig. 5.19 indicate that, relative to the $k = 0$ solution, the concentration magnitudes do not increase in the transverse direction for $(y = 5)$. For both $k = 20$ and $k = 50$, the peak concentrations are less than the results obtained for $k = 0$. For large distances or large time $(t = 50)$, the increase in concentration levels is not as large in the longitudinal direction $(y = 0)$. In the transverse direction $(y = 2)$ and $(y = 5)$, however, five- and threefold increases are observed, respectively.

Thus, for early times and for the case of an asymptotically varying dispersion coefficient $(k = 0$ to $k = 50)$, significant concentration increases are expected in the longitudinal direction, whereas when the dispersion coefficient does not asymptotically vary, concentration magnitudes do not show as much of an increase, or can even be reduced as the results in the transverse direction indicate. On the other hand, for large times over the same variation range in the dispersion coefficient, the increase in concentration magnitudes in the longitudinal direction is not significant, whereas the increase in the transverse direction becomes significant. The reversal of the increase in peak concentrations in the transverse direction, for large k values and for small times, and the increase of transverse concentration for large times are trends which have been observed repeatedly for other solutions as well. In all cases, for $(a^2 = 1)$, the travel distance of the peak concentration was not altered.

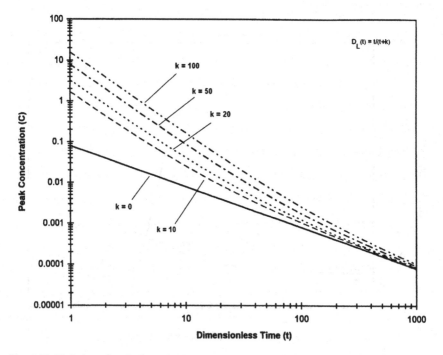

Fig. 5.20 Variation of peak dimensionless concentration profiles as a function of time and k at $y = 0$ for an instantaneous point source $(a^2 = 1)$

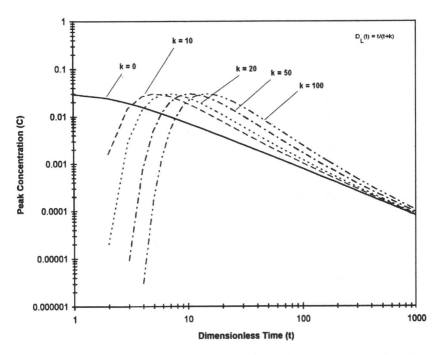

Fig. 5.21 Variation of peak dimensionless concentration profiles as a function of time and k at $y = 2$ for an instantaneous point source ($a^2 = 1$)

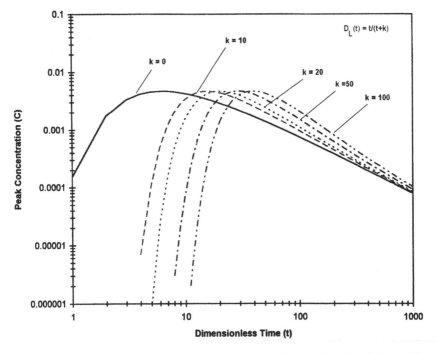

Fig. 5.22 Variation of peak dimensionless concentration profiles as a function of time and k at $y = 5$ for an instantaneous point source ($a^2 = 1$)

This observation is illustrated in Figs. 5.20–5.22 in which the peak concentration is plotted as a function of dimensionless time for $(y = 0; \; y = 2$ and $y = 5)$. From Fig. 5.20 one can observe that peak concentrations are higher for early times in the longitudinal direction and that as the solution time increases for all "k" values the peak concentration asymptotically reduces to the level of peak concentration of case $k = 0$. In the transverse direction, the variation of peak concentration with time shows a different trend. The arrival of higher peak concentrations in the transverse direction, such as those at $(y = 2$ and $y = 5)$, does not necessarily occur at early times but may occur at much later stages of the advective–dispersive transport period. The other observation that is evident in Figs. 5.20–5.22 is that, for the case considered here, the time dependent nature of the dispersion coefficient does not affect the concentration distribution solution at very large times.

Similarly, numerical results obtained for the case $(a^2 = 6)$ are given in Figs. 5.23–5.28. Conclusions derived for this case follow the same pattern discussed above. From these results it can be seen that the relation between time dependent dispersivity values and contaminant distribution in the longitudinal and transverse directions is complex. The general trend observed is the reversal of the advective–dispersive expansion patterns as k or y increases as a function of time. For an asymptotically varying dispersion coefficient, the general pattern is that concentration magnitudes increase in the longitudinal direction during early times. However, similar increases are expected in the transverse direction only in later solution periods and thus at large distances.

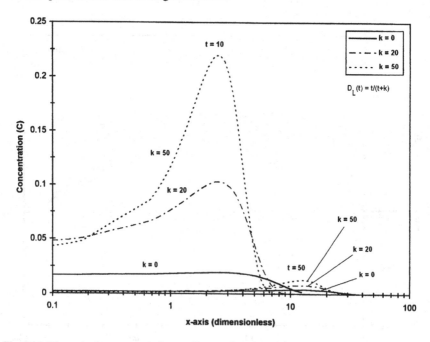

Fig. 5.23 Dimensionless concentration profiles as a function of time and x-coordinate at $y = 0$ for an instantaneous point source $(a^2 = 6)$

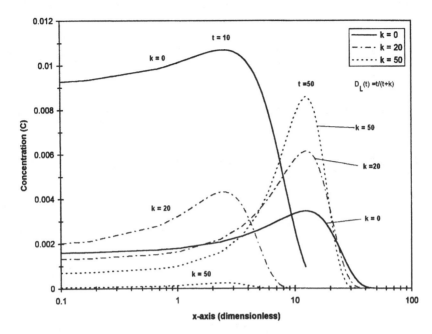

Fig. 5.24 Dimensionless concentration profiles as a function of time and x-coordinate at $y = 2$ for an instantaneous point source ($a^2 = 6$)

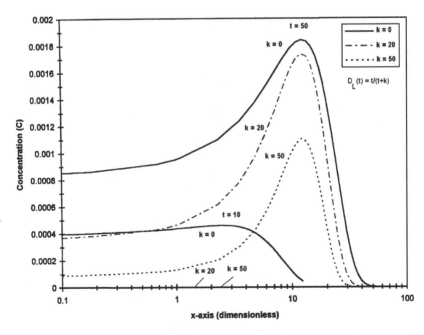

Fig. 5.25 Dimensionless concentration profiles as a function of time and x-coordinate at $y = 5$ for an instantaneous point source ($a^2 = 6$)

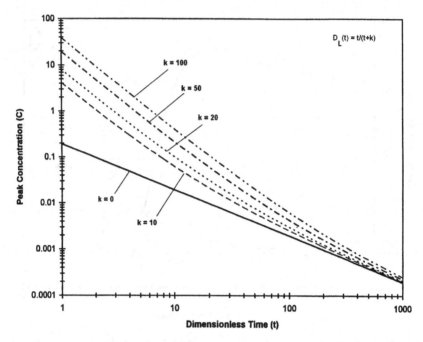

Fig. 5.26 Variation of peak dimensionless concentration profiles as a function of time and k at $y = 0$ for an instantaneous point source ($a^2 = 6$)

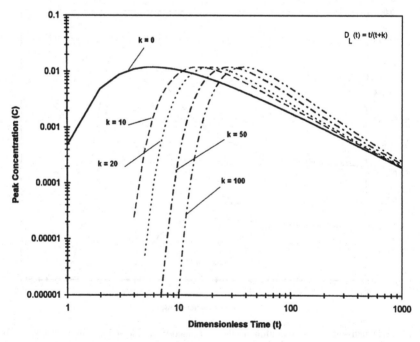

Fig. 5.27 Variation of peak dimensionless concentration profiles as a function of time and k at $y = 2$ for an instantaneous point source ($a^2 = 6$)

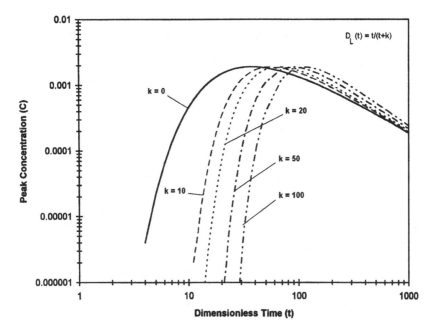

Fig. 5.28 Variation of peak dimensionless concentration profiles as a function of time and k at $y = 5$ for an instantaneous point source ($a^2 = 6$)

The numerical results for an instantaneous point source problem, based on the exponentially varying dispersivity coefficient, showed similar trends in our analysis. These results are not included here due to space limitations. Results for other problems may be obtained in a similar manner using the analytical solutions given above, which are included in the ACTS software.

In this section, general analytical solutions for the two-dimensional advection–dispersion equation with time dependent dispersion coefficients are developed. These solutions can be employed to obtain particular solutions for several time dependent dispersion coefficient functions, and also for various injection and initial concentration distributions in an aquifer. It is analytically shown that the point pulse initial distribution solution and the instantaneous point injection case tend to similar solutions. As special situations, analytical solutions for instantaneous line injection and continuous line injection cases are also given. It is shown that both of these cases yield solutions similar to a one-dimensional problem with instantaneous point injection and continuous point injection as expected. For these cases, it is shown that the analytical solutions presented for the four special dispersion coefficient functions yield the same solutions as those given by Basha and Elhabel (1993).

In the experimental and analytical work conducted by researchers it has been shown that there are two important scales in the analysis of the effects of dispersion: (i) the smaller time scale in which the dispersivity grows with time and distance; and, (ii) the larger time scales in which the dispersion coefficient becomes constant.

These studies also indicate that both the longitudinal and transverse dispersions of contaminants will be influenced during small time scales in which the variability of dispersivity is significant, and that the asymptotic behavior of the dispersion coefficient can be reached only for unrealistically large time or distances, and thus the so-called "pre-asymptotic" period is important in most practical applications.

Based on the results of the case studies discussed above, for the "pre-asymptotic" period, it can be concluded that the effect of the time dependent dispersion coefficient on the contaminant dispersion problem is not the same in both the longitudinal and transverse directions. As time dependent dispersion coefficients increase, the concentration magnitudes in the longitudinal direction in the "pre-asymptotic" period increase. Similar increases are not observed in the transverse direction during the same periods. Instead, comparable increases in concentration levels in the transverse direction occur at much larger times. Thus, scale dependence effects on contaminant dispersion in longitudinal and transverse directions do not follow the same pattern. Again, for the case studies discussed above, for very large times, the analytical solutions indicate that the time dependent nature of the dispersion does not significantly influence the contaminant migration pattern in both longitudinal and transverse directions.

The analytical solutions discussed here are benchmark solutions for scale dependent dispersivity problems for contaminant transport analysis in two-dimensional domains. These solutions may be used to analyze problems in which scale dependence is of concern. These solutions may also be used to provide tools to evaluate field data in which scale dependence of the dispersion coefficient is expected to influence contaminant migration patterns in an aquifer.

5.6 Unsaturated Groundwater Pathway Models

When contaminants are released to the soil surface or near the soil surface above the water table, the contaminant plume may migrate through the unsaturated zone and reach the saturated water table aquifer (unconfined aquifer). In such situations it is important to include the unsaturated zone in the analysis of contaminant transformation and transport. A schematic diagram of the contaminant migration in the unsaturated zone is shown in Fig. 5.29.

5.6.1 Marino Model

The first model included to the ACTS software which deals with the unsaturated zone is identified as the Marino model (Marino 1974). In this model the transport of contaminants in the unsaturated zone is treated as a one-dimensional problem. Similar to the saturated zone analysis, important transformation and transport

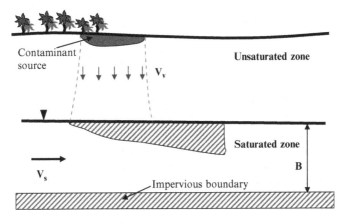

Fig. 5.29 Definition sketch for an unsaturated aquifer

mechanisms that may be considered in the analytical model include advection and dispersion in the vertical direction, linear adsorption, and first-order decay of the contaminant. With these components, the transport equation can be given as,

$$R_v \frac{\partial C}{\partial t} + v_v \frac{\partial C}{\partial z} = D_v \frac{\partial^2 C}{\partial z^2} - \lambda_v R_v C \qquad (5.164)$$

in which C is the dissolved phase contaminant concentration in the unsaturated zone $[ML^{-3}]$; D_v is the hydrodynamic dispersion coefficient in the vertical direction in the unsaturated zone $[L^2 T^{-1}]$; λ_v is the first-order degradation rate within the unsaturated zone $[T^{-1}]$; R_v is the unsaturated zone retardation factor [dimensionless]; V_v is the steady-state unsaturated zone seepage velocity $[LT^{-1}]$; t is time; z is the vertical coordinate, which is selected to be positive downwards. In the unsaturated advection–dispersion-reaction models the retardation factor in Eq. (5.164) is computed using Eq. (5.165):

$$R_v = 1 + \frac{\rho_b K_d}{\theta S_w} \qquad (5.165)$$

in which ρ_b is the bulk density of the unsaturated zone soil $[ML^{-3}]$; K_d is the distribution coefficient for the unsaturated zone $[L^3 M^{-1}]$; θ is the porosity of the unsaturated zone [dimensionless]; S_w is the water saturation within the unsaturated zone [dimensionless]. The overall first-order degradation rate λ_v, which is calculated using Eq. (5.29), includes the effects of both biodegradation and chemical hydrolysis reactions.

The solution of the above differential equation requires two boundary conditions. The first boundary condition describes the source concentration and may be given as:

$$C(0, t) = C_o \tag{5.166}$$

or

$$C(0, t) = C_o \exp(-\Lambda t) \tag{5.167}$$

in which Λ is the decay rate of the source concentration $[T^{-1}]$; C_o is the initial (or steady-state) concentration at the top of the unsaturated zone $[ML^{-3}]$. Note that Eq. (5.166) represents a constant source concentration condition and Eq. (5.167) represents an exponentially decaying source boundary concentration. The second boundary condition, applied at a large distance from the source in the downstream direction (z-axis) may be given as,

$$C(\infty, t) = 0 \tag{5.168}$$

The background concentration of the contaminant in the unsaturated zone is assumed to be negligible. Therefore, the initial condition is defined as,

$$C(z, 0) = 0 \tag{5.169}$$

The analytical solution for the above system of equations is given by various authors including Marino (1974). Using the constant concentration boundary condition, Eq. (5.166), the solution can be expressed as:

$$\frac{C}{C_o} = \frac{1}{2} \exp\left[\frac{(v_v - \Gamma)z}{2D_v}\right] erfc\left[\frac{R_v z - \Gamma t}{2\sqrt{D_v R_v t}}\right]$$
$$+ \frac{1}{2} \exp\left[\frac{(v_v + \Gamma)z}{2D_v}\right] erfc\left[\frac{R_v z + \Gamma t}{2\sqrt{D_v R_v t}}\right] \tag{5.170}$$

Using the exponentially decaying concentration boundary condition, the solution to Eq. (5.164) can be given as,

$$\frac{C}{C_o} = \frac{1}{2} \exp(-\Lambda t)\left\{\exp\left[\frac{(v_v - \Gamma)z}{2D_v}\right] erfc\left[\frac{R_v z - \Gamma_1 t}{2\sqrt{D_v R_v t}}\right]\right.$$
$$\left. + \exp\left[\frac{(v_v + \Gamma_1)z}{2D_v}\right] erfc\left[\frac{R_v z + \Gamma_1 t}{2\sqrt{D_v R_v t}}\right]\right\} \tag{5.171}$$

where Γ and Γ_1 are given by,

$$\Gamma = (v_v^2 + 4D_v \lambda_v)^{\frac{1}{2}} \tag{5.172}$$

$$\Gamma = (v_v^2 + 4D_v(\lambda_v - \Lambda R))^{\frac{1}{2}} \tag{5.173}$$

Assumptions and Limitations: The more important assumptions on which the unsaturated zone transport model is based are:

i. The flow field within the unsaturated zone is at steady state.
ii. The seepage velocity and other model parameters (e.g., the diffusion coefficient, partition coefficient) are uniform and constant (i.e., the unconfined aquifer is homogeneous and isotropic).
iii. Transport is assumed to be strictly one-dimensional. Lateral and transverse advection and diffusion are neglected.
iv. Decay of the solute may be described by a first-order decay constant. The daughter products of chemicals are neglected.

5.6.2 Jury Model

The second unsaturated zone model included in the ACTS software is the Jury model. This model may be used in vapor and solute transport analysis in the unsaturated zone and has been discussed in Chapter 4. The reader is referred to that section for further details of this application. The Jury model (Jury et al. 1990), is an unsaturated zone model that may be used to estimate both volatilization from soil and time dependent concentration profiles within the unsaturated zone. The mathematical model used in this case is similar to that used in other applications in this chapter and is repeated below,

$$R_v \frac{\partial C}{\partial t} + v_v \frac{\partial C}{\partial z} = D_v \frac{\partial^2 C}{\partial z^2} - \lambda_v R_v C \tag{5.174}$$

The parameters of this model are as defined in the Marino model. In this case, although the solution domain is infinite the soil contamination zone is finite $(0 \leq z \leq L)$ and the z-axis is oriented as positive downward from the soil surface. The initial and boundary conditions are defined as given in Eq. (5.175).

$$
\begin{aligned}
C &= C_o; & 0 < z < L;\ t = 0 \\
C &= 0; & z \geq L;\ t = 0 \\
D_v \frac{\partial C}{\partial z} + v_v C &= H_e C; & z = 0;\ t \geq 0 \\
C &= 0; & z \Rightarrow \infty;\ t \geq 0
\end{aligned}
\tag{5.175}
$$

The first equation above implies that, as an initial condition, the contaminant is uniformly incorporated in the soil to a depth L. The second equation above implies that the contaminant concentration below depth L is zero, or that the soil is clean (see Fig. 4.15). The third equation above defines the upper boundary condition, which indicates that the contaminant vapor is released to the atmosphere into a stagnant air boundary layer, in which the contaminant concentration of the air is assumed to be zero. This release reduces the contaminant concentration in the soil gradually. Here H_e is the mass transfer coefficient, estimated as,

$$H_e = \frac{hH}{(\rho_b K_d + \theta_w + \theta_a H)} \tag{5.176}$$

in which $h [LT^{-1}]$ is a boundary layer transfer coefficient estimated as $\left(h = D_g^a/d\right)$, $D_g^a [L^2 T^{-1}]$ is a chemical specific gaseous diffusion coefficient in air, $d [L]$ is the stagnant air boundary layer thickness, H is the dimensionless Henry's law constant, θ_w is the volumetric water content, θ_a is the air porosity, K_d is the chemical specific soil-water partition coefficient $(K_d = K_{oc} f_{oc})$ and ρ_b is the bulk density of soil. The lower boundary condition of the infinite domain is assumed to be zero as shown in the last equation above.

In Eq. (5.174), the total soil concentration C is assumed to be distributed between the solid, aqueous and vapor phases. It is estimated using,

$$C = \rho_b C_s + \theta_w C_l + \theta_a C_g \tag{5.177}$$

in which $(C_s; C_l; C_g)$ are the adsorbed phase soil concentration, aqueous phase concentration and gas phase concentration respectively. The three concentrations are related to each other by the partition coefficients as follows,

$$\begin{aligned} C_s &= K_d C_l \\ C_g &= H C_l \end{aligned} \tag{5.178}$$

The effective diffusion coefficient in Eq. (5.174) is estimated as,

$$D_v = \frac{\left[\left(\theta_a^{10/3} D_g^a H + \theta_a^{10/3} D_l^w\right)/\theta^2\right]}{[\rho_b f_{oc} K_{oc} + \theta_w + \theta_a H]} \tag{5.179}$$

in which θ is the soil porosity, D_l^w is the chemical specific liquid diffusion coefficient in water and f_{oc} is the fraction of organic carbon content. The effective contaminant velocity in the soil is estimated by,

$$v_v = \frac{J_w}{[\rho_b f_{oc} K_{oc} + \theta_w + \theta_a H]} \tag{5.180}$$

in which $J_w [LT^{-1}]$ is the volumetric soil–water flux, or the percolation rate when J_w is positive.

The analytical solution to this model is given as (Jury et al. 1990),

$$
\begin{aligned}
C = \frac{1}{2}C_o \exp(-\lambda t) &\left\{ erfc\left[\frac{(z - L - v_v t)}{(4D_e t)^{1/2}}\right] - erfc\left[\frac{(z - v_v t)}{(4D_e t)^{1/2}}\right] \right\} \\
&+ \left(1 + \frac{v_v}{H_e}\right) \exp\left(\frac{v_v z}{D_e}\right)\left[erfc\left[\frac{(z + L + v_v t)}{(4D_e t)^{1/2}}\right] - erfc\left[\frac{(z + v_v t)}{(4D_e t)^{1/2}}\right]\right] \\
&+ \left(2 + \frac{v_v}{H_e}\right) \exp\left(\frac{(H_e(H_e - v_v)t + (H_e - v_v)z)}{D_e}\right) \\
&\bullet \left[erfc\left[\frac{(z + (2H_e + v_v)t)}{(4D_e t)^{1/2}}\right] - \exp\left(\frac{H_e L}{D_e}\right) erfc\left[\frac{(z + L + (2H_e + v_v)t)}{(4D_e t)^{1/2}}\right]\right]
\end{aligned}
$$

$$(5.181)$$

Equation (5.181) represents the time dependent solution of the concentration in the soil which is included in the ACTS software.

5.7 Applications

The environmental pathway models discussed in this chapter cover a wide range of saturated and unsaturated groundwater pathway transformation and transport models. Providing applications for each of these cases would be an almost impossible task due to the multitude of cases that can be covered using these models. In this section, several applications are selected and solved to demonstrate the use of the important features of the ACTS software. As the reader becomes familiar with the ACTS software, he or she will recognize that the features and procedures discussed below are standardized for all other pathway applications within the ACTS software. These procedures can be repeated in other studies that involve other environmental pathways to extend the analysis to a more sophisticated level. Thus, the purpose here is to introduce the reader to some applications in groundwater pathway transformation and transport analysis using ACTS software, and in so doing help familiarize the reader with the important features of this software. In these applications we will be using the original version of the graphics package as opposed to the new version that was utilized in the applications discussed in Chapter 4. This selection can be done using the "Options" pull down menu when the ACTS software is started (see Appendix 3).

Example 1: Contamination of an aquifer has occurred due to a spill of 2,4-dinitrotoluene over an area of 50 m^2 (10 m by 5 m) at a concentration of 1,000 mg/L. The aquifer has a porosity of 35% and a bulk soil density of 1.6 g/cm^3. The effective aquifer thickness is 10 m and the infiltration rate is 0.0005 m/day. It is estimated that the Darcy velocity in the aquifer is about 0.8 m/day with an estimated range in between 0.2 and 1.95 m/day, and that the longitudinal dispersion coefficient is about 1.6 m²/day with an estimated range between 0.8 and 2.50 m²/day. The distribution coefficient of 2,4-dinitrotoluene for this aquifer has been measured to be 2.5 mL/g, 2,4-dinitrotoluene biodegrades through a first-order reaction rate, and has a half life of 40 days. After one year (360 days) of extensive soil excavation efforts, the contaminant source area is completely removed from the aquifer. Estimate the extent of the contaminant plume in the aquifer remaining and analyze the migration pattern.

Solution: In this case we will use a two-dimensional aquifer model (Fig. 5.6). All of the data given in the problem above are defined as deterministic values, except for the Darcy velocity and the dispersion coefficients. We will perform uncertainty analysis on those variables later on to evaluate the effects of the uncertainty on the results. First let's solve the problem using a deterministic analysis. Selecting the two-dimensional model we want to work with, we enter the necessary data using the three folders available for this model (Figs. 5.30–5.32. As seen in Fig. 5.30, the solution domain needs to be identified by entering the minimum and maximum (x, y) coordinates and the discretization step size. In this case the longitudinal aquifer length is chosen as $(0 \leq x \leq 200$ m$)$, and the computational step length in the x-direction is selected as 10 m. In the y-direction the aquifer length is selected as 50 m, and the computation step length in the y-direction is selected as 5 m. The maximum simulation time is selected as 700 days with a time step of 10 days. All of these selections will be made by the user and can be changed if too small or too large values are selected initially. In Fig. 5.31 the boundary

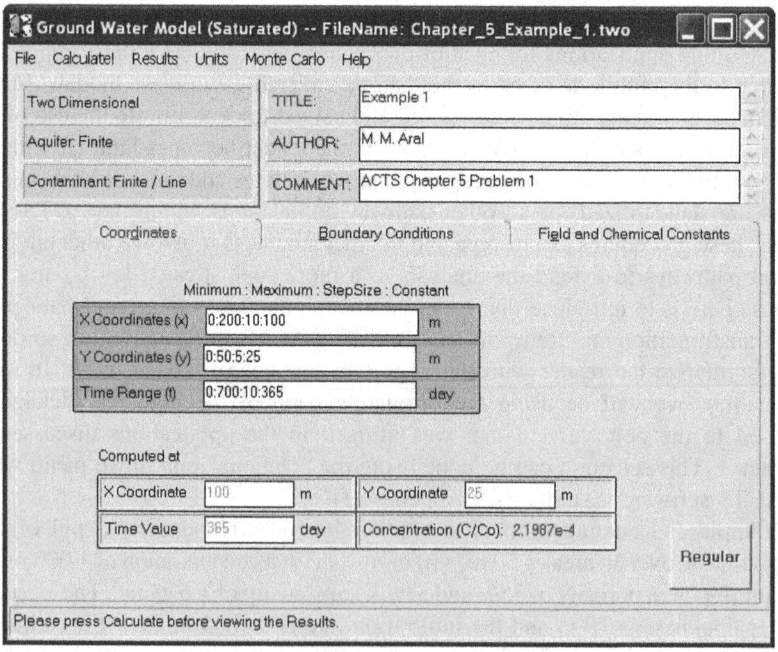

Fig. 5.30 Example 1 coordinate data entry folder

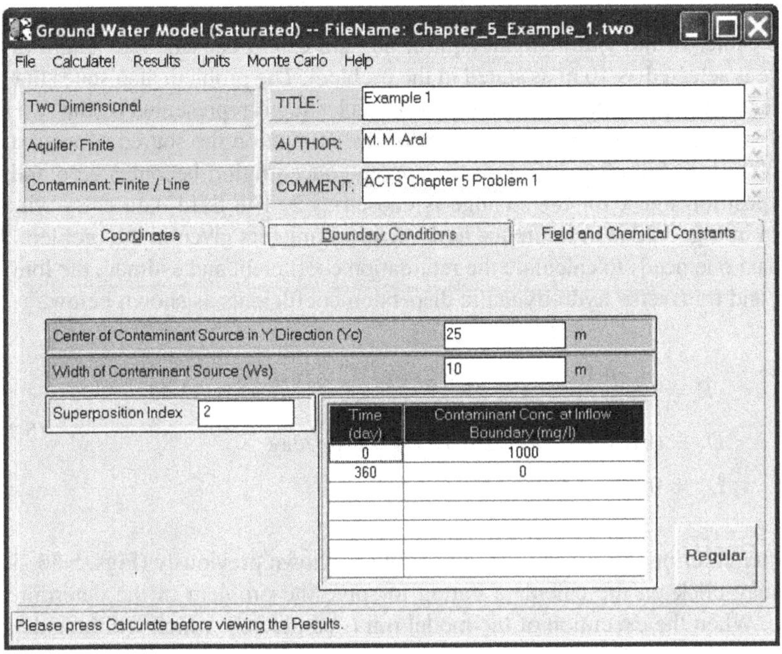

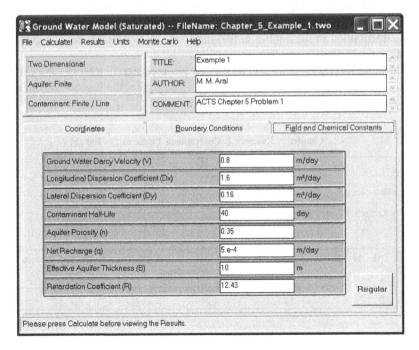

Fig. 5.31 Example 1: Boundary condition data entry folder

Fig. 5.32 Example 1: Aquifer parameter data entry folder

condition data is shown. The contaminant source center is selected as 25 m, which is the center of the y-directional aquifer domain extent chosen. The length of the source is selected as 10 m as stated in the problem. The problem also states that the aquifer source is cleaned after 360 days. This cleanup is represented by the two step superposition data entered. At the start of the simulation the source concentration is 1,000 mg/L but after 360 days the source concentration becomes zero and the superposition index for this change is chosen as 2. The third data entry window shown in Fig. 5.32 is in reference to the field parameters given in the problem. For this data one needs to calculate the retardation coefficient and estimate the longitudinal and transverse hydrodynamic dispersion coefficients as shown below,

$$R = 1 + \frac{\rho_b K_d}{n} = 1 + \frac{(1.6 \text{ g/cm}^3)(2.5 \text{ mL/g})}{0.35} = 12.43$$

$$D_x = 0.1 L v_x = 0.1(200)(0.8) = 1.6 \text{ m}^2/\text{day} \qquad (5.182)$$

$$D_y = 0.1 D_x = 0.16 \text{ m}^2/\text{day}$$

After entering these data in the three folders shown previously (Figs. 5.30–5.32) one may click on the calculate button to solve the problem in the deterministic mode. When the execution of the model run is completed, which will be indicated by the red bar at the bottom of the data entry folders, the results may be analyzed.

The numerical results obtained can be viewed in graphical, text or spreadsheet formats. Numerous types of outputs can be viewed by selecting the graphical option. In Figs. 5.33 and 5.34, two plots are given which show the contaminant plume distribution in the aquifer at two different solution times, i.e. $t = 200$ and 500 days.

The numerical results may also be viewed as a concentration profile at a certain point in the aquifer as shown in Fig. 5.35. In this figure, the numerical results at $(y = 25 \text{ m})$ are shown as a function of (x, t).

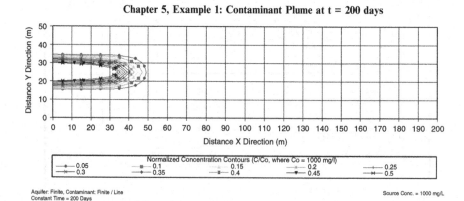

Fig. 5.33 Example 1: Contaminant plume distribution in the aquifer at $t = 200$ days

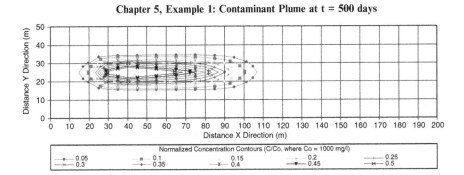

Fig. 5.34 Example 1: Contaminant plume distribution in the aquifer at t = 500 days

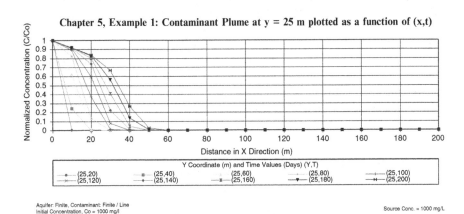

Fig. 5.35 Example 1: Contaminant plume distribution in the aquifer at y = 25 m

These results represent the deterministic solution of Example 1. The reader should also recognize the effect of the changing boundary condition at the source as shown in Fig. 5.34 where the source concentration is zero after the clean-up operation at time 360 days. Further, the user should experiment with the alternatives that are available in the graphics module to display the results obtained. These alternatives include changing the axis scale, title, number of concentration profiles plotted, left and center footnote insertions, and color and font. Those options that are available to the user can be learned only through practice and through the repeated use of the functions of this module.

As defined in the problem, there are uncertainties associated with the values of certain parameters. It is estimated that the Darcy velocity in the aquifer is about 0.8 m/day, with an estimated range in between 0.2 and 1.95 m/day, and that the longitudinal dispersion coefficient is about 1.6 m²/day with an estimated range in between 0.8 and 2.50 m²/day. Accordingly the transverse dispersion coefficient varies in the range of 0.08–0.25 m²/day. Given the uncertainty in these variable parameters, one may analyze its effects on the numerical results. To evaluate uncertainty, a two stage Monte Carlo analysis will be conducted for the solution at $(x = 100$ m; $y = 25$ m; $t = 365$ days$)$. To begin, enter the Monte Carlo window by selecting the "Generate" option under that Monte Carlo pull down menu. As discussed in Chapter 4, certain parameters need to be specified to generate the probability density functions of the three uncertain parameters. These selections are shown in Fig. 5.36, which indicates that lognormal representations of the three parameters are chosen and that 5,000 random variables are generated for each case. The resulting distributions are shown in Figs. 5.37 and 5.38 for the velocity and longitudinal dispersion parameters. It should also be noticed that in this analysis a two stage Monte Carlo approach is initiated since the complete parameter sets are chosen for the random variables selected to evaluate the mathematical model. With this information we will return to the main model window by closing the Monte Carlo window. As can be seen in Fig. 5.39, the Darcy velocity, longitudinal and lateral dispersion coefficient boxes display "Monte Carlo". Using this new data set Monte Carlo analysis can be completed by clicking on the calculate button. When the computation is complete the results can be viewed as a probability density or in some other formats at $(x = 100$ m; $y = 25$ m; $t = 365$ days$)$. The probability density functions obtained for the concentration at these coordinates are shown in Fig. 5.40, which indicates that the variability effects are negligible over the ranges considered.

Monte Carlo Simulation (Two Dimensional)

File Generate Options Help

AQUIFER: Finite CONTAMINANT: Finite / Line

Input Parameters

Variants	Mean	Minimum	Maximum	Variance	Number of Terms	Distribution Type
Ground Water Darcy	0.8	0.2	1.95	0.02	5000	Lognormal
Longitudinal Dispersion	1.6	0.8	2.5	0.03	5000	Lognormal
Lateral Dispersion	0.16	0.08	0.25	0.03	5000	Lognormal

Simulation Results

Variants	Arithmetic Mean	Geometric Mean	Median	Minimum	Maximum	Standard Deviation
Ground Water Darcy	0.8003	0.78805	0.78829	0.44591	1.4611	0.14201
Longitudinal Dispersion	1.6003	1.591	1.5913	1.1207	2.3262	0.17366
Lateral Dispersion	0.14456	0.13737	0.13554	0.08001	0.25	0.04648

Select Exit from File menu to use the selected values in the model.

Fig. 5.36 Example 1: Monte Carlo data entry window

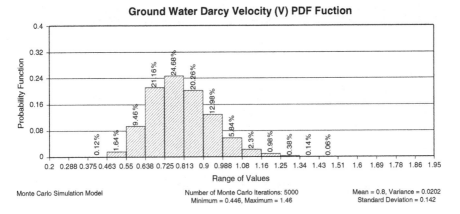

Fig. 5.37 Example 1: Probability density function for Darcy velocity

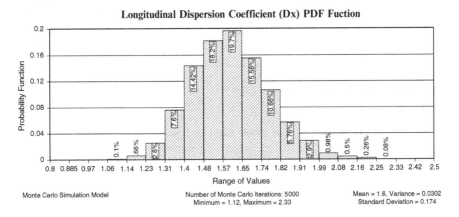

Fig. 5.38 Example 1: Probability density function for longitudinal dispersivity

Similarly, cumulative or complementary cumulative plots may also be generated to facilitate the analysis of the results.

Similar to the previous presentations numerous changes on these figures can be made. It is recommended that the user should learn the functions of these options through practice.

Example 2: In this application the problem given in Example 1 will be analyzed using variable dispersivity models with some changes to the input data. The following data is given for this problem: contamination of an aquifer has occurred due to a spill of 2,4-dinitrotoluene as a point source at a concentration of 1,000 mg/m/day. The aquifer has a porosity of 35% and a bulk soil density of 1.6 g/cm^3. It is estimated that the Darcy velocity in the aquifer is 0.8 m/day in the longitudinal direction and that the Darcy velocity in the transverse direction is given as 0.2 m/day.

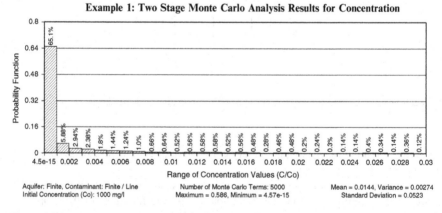

Fig. 5.39 Example 1: Monte Carlo data entry window

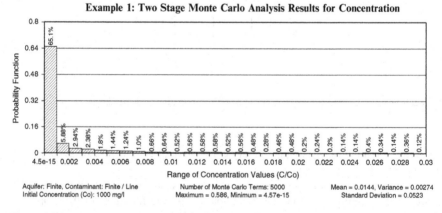

Fig. 5.40 Example 1: Probability density function obtained for concentrations

The longitudinal dispersion coefficient is 1.6 m²/day, and the maximum dispersion coefficient is given as 2.5 m²/day. The distribution coefficient of 2,4-dinitrotoluene for this aquifer has been measured to be 2.5 mL/g. 2,4-dinitrotoluene biodegrades through a first-order reaction rate and has a half life of 40 days. Estimate the extent of the contaminant plume in the aquifer using an asymptotic dispersion parameter model with a dispersion ratio coefficient of a = 0.5 and a mean travel time of k = 100 days.

Solution: In this case we will use a continuous point source model with the data entry folders shown in Figs. 5.41–5.43, which indicate a two-dimensional velocity field. In this application the uncertainty analysis is not considered. The contaminant source is located at x = 0 and y = 0 and the source is continuous. The numerical results obtained for this problem are given in Fig. 5.44.

Example 3: In this example a deterministic analysis of a multi-species application is discussed for a one-dimensional contaminant transport problem. The hypothetical chemical C1 is introduced to the aquifer of 20.0 m length at a constant rate of 100 mg/L. The chemical C1 degrades into byproducts by the sequence C1 → C2 → C3 → C4 at the reaction rates $k_1 = 0.075$ day^{-1}, $k_2 = 0.05$ day^{-1}, $k_3 = 0.02$ day^{-1}, $k_4 = 0.045$ day^{-1}. The Darcy velocity in the aquifer is 0.21 m/day, the longitudinal hydrodynamic dispersion coefficient is 0.6 m^2/day and the aquifer porosity is 0.35. Due to the limitations of the analytical solution, as discussed in Section 5.4, the retardation coefficient is assumed to be a constant $R = 1$ for all species. The effective yield factors of the chemicals are also assumed to be 1 to simplify the problem. The task is to determine the formation and migration of the chemical species sequence C1 → C2 → C3 → C4 in the aquifer.

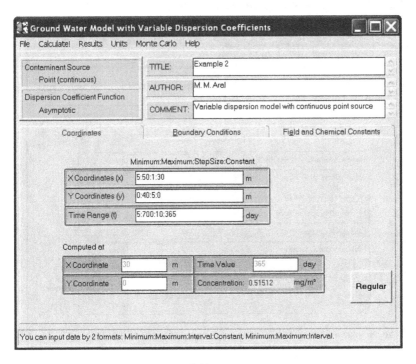

Fig. 5.41 Example 2: coordinate data entry folder

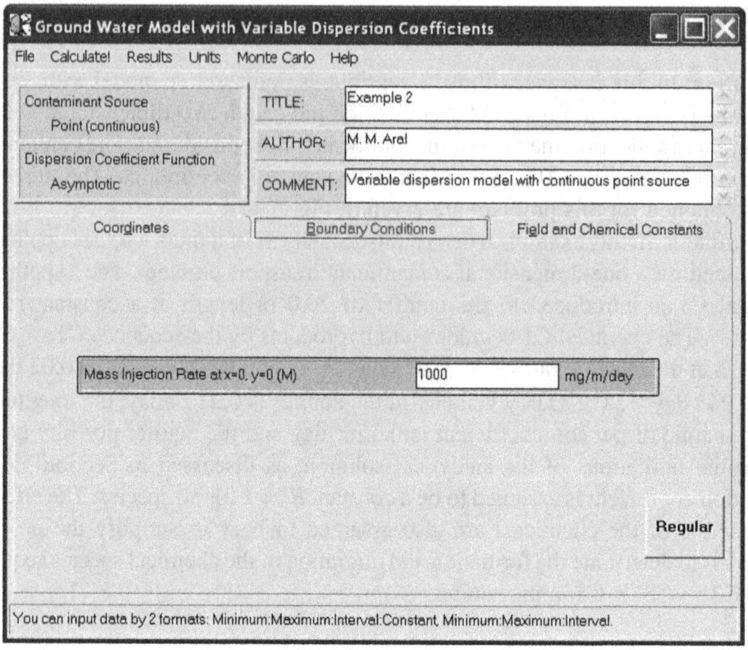

Fig. 5.42 Example 2: boundary condition data entry folder

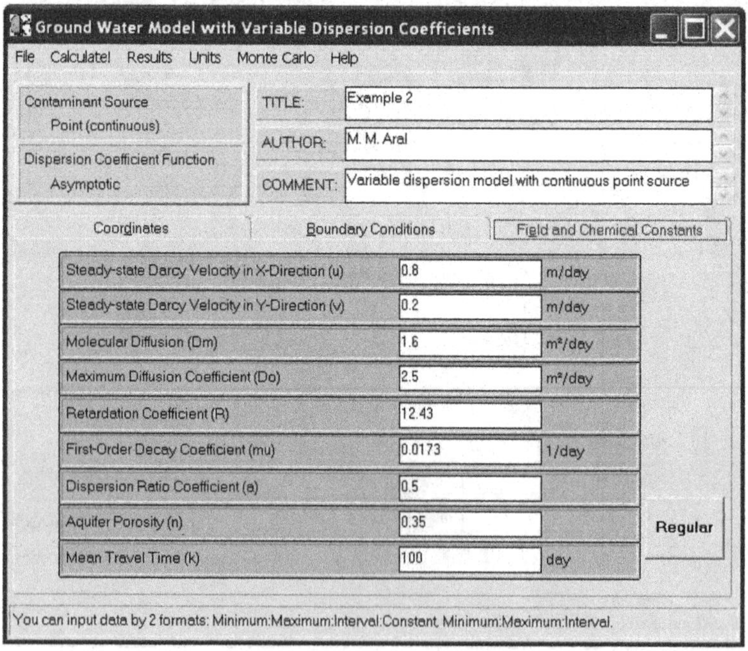

Fig. 5.43 Example 2: Aquifer parameter data entry folder

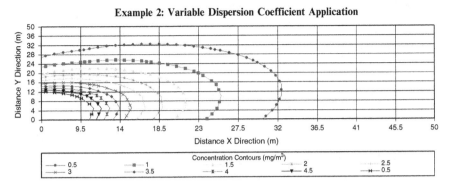

Fig. 5.44 Contaminant plume distribution in the aquifer for a two dimensional velocity field

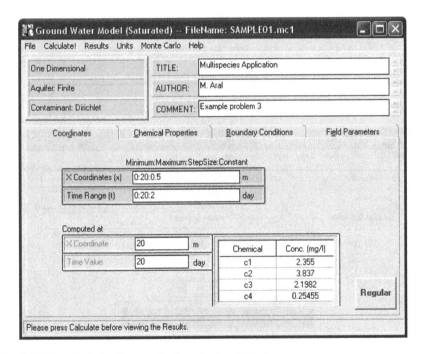

Fig. 5.45 Example 3: Aquifer domain discretization folder

Solution: The 20.0 m length aquifer is discretized both spatially and temporarily, and the other input parameters of the problem are entered into the ACTS software as shown in Figs. 5.45–5.48.

Once the data is entered to the model by clicking on the "Calculate" button the computations are done. The results are displayed at the end of the aquifer 20.0 m,

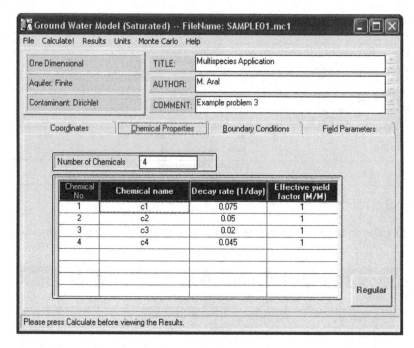

Fig. 5.46 Example 3: Chemical properties data entry folder

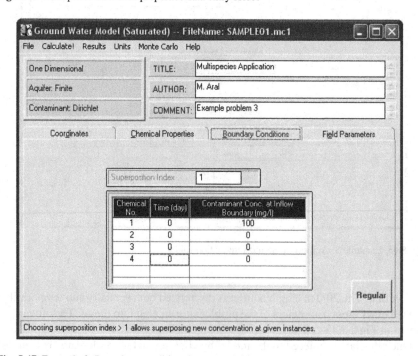

Fig. 5.47 Example 3: Boundary condition data entry folder

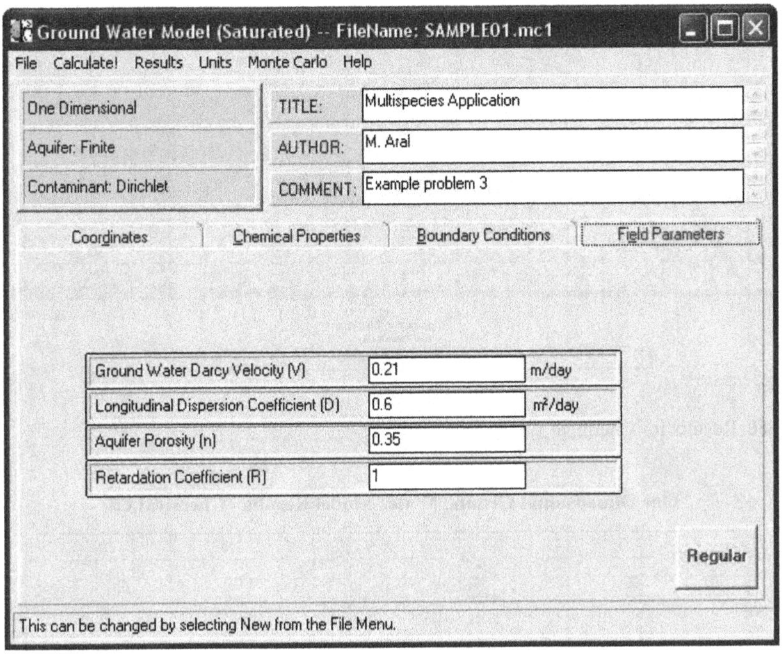

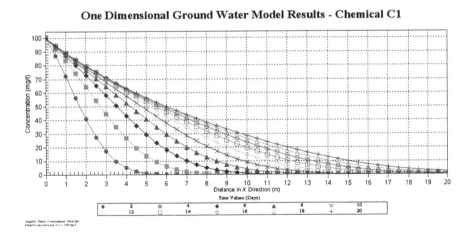

Fig. 5.48 Example 3: Aquifer properties data entry folder

Fig. 5.49 Results for Chemical C1

and at time 20.0 days, as shown in the lower grid of Fig. 5.45. This outcome should be apparent to the user by now and stems from the fact that we have not specified as specific spatial point in the aquifer and time in the upper grid and thus

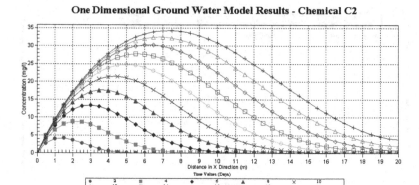

Fig. 5.50 Results for Chemical C2

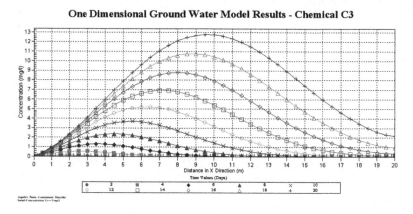

Fig. 5.51 Results for Chemical C3

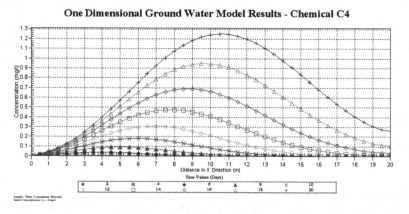

Fig. 5.52 Results for Chemical C4

end of domain and final time is selected as default values in the ACTS software for the X-coordinate and Time grid as shown in Fig. 5.45. In that case the results are displayed at (x_{max}, t_{max}) as a default. The results may also be plotted using the graphics package of the ACTS software. These results are shown in Figs. 5.49–5.52 for the chemical sequence C1 → C2 → C3 → C4 and complete the analysis of this problem. The user should also recognize that these results may be displayed as breakthrough plots if desired.

References

Anderson BA, Maslia ML et al (2007) Probabilistic analysis of pesticide transport in shallow groundwater at the Otland Island Education Center, Otland Island, Georgia. Atlanta, Agency for Toxic Substances and Disease Registry (ATSDR): 48 p

Anderson MP (1984) Movement of contaminants in groundwater: groundwater transport – advection and dispersion. National Academy Press, Washington D.C., pp 37–45

Barry DA, Sposito G (1989) Analytical solution of a convection-dispersion model with time-dependent transport-coefficients. Water Resour Res 25(12):2407–2416

Basha HA, Elhabel FS (1993) Analytical solution of the one-dimensional time-dependent transport-equation. Water Resour Res 29(9):3209–3214

Bear J (1972) Dynamics of fluids in porous media. American Elsevier, New York, 764 pp

Bear J (1979) Hydraulics of groundwater. McGraw Hill, New York, 569 pp

Charbeneau RJ (2000) Groundwater hydraulics and pollutant transport. Prentice Hall, Upper Saddle River, NJ, 593 pp

Cherry JA, Gillham RW et al (1984) Contaminants in groundwater – chemical processes. National Academy Press, Washington D.C., pp 46–64

Clement PT (2001) Generalized solution to multispecies transport equations coupled with a first-order reaction network. Water Resour Res 37(1):157–163

Dagan G (1986) Statistical theory of groundwater flow and transport: pore to laboratory, laboratory to formation, and formation to regional scale. Water Resour Res 22(9):120–134

Fetter CW (1999) Contaminant hydrogeology. Prentice Hall, Upper Saddle River, NJ

Freeze RA, Cherry JA (1979) Groundwater. Prentice Hall, Englewood Cliffs, NJ, 604 pp

Fried JJ (1975) Groundwater pollution. American Elsevier, New York, 930 pp

Gelhar LW, Gutjahr AL et al (1979) Stochastic analysis of macrodispersion in a stratified aquifer. Water Resour Res 15(6):1387–1397

Gunduz O, Aral MM (2005) A Dirac-delta function notation for source/sink terms in groundwater flow. J Hydrol Eng 10(5):420–427

Haberman R (1987) Elementary applied partial differential equations. Prentice-Hall, Englewood Cliffs, NJ, 547 p

Huyakorn PS, Ungs MJ et al (1987) A three-dimensional analytical method for predicting leachate migration. Ground Water 25(5):588–598

Jury WA, Russo D et al (1990) Evaluation of organic chemicals residing below the soil surface. Water Resour Res 26(1):13–20

Marino AM (1974) Distribution of contaminants in porous media flow. Water Resour Res 10 (5):1013–1018

Maslia ML, Aral MM (2004) ACTS – a multimedia environmental fate and transport analysis system. Practice Period Hazard Toxic Radioact Waste Manage – ASCE 8(3):1–15

Pickens JF, Grisak GE (1981) Modeling of scale-dependent dispersion in hydrogeologic systems. Water Resour Res 17(6):1701–1711

Pickens JF, Grisak GE (1987) Modeling of scale-dependent dispersion in hydrogeologic systems – reply. Water Resour Res 23(3):523–523

Schnoor JL (1996) Environmental modeling: fate and transport of pollutants in water, air, and soil. Wiley, New York, 682 pp

Sun Y, Buscheck TA et al (2007) Modeling reactive transport using exact solutions for first-order reaction networks. Transport Porous Med DOI. doi:10.1007/s 11242-007-9121-8

Wexler EJ (1989) Analytical solutions for one-, two-, and three-dimensional solute transport in groundwater systems with uniform flow. Open File Report 89-56. Tallahassee, FL, USGS

Yates SR (1992) An analytical solution for one-dimensional transport in porous-media with an exponential dispersion function. Water Resour Res 28(8):2149–2154

Chapter 6
Surface Water Pathway Analysis

When you work on the science of the motion of water, remember to include its application and use, so that the science will be useful.
Leonardo Da Vinci

When surface water pathway exposure and health effects are of concern, the transformation and transport analysis may include several aquatic environments such as impoundments, rivers, streams, reservoirs, wetlands, lakes, estuaries and open oceans. Among these, fresh water aquatic environments are of significant importance since they are the primary drinking water source for many communities. Estuaries and oceans may be of concern due to other exposure pathways, such as recreational activities, that are important in coastal areas. Water quality in surface water environments is regulated and routinely monitored by federal and state agencies. In the United States, the first surface water regulatory standard for an aquatic environment appeared with the enactment of the Public Health Service Act of 1912. This regulation was followed by the Clean Water Act (CWA) of 1972, in which jurisdiction for oversight of water quality in rivers, lakes, estuaries, and wetlands was established and the regulations on wastewater effluents were identified. The Safe Drinking Water Act (SWDA) enacted in 1974 further included regulations for the tap water quality in all community water systems, defined as those systems which have 15 or more outlets or serve 25 or more customers. In this act, the USEPA was required to set the national standards for drinking water quality within the United States and its territories.

In regulatory terms, surface water quality refers to the amount of pollution that can be present in a surface water body and still meet the current water quality standards set by the regulatory agencies. In the United States, these regulatory guidelines are established by the USEPA based on the National Pollutant Discharge Elimination System (NPDES) rules for wastewater effluents that are discharged to surface water systems. In essence, under the NPDES rule, the Maximum Contaminant Level (MCL) standards and the Maximum Contaminant Level Goals (MCLGs) are somewhat relaxed in terms of the level of effluent concentrations

M.M. Aral, *Environmental Modeling and Health Risk Analysis (ACTS/RISK)*, 275
DOI 10.1007/978-90-481-8608-2_6, © Springer Science+Business Media B.V. 2010

that may be discharged to aquatic environments. This relation stems from the potential dilution of the contaminant concentrations as the contaminant plume moves in the downstream direction. The MCL standards are based on health effects criteria and the MCLG, which are unenforceable goals, are based on the more stringent criteria of no known or anticipated health effects, regardless of the technological feasibility or cost of achieving that standard.

In spite of the stringent regulations established by these rules, in the 1990s a water quality inventory identified that almost 40% of all rivers and 45% of all lakes in the U.S. were polluted. More than 95% of the water tested near four population centers in the Great Lakes between 2001 and 2002 contained unsafe levels of mercury and pesticides, according to USEPA and National Wildlife Federation studies (NWLF 2009; USEPA 2009). Since 1972, billions of dollars have been invested in building and upgrading the sewage treatment facilities and in the appropriate design of effluent discharge facilities. This spending is all part of a significant effort to provide clean water to populations. Currently, more than 30,000 major industrial dischargers pre-treat their wastewater before it enters local sewers in the USA. By the year 2000, only 75% of toxic discharges, including heavy metals and PCBs, were prevented from entering the clean water stream. This percentage implies that there is still a lot of work to be done if the goal of providing clean and safe water to the general public for consumption and recreational purposes is to be achieved.

Whether to evaluate the conditions at an effluent discharge site or to evaluate the performance of an engineering design of an effluent discharge system, the most important and effective scientific methods available to engineers are physical and mathematical modeling tools. In this chapter, we will focus our attention on mathematical modeling tools, although we will also use extensively the interpretations and the outcome of physically based modeling studies. The purpose of an aquatic transformation and transport modeling study is to provide estimates of pollutant concentrations in a surface water body, based on the appropriate characterization of the mixing conditions in the near field or far field from of an effluent discharge point. The transformation and transport processes in surface water bodies are mostly dominated by the advective transport process, except possibly for impoundments or in reservoirs and lakes where diffusion processes may become a dominant dilution and dispersion mechanism. A wide variety of empirical and physically based mathematical models are described in the literature to evaluate the transport of pollutants in this pathway and to characterize the interaction of pollutants with the sediments that constitute one of the boundaries of surface water domain. These models may range from simple algebraic empirical models to sophisticated multidimensional transformation and transport models, the solutions of which are based on numerical methods. Thus, considerable familiarity with the scientific subject matter and the site specific conditions is required for both the appropriate selection and application of these models.

The emphasis in this chapter is on the use of simple empirical or analytical models that are used to simulate the transformation and transport of pollutants in a surface water body and the deposition of pollutants on shoreline and river sediments from both routine and accidental releases of liquid effluents. These simpler models, which may be used to provide a link between the effluent release and direct or indirect pathways to humans for exposure analysis, may yield conservative outcomes, which are desirable for health effects analysis. In this section, the theory background, limitations and applications of several surface water pathway models are reviewed. The examples included cover a range of applications in which these models are used in deterministic as well as stochastic analysis mode in finding solutions of the environmental migration of pollutants.

In surface water pathway analysis there are four distinct transformation and transport processes that control the spread and transport of contaminants. These are, in no particular order: (i) Advective and diffusive transport in water bodies and transport with sediment movement; (ii) intermedia transfer characterized by processes such as adsorption, desorption, precipitation, dissolution and volatilization; (iii) chemical decay, mostly characterized by the half-life of chemicals; and, (iv) chemical transformation which may yield daughter products. Beyond these four processes, another important aspect of aquatic transformation and transport modeling is the characterization of source conditions. Various effluent source discharge conditions that may be considered include: (i) direct discharge from point sources; (ii) dry and wet deposition from the atmosphere; (iii) runoff and soil erosion from land surfaces; and, (iv) seepage to or from groundwater. Based on these pathway specific conditions, several combinations of these transport processes and source conditions can be considered by using the models included in the surface water module of the ACTS software. The purpose of this chapter is to provide a review of these topics and to include applications that would demonstrate the use of these models. A more detailed discussion of surface water related engineering concepts can be found in the following literature (Chow 1959; USNRC 1978; Fisher et al. 1979; Singh 1995; Chaudhry 2007). In Fig. 6.1, the opening window of the surface water module is shown which includes the standard operational pull down menus of the ACTS software. More detailed use of these menus is discussed in Appendix 3.

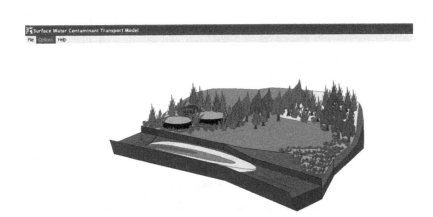

Fig. 6.1 Opening window of the ACTS software surface water pathway module

6.1 Definitions and Governing Principles

The definition sketch of an open channel cross section and the representation of the energy levels that are associated with the flow regime are shown in Fig. 6.2. There are several observations that can be made according to the conditions described in this figure. It is important to notice that in a typical open channel cross-section, the slope of the energy line, the slope of the water surface and the slope of the channel bottom may not be the same. By definition, this state implies that the flow condition shown in Fig. 6.2 is for a non-uniform flow in which the water depth, $(y_1; y; y_2)$ and the flow velocity in the channel, $(V_1; V; V_2)$ are variable. As indicated in this figure, the elevation of the channel base (z_1, z_2) is also variable. This condition results in the channel slope, S_o and the energy line slope, S_e which are not equal to each other under non-uniform flow conditions. The total energy level in an open channel is the sum of the three energy levels that are identified in open channel literature as "heads": the elevation head (z), the pressure head, $\left(\frac{P}{\gamma} = y\right)$ and the velocity head, $\left(\frac{V^2}{2g}\right)$. The choice of the term "head" is associated with the observation that the dimensions of these three energy levels are in length units $[L]$, i.e. in this case a measure of energy is made in length units. As shown in Fig. 6.2, potential energy is defined as the sum of the elevation head and the pressure head which coincides with the water surface elevation in open channels. Thus the water surface represents the position of the potential energy in the channel, which is identified as the hydraulic grade line (HGL) in open channel terminology. The sum of the potential energy in the channel and the kinetic energy that is represented by the velocity head gives us the total energy in the system. This sum is indicated by the energy line (EL) in Fig. 6.2. As the flow develops in the downstream direction, the primary loss of energy in the system is due to friction losses. This is represented in terms of shear stresses that are experienced at the boundaries of the flow domain. This energy loss is indicated by the head loss term (h_L) in (Fig. 6.2). Thus, the head loss term in an open channel is a function of the friction that characterizes the resistance to flow in the channel. One should also notice that the free surface is also characterized as a flow boundary where the pressure is atmospheric, or zero

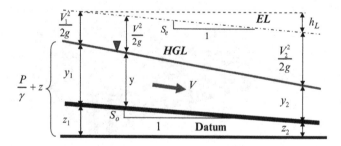

Fig. 6.2 Energy diagram for an open channel segment

pressure in gauge scale. This conceptualization would be the starting point of the analysis of flow conditions in an open channel environment.

The gradually varied and rapidly varied conditions are a sub-classification of non-uniform flow regimes in which there is a gradual or rapid change of water depth over short longitudinal distances (Fig. 6.3). Under gradually varied flow conditions the pressure distribution in the vertical direction in an open channel is considered to be hydrostatic. This is an assumption which cannot be made under rapidly varying flow conditions. In Fig. 6.3, a rapidly changing channel base is considered in order to characterize the three regimes in a definition sketch. This conceptualization also leads one to the definition of the uniform flow condition in an open channel, which is the condition in which the energy line is parallel to the hydraulic grade line, both of which are also parallel to the channel base. The uniform flow condition implies that the slopes of all of these characteristic energy levels have the same slope in the flow direction. Thus, the uniform flow condition represents a condition of equilibrium between the driving gravitational forces and the resisting shear forces. The governing equations for this flow regime are derived based on this assumption. In open channels uniform flow conditions are usually expected to occur over long distances over which there is no change in the channel slope and the friction condition at the channel boundaries. In this flow regime, the pressure distribution in the vertical direction is also hydrostatic. As stated earlier, hydrostatic pressure distributions are also used for gradually varying flows. A hydrostatic pressure distribution implies that the pressure in the vertical direction in a aquatic environment changes linearly as a function of depth and the specific weight of the fluid in the channel$(P = \gamma y)$. Under uniform flow conditions, since the flow velocity is constant over a longitudinal stretch, this condition also implies that the acceleration of fluid particles in the flow domain is zero. This result does not occur for varied flow conditions, and especially for rapidly varied flow conditions. If there is acceleration of fluid particles in the flow domain, the assumption of the pressure distribution being hydrostatic in the vertical direction no longer holds (Bird et al. 2002). All of these definitions are important concepts and constitute the basis of the terminology that we will use in

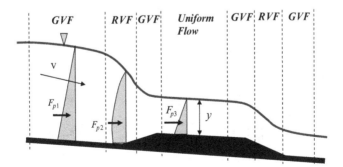

Fig. 6.3 Definition sketch for expected flow regimes according to water surface profile in an open channel section with a raised channel base

this chapter in to identify the other processes that are of importance in the field of transformation and transport analysis of contaminants in aquatic environments.

6.1.1 Uniform Flow in Open Channels

The governing equations for uniform flow in open channels can be derived based on the assumption that gravitational forces are balanced by friction forces in the flow domain. This flow regime may occur in open channels over a long reach if the slope of the channel and the friction coefficient are not changing. The definition sketch of this regime is shown for a representative elementary volume in the longitudinal and the cross sectional direction in Fig. 6.4.

In this figure, the gravitational forces are characterized in terms of the component of the weight force W $[MLT^{-2}]$ in the flow direction and the friction force $F_{\tau_w}[MLT^{-2}]$ in which the term $\tau_w[ML^{-1}T^{-2}]$ represents the wall friction experienced across the channel boundary indicated as the wetted perimeter $P[L]$. The cross section area of the channel is given as $A[L^2]$.

The uniform flow equations which are based on the conditions described in Fig. 6.4 can be derived using Newton's second law, in which the acceleration of the fluid particles is considered to be zero,

$$\sum F = m\vec{a} = 0 \tag{6.1}$$

As shown in Fig. 6.4, since the pressure forces are equal and in opposite directions, there are two forces that characterize the flow in the REV. These are the component of the weight force in the flow direction and the resistance force in the opposite direction. The definition for these two forces is given below,

$$W \sin \theta = \gamma A L \sin \theta \tag{6.2}$$

$$F_{\tau_w} = \tau_w P L \tag{6.3}$$

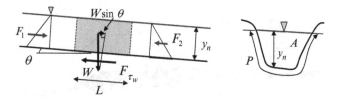

Fig. 6.4 Definition sketch for uniform flow conditions along a longitudinal and cross-section for a representative elementary volume

in which

$$\sin \theta = \frac{h_L}{L} = S_o \tag{6.4}$$

τ_w is the wall shear stress $[ML^{-1}T^{-2}]$ and P is the wetted perimeter. For the equilibrium conditions defined in Eq. (6.1) and considering all of the forces acting on the REV, one can write,

$$F_1 + W \sin \theta - F_2 - F_{\tau_w} = 0 \tag{6.5}$$

or

$$\gamma A L \sin \theta = \tau_w P L \tag{6.6}$$

which leads to the definition of the wall shear stress in terms of the parameters of the flow domain as defined in Fig. 6.4,

$$\tau_w = \gamma \frac{A}{P} \sin \theta = \gamma R_h \sin \theta = \gamma R_h \frac{h_L}{L} = \gamma R_h S_o \tag{6.7}$$

here $R_h = A/P$ is the hydraulic radius, which is defined as the ratio of the cross section area of the channel to the wetted perimeter $[L]$.

Using the standard definition of wall shear stress in terms of the Darcy–Weisbach friction factor that is given as (Chow 1959; Chaudhry 2007),

$$\tau_w = \frac{f \rho V^2}{8} \tag{6.8}$$

one can arrive at the definition of uniform flow velocity under uniform flow conditions,

$$\frac{f \rho V^2}{8} = \gamma R_h S_o \tag{6.9}$$

or

$$V = \sqrt{\frac{8\gamma}{f\rho} R_h S_o} = \left(\frac{8g}{f}\right)^{1/2} (R_h)^{1/2} (S_o)^{1/2} \tag{6.10}$$

which is identified as the Chezy equation in open channel flow literature, in which the Chezy constant is defined as (Chow 1959),

$$C = \left(\frac{8g}{f}\right)^{1/2} \tag{6.11}$$

This gives the uniform velocity in an open channel as,

$$V = C(R_h)^{1/2}(S_o)^{1/2} \tag{6.12}$$

In the open channel literature the use of the Manning coefficient is more common than the use of the Chezy constant since the tabulated values of Manning's constant are more commonly available. The relationship between the Chezy constant and the Manning's constant are given as,

$$\left. \begin{array}{ll} C = \dfrac{1.49}{n} R_h^{1/6} & \text{in British units} \\[3mm] C = \dfrac{1}{n} R_h^{1/6} & \text{in Standard International (SI) units} \end{array} \right\} \tag{6.13}$$

in which n is the Manning's roughness coefficient (Table 6.1). Accordingly the uniform flow velocity in British units is given as (Chow 1959; Chaudhry 2007),

$$V = \frac{1.49}{n}(R_h)^{2/3}(S_o)^{1/2} \tag{6.14}$$

and this velocity for SI units can be defined as,

$$V = \frac{1}{n}(R_h)^{2/3}(S_o)^{1/2} \tag{6.15}$$

Depending on the unit system considered, these two equations are used in most applications to characterize the flow field, the characterization of which is

Table 6.1 Manning's roughness coefficient for various Channels

Channel type	Channel roughness condition	Manning's coefficient n
Artificial channels	Finished cement	0.012
	Unfinished cement	0.014
	Brick work	0.015
	Rubble masonry	0.025
	Smooth dirt	0.022
	Gravel	0.025
	With weeds	0.030
	With cobbles	0.035
Natural channels	Clean and straight	0.030
	Most rivers	0.035
	Rivers with deep pools	0.040
	Rivers with irregular sides	0.045
	Dense side growth	0.080
Flood plains	Farmland	0.035
	Small brushes	0.125
	With trees	0.150

necessary in transformation and transport analysis that will follow. The Manning's roughness coefficient for typical open channel characterizations is given in Table 6.1. A more detailed discussion of these topics can be found in the following references (Chow 1959; Chaudhry 2007).

6.1.2 Mixing Models in Open Channels

The primary mechanism that controls water quality in surface water environments for which the NPDES rules are based is the potential for mixing that occurs at different scales through various physical mixing processes. The primary mixing processes that contribute to dilution in aquatic environments are turbulent, dispersive and diffusive mixing processes. These mixing processes simultaneously occur in aquatic media as the contaminant is transported in the downstream direction due to the advective process. Advective transport refers to the bulk movement of the contaminant in the flow direction under mean velocities. Dilution is achieved by the entrainment of the ambient environment into the plume released. Turbulent and dispersive mixing contributes to this process, which is usually analyzed as a lumped process, which refers to the smearing of pollutant plumes due to fluctuating flow velocities in both magnitude and direction. Diffusion refers to random migration of an ensemble of particles from regions of high concentration to regions of low concentration. This process is characterized by the use of Fick's law as discussed in Chapters 2 and 3. The relative importance of the process of advection and mixing on water quality parameters is determined by the use of the dimensionless Peclet number, which represents the ratio of the parameters that characterize advection and mixing as shown below.

$$P_e = \frac{\text{Advective processes}}{\text{Diff. - Disp. processes}} = \frac{VL}{D} \tag{6.16}$$

where L is a characteristic length, V is the mean velocity in the channel and D is the diffusion dispersion mixing coefficient. If P_e is large, then advective processes dominate the mixing process. If P_e is small, then mixing processes dominate the water quality conditions at a site. The mathematical models that are used in the analysis of transformation and transport conditions in surface water environments again are based on the advection–diffusion-reaction equation that is used in other pathways and covered in detail in Chapter 3. The difference in this case lies in the characterization of the flow field, basic principles of which are summarized above. The definition of the mixing coefficient is what we will discuss next to complete the introduction of the basic concepts that is necessary for the aquatic pathway analysis.

The mixing coefficient for surface water problems is readily characterized in terms of wall shear stresses and the shear velocity. These two terms are related as shown below, for which Newton's second law is utilized to define the equilibrium between the wall shear forces and mass times acceleration, which is defined in terms of the acceleration of a largest eddy swirl that may occur in an open channel,

$$\left.\begin{array}{l} F_{\tau_w} = m\,\vec{a}_{eddy} \\ \tau_w LW = \rho LW y \dfrac{u_*}{t_{eddy}} \end{array}\right\} \qquad (6.17)$$

in which t_{eddy} is the eddy turnaround time for an open channel cross-section of depth y, $t_{eddy} \approx y/u_*$. Thus the relationship between wall shear stress and shear velocity can be given as,

$$u_* = \sqrt{\dfrac{\tau_w}{\rho}} \qquad (6.18)$$

in which u_* is the shear velocity $[LT^{-1}]$, τ_w is the wall shear stress as defined earlier, L and W are the length and the width of the channel section (Fig. 6.5) and ρ is the density of the fluid. For steady uniform flow as discussed earlier, the wall shear stress can also be evaluated as given in Eq. (6.7). According to Eq. (6.18), shear velocity can also be written as,

$$u_* = \sqrt{\dfrac{\gamma R_h S_o}{\rho}} = \sqrt{g R_h S_o} \qquad (6.19)$$

It is important to note that if the flow in the channel is not uniform, Eq. (6.19) can still be used with the replacement of the channel slope by the energy slope.

In essence shear velocity is a representation of the eddy mixing process, which is characterized by a velocity profile influenced by wall shear stresses or the shear stress distribution in the vertical direction (perpendicular to the flow direction) the transverse direction (Fig. 6.5). For example, the vertical dispersion coefficient for an infinitely wide open channel is based on the logarithmic velocity distribution assumption in an open channel, (Fisher et al. 1979). This empirical model can be given as shown in Eq. (6.20), in which z is the vertical direction,

$$\left.\begin{array}{l} D_z \approx u_* y \\ D_z = 0.067 u_* y = 0.067 y \sqrt{g R_h S_o} \end{array}\right\} \qquad (6.20)$$

The transverse mixing coefficient cannot be derived based on the analytic model of a velocity profile in the transverse direction, but this model is usually given in terms of a similar form to Eq. (6.20) (Fisher et al. 1979) and the y direction is used to characterize the transverse direction,

$$\left.\begin{array}{l} D_y \approx u_* y \\ D_y = C_{trans} u_* y = C_{trans} y \sqrt{g R_h S_o} \end{array}\right\} \qquad (6.21)$$

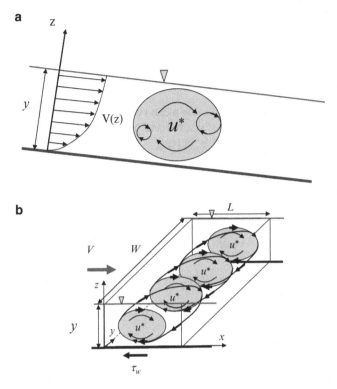

Fig. 6.5 Velocity profile in an open channel (**a**) and Eddy mixing concepts (**b**). (**a**) Conceptualization of Eddy mixing in a channel cross-section. (**b**) Conceptualization of Eddy mixing in transverse direction

in which C_{trans} is a dimensionless coefficient which is a function of the channel type, channel cross-section and channel roughness. Typical laboratory studies indicate that this coefficient is on the order of two times larger than the vertical dispersion coefficient, $(0.3 < C_{trans} < 0.9)$. The longitudinal mixing coefficient is generally expected to be much larger than the vertical mixing coefficient. In an analysis that is similar to the derivation of the vertical mixing coefficient, the following empirical model is used in the literature for the longitudinal mixing coefficient, (Elder 1959).

$$D_x = 5.93 u_* y = 5.93 y \sqrt{g R_h S_o} \qquad (6.22)$$

Field studies conducted on natural rivers and streams indicate that the longitudinal dispersion in natural streams is always greater than that predicted by Eq. (6.22), (Fisher et al. 1979). The other suggested empirical model which is commonly used in the literature that reflects this observation is the Elder equation (Elder 1959),

$$D_x = 0.011 \frac{V^2 W^2}{u_* y} = 0.011 \frac{V^2 W^2}{y\sqrt{gR_h S_o}} \tag{6.23}$$

in which W is the channel width, V is the mean velocity in the channel, and the other parameters are as defined earlier (Figs. 6.4 and 6.5).

Consider a river width $W = 30$ m, depth $y = 2$ m, and average uniform velocity $V = 0.5$ m/s. Assuming that the channel slope is $S_o = 1.0 \times 10^{-4}$, the longitudinal, transverse and vertical dispersion coefficients can be estimated based on Eqs. (6.20), (6.21) and (6.23) as follows,

$$D_x = 0.011 \frac{V^2 W^2}{u^* y} = 0.011 \frac{V^2 W^2}{y\sqrt{gR_h S_o}} = 0.011 \frac{(0.5)^2 (30)^2}{2\sqrt{(9.81)(1.77)(10^{-4})}} = 29.5\,\text{m}^2/\text{s}$$

$$D_y = 0.3 u_* y = 0.3 y\sqrt{gR_h S_o} = 0.3(2)\sqrt{(9.81)(1.77)(10^{-4})} = 0.025\,\text{m}^2/\text{s}$$

$$D_z = 0.067 u_* y = 0.067 y\sqrt{gR_h S_o} = 0.067(2)\sqrt{(9.81)(1.77)(10^{-4})} = 0.006\,\text{m}^2/\text{s}$$

In the transformation and transport analysis of effluents in aquatic pathways, these parameters will be used in models that are based on conservation of mass principles and thus need to be completed ahead of the contaminant transport analysis step.

6.1.3 Elementary Transport Models in the Aquatic Pathway

The derivation of mass balance based contaminant transformation and transport models for open channels follows the same principles discussed in detail in Chapter 3. However, since advective transport can be a dominant transport mechanism in open channels, it is possible to define simpler plug flow type models for contaminant transformation and transport analysis in open channels or for all aquatic pathways in general. Considering the channel cross-section as shown in Fig. 6.6, the mass balance equation may be written as follows, including a reaction term,

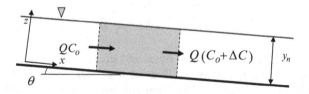

Fig. 6.6 Definition sketch of an open channel cross-section for mass balance analysis

$$\text{Mass Accumulation} = \sum \text{Mass Influx} - \sum \text{Mass Outflux}$$
$$\pm \sum \text{Reactions} \tag{6.24}$$

Utilizing the REV approach described in Chapter 3 and considering advective transport while ignoring longitudinal mixing effects, the mass balance analysis will yield the following model,

$$\frac{\partial C}{\partial t} + V \frac{\partial C}{\partial x} = -\lambda C \tag{6.25}$$

in which V is the mean longitudinal velocity in the channel, and the reaction term is considered to be a first order decay term with a decay rate λ.

This model can also be analyzed as a box model, ignoring the spatial variations which yields,

$$\frac{dC}{dt} = VC_0 - VC - \lambda C \tag{6.26}$$

in which C_o is the inflow concentration into the box and C is the outflow concentration. The steady state solution $\left(\frac{dC}{dt} = 0\right)$ in this case implies,

$$C_{t \Rightarrow \infty} = \frac{VC_o}{V + \lambda} \tag{6.27}$$

The change of the time dependent concentration in the box between the initial concentration and the infinite time concentration can be given by the analytical solution of Eq. (6.26),

$$C(t) = C_\infty + (C_o - C_\infty) \exp(-(V + \lambda)t) \tag{6.28}$$

For a steady state application over a longitudinal reach the mathematical model given in Eq. (6.25) reduces to,

$$V \frac{dC}{dx} = -\lambda C \tag{6.29}$$

In this case spatial variation of concentration in the channel can be analyzed at infinite time considering the first order decay of the chemical in the channel. The analytical solution of this model can be given as,

$$C = C_o \exp(-\lambda x / V) \tag{6.30}$$

in which C_o is the contaminant concentration at the entrance of the channel.

The steady state solution without the reaction term implies that inflow concentration is equal to outflow concentration if there is a single reach in the system. If

Fig. 6.7 Definition sketch for a multi reach channel system

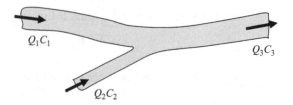

there is more than one reach in the stream (Fig. 6.7), then cumulative inflows and outflows can be considered for the mass balance analysis in steady state analysis.

In this case the continuity and mass balance equations will yield,

$$\left.\begin{array}{c} Q_1 + Q_2 = Q_3 \\ \\ Q_1 C_1 + Q_2 C_2 = Q_3 C_3 \end{array}\right\} \tag{6.31}$$

This analysis can also be extended to specific cases that may involve variable discharges, variable cross-sections and the decay term. For simplicity, if we assume that the variation in these three parameters can be expressed in terms of an exponential function,

$$\begin{array}{c} Q = Q_o \exp(qx) \\ A = A_o \exp(ax) \\ R = \lambda_o \exp(-\lambda t) \end{array} \tag{6.32}$$

in which it is implied that the discharge and the cross-section area are increased exponentially in the longitudinal direction, and the decay is characterized as a first order term. In terms of discharge, the mathematical model of this problem can be given as,

$$\frac{\partial C}{\partial t} = -\frac{C}{A}\frac{\partial Q}{\partial x} - \frac{Q}{A}\frac{\partial C}{\partial x} - \frac{\partial Q \partial C}{A \partial^2 x} - \sum R \tag{6.33}$$

where Q is the discharge in the channel, A is the cross-section area of the channel and R again represents the reaction terms, all of which are assumed to be varying exponentially as shown in Eq. (6.32). This is a nonlinear differential equation, for which analytical solutions cannot be given and which may require an iterative numerical analysis. However, this model can be simplified if we make an order of magnitude analysis and assume that the contribution of the terms which includes the product of the two gradient terms, the fourth term in Eq. (6.33), may be negligible. With this assumption the mathematical model reduces to,

$$\frac{\partial C}{\partial t} = -\frac{1}{A}\frac{\partial(CQ)}{\partial x} - \lambda C \tag{6.34}$$

and the steady state model for this case reduces to,

$$\frac{1}{A}\frac{d(CQ)}{dx} = -\lambda C \tag{6.35}$$

the analytical solution of this mathematical model for an exponentially varying Q and A can be given as,

$$C = C_o \exp\left[-qx - \frac{\lambda A_o}{(a-q)Q_o}e^{(a-q)x} + \frac{\lambda_o A_o}{(a-q)Q_o}\right] \tag{6.36}$$

All models that are discussed above constitute the simpler applications of the conservation of mass principles in rivers and streams which may yield useful applications for cases in which plug flow or box analysis can be justified.

The conservation of mass principle equations for the general one dimensional transformation and transport problem in a semi infinite river reach with first order decay and longitudinal diffusion effects can be given as:

$$\frac{\partial C}{\partial t} + V\frac{\partial C}{\partial x} - D_x\frac{\partial^2 C}{\partial x^2} + \lambda C = 0 \tag{6.37}$$

The boundary conditions for this problem can be given as,

$$\begin{aligned}C = C_0 &\quad \text{at} \quad x = 0; \quad t \geqslant 0 \\ C, \frac{\partial C}{\partial x} = 0 &\quad \text{at} \quad x \Rightarrow \infty; \quad t \geqslant 0\end{aligned} \tag{6.38}$$

The initial condition can be defined as,

$$C = 0; \quad 0 < x < \infty; \quad t = 0 \tag{6.39}$$

which implies a clean river reach. Again, the parameters V and D_x in Eq. (6.37) are the mean velocity in the channel in the longitudinal direction and the longitudinal mixing coefficient given by Eq. (6.23), respectively. The analytical solution of this problem was given earlier and is repeated below,

$$C(x,t) = \frac{C_0}{2}\left\{\exp\left[\frac{x(V-U)}{2D_x}\right]erfc\left[\frac{x-Ut}{2\sqrt{D_xt}}\right] + \exp\left[\frac{x(V+U)}{2D_x}\right]erfc\left[\frac{x+Ut}{2\sqrt{D_xt}}\right]\right\}$$
$$\text{where} \quad U = \sqrt{V^2 + 4\lambda D_x}. \tag{6.40}$$

The governing equation and the boundary conditions of a semi-infinite river reach in which the contaminant is introduced as a flux can be given as,

$$\frac{\partial C}{\partial t} + V\frac{\partial C}{\partial x} - D_x\frac{\partial^2 C}{\partial x^2} + \lambda C = 0 \tag{6.41}$$

The boundary conditions for this case are defined as,

$$VC + D_x\frac{\partial C}{\partial x} = VC_0 \quad \text{at} \quad x = 0; \quad t \geqslant 0$$

$$\frac{\partial C}{\partial x} = 0 \quad \text{at} \quad x \Rightarrow \infty; \quad t \geqslant 0 \tag{6.42}$$

The initial condition representing an uncontaminated open channel reach is given as,

$$C = 0; \quad 0 < x < \infty; \quad t = 0 \tag{6.43}$$

The analytical solution of this problem is given as,

$$C(x,t) = \frac{C_0 V^2}{4\lambda D_x}\left\{ 2\exp\left[\frac{xV}{D_x} - \lambda t\right]erfc\left[\frac{x + Ut}{2\sqrt{D_x t}}\right]\right.$$
$$+ \left(\frac{U}{V} - 1\right)\exp\left[\frac{x(V + U)}{2D_x}\right]erfc\left[\frac{x - Ut}{2\sqrt{D_x t}}\right]$$
$$\left. - \left(\frac{U}{V} + 1\right)\exp\left[\frac{x(V + U)}{2D_x}\right]erfc\left[\frac{x + Ut}{2\sqrt{D_x t}}\right]\right\} \tag{6.44}$$

where $U = \sqrt{V^2 + 4\lambda D_x}$.

The last two models given above are a more realistic representation of the longitudinal transformation and transport process that may be observed in a river reach. These models are included in the ACTS software as will be discussed later on.

6.1.4 Elementary Transport Models for Small Lakes and Impoundments

The mass balance based elementary models for lakes and impoundments can be derived in a similar manner as discussed in the section above. For these models, simplicity implies that we will be neglecting the spatial variations in a lake and consider the lake to be a completely mixed system. The conservation of mass based mathematical model for this case can be given as (Fig. 6.8)

Fig. 6.8 Definition sketch for
a small lake

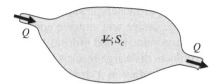

$$\mathcal{V}\frac{\partial C}{\partial t} = -\frac{\mathcal{V}}{t_r}(C) - \lambda C \mathcal{V} + S_c \qquad (6.45)$$

in which $\mathcal{V}$ is the volume of the lake, t_r is defined as the residence time of the fluid in the lake $(t_r = \mathcal{V}/Q)$, Q is the steady state throughput to the lake, λ is the first order decay rate and S_c represents a constant source of contamination as it is introduced to the lake. The analytical solution of this problem is,

$$C(t) = \frac{S_c}{\beta \mathcal{V}}\left(1 - e^{-\beta t}\right) + C_o e^{-\beta t} \qquad \beta = \left(\frac{1}{t_r} + \lambda\right) \qquad (6.46)$$

The steady state solution will be,

$$C(t = \infty) = \frac{S_c}{\beta \mathcal{V}} \qquad \beta = \left(\frac{1}{t_r} + k\right) \qquad (6.47)$$

If there is no external contamination source, the ambient contaminant concentration in the lake, C_o will decay as given in Eq. (6.48). The concentration C_o is considered to be the initial pollution level in the lake.

$$C(t) = C_o e^{-\beta t} \qquad \beta = \left(\frac{1}{t_r} + k\right) \qquad (6.48)$$

These are all relevant and simple applications of the mass balance concept to a surface impoundment or small lake system. These models are also included in the ACTS software system as will be discussed in the following sections.

6.2   Surface Water Pathway Models

The mass conservation based analysis discussed above differs from the models used to describe the initial mixing processes in an effluent discharge point. It is expected that the dissipation and the dilution process take place immediately at the discharge point where the effluent is subject to efflux discharge and ambient conditions. Traditionally, the mixing of effluents and receiving waters is analyzed in terms of "near field" and "far field" mixing processes. The near field analysis is based on empirical models characterized by the specific characteristics of the outfall discharge design and ambient conditions. In the near field region, factors such as discharge velocity and effluent buoyancy effects dominate effluent mixing. Once the effluent discharge reaches a point of neutral buoyancy, either by being trapped below the surface as a stratified layer or by impacting the surface or the bottom of the channel, the mixing in the channel is controlled by ambient currents in the channel in the vicinity of the effluent discharge point and more importantly by the turbulence in the channel at the discharge point. The end of the near field mixing region marks the beginning of the far field mixing region, and the effluent dilution in the near field is identified as the "initial dilution." The buoyancy affected spreading, or density effects, also occur at the onset of the far field as the lighter effluent plume spreads out in a layer of neutral buoyancy while entraining the usually heavier, ambient fluid. In the far field zone, dilution is characterized by environmental factors only. Thus, in this zone the effluent discharge design effects are considered to be negligible. Figure 6.9 shows the characterization of the near field and far field regions for a buoyant discharge. It is also important to point out that the beginning of the far field may be characterized as the intermediate near field, where the effect of residual near field fluxes may be observed. Four important mixing processes can be highlighted as they characterize the near field and far field mixing conditions.

Discharge Buoyancy and Ambient Stratification: The buoyancy effect of an effluent discharge is caused by the difference in the effluent and ambient fluid densities. This density difference is mainly due to dissolved solids, such as salt in saltwater, or suspended solids, such as sediment in fresh water or temperature effects.

The buoyancy effect is one of the more important forces that influence the mixing of an effluent discharge into a receiving water body. A fluid that is discharged into another fluid of dissimilar density will rise or sink until it becomes neutrally buoyant. Positively buoyant discharges will rise regardless of discharge momentum conditions until they reach the surface or entrain enough ambient water to become in equilibrium with the density of the receiving water body. Negatively

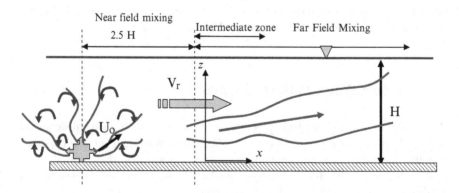

Fig. 6.9 Near field and far field mixing zones

buoyant discharges will sink and attach to the bottom. Such conditions are caused by discharging heavy brine solutions or by discharging into colder waters at or near freezing. This latter case is due to the temperature dependent nature of the density of liquids as they freeze. For example, fresh water has its maximum density at approximately 4°C. Warm effluent discharged to ambient water colder than 4°C will mix and attach to the bottom unless discharge momentum and turbulent mixing cause mixing and elevates it.

Ambient density difference effects can be described in one of three general stratification categories. The density stratification profile (a plot of density versus depth) can be uniform, constant gradient, or uniform with a sharp density increase at specific depths. The last condition is known as a pycnocline (or thermocline), which can be caused by temperature effects, subsurface currents, or tidal effects in estuaries known as salt wedges.

Transverse and Vertical Mixing: The mixing of effluent within receiving water occurs in both the transverse and vertical directions due to turbulence effects away from the point of discharge. In shallow channels, turbulence along the channel bottom and in the water column promotes rapid and uniform vertical mixing. Transverse mixing in rivers occurs over relatively longer distances and is dependent on factors such as ambient velocity in the channel, channel morphology, and the effects of bottom roughness and rapids.

In tidal systems, vertical mixing may be limited due to water column stratification. For example, a positively buoyant plume may rise to the surface of the receiving water, or become trapped below the surface, or oscillate between these two conditions depending on the strength of tidal currents and the ambient vertical density stratification. Once neutral buoyancy is achieved, subsequent lateral and vertical mixing will be driven by ambient currents and turbulence, including the effects of current interactions with the bottom of the water body as well as currents created due to wind effects and turbulence at the surface.

Current Interactions: The direction and magnitude of the ambient current can have an important effect on effluent mixing. Typically, these processes have a greater influence in the far field region, although they may also impact the near field processes if they are strong relative to the discharge momentum created by the effluent discharge mechanism. Similar to the density, velocity can also be stratified in a water body. This situation is particularly true in rivers where the friction drag effect of shorelines and the channel bottom may cause significant differences in the velocity profile both in the vertical and in the transverse directions across the stream.

Tidal effects have also been found to cause reversing currents, which can occur near the bottom or across an entire water cross-section. For example, in an estuary, an outward velocity may occur in a fresh water layer at the surface while an inward velocity may be observed in an underlying salt wedge that may exist along the bottom. Discharge into the saline bottom layer may result in effluent being trapped near the bottom and actually being carried upstream due to the reversed flow patterns in the salt wedge.

Tidal Effects on Mixing: Tidal cycles may have significant effects on far field mixing processes. Re-entrainment of previously discharged effluent may cause accumulation of pollutants in the discharge zone. This accumulation commonly occurs in bays with extremely long residence, or flushing, times. Effluent cannot be entirely swept away during a single tidal cycle, and some is perpetually left behind in the initial mixing zone. Such effects illustrate the importance of physical scale models that provide a better picture of the near field and far field mixing process that cannot be predicted by mathematical models.

Based on all of these conditions, the most important aspect of an effluent discharge into an aquatic environment is the engineering control or maximization of the initial dilution that can be accomplished at a discharge point. The models that are used in this analysis, which are mostly empirical models, constitute an important component of the dilution analysis of effluents in the environment.

The water quality based NPDES permit limit is the most common regulatory control utilized in recent years by states that have adopted ambient criteria for toxic pollutants. In issuing these permits, the USEPA and many states acknowledge that the designated uses of a water body can be maintained without requiring effluents to meet fully the water quality criteria at the point of discharge. Thus, allowances are made which consider the initial mixing and dilution that take place in the vicinity of an effluent discharge. The near field mixing and the dilution effects are mostly associated with the turbulent mixing produced from a discharge momentum exchange process in the case of effluent jet discharges or from discharge buoyancy mixing in the case of density differences between the effluent density and the ambient water body. The initial dilution process is assumed to occur over a short period of time and over short distances. The dilution concept can be illustrated by the following simplified algebraic expression:

$$S = \frac{C_o}{C} \tag{6.49}$$

in which C $[ML^{-3}]$ is the allowable effluent concentration or the concentration at a point in the mixing field, C_o $[ML^{-3}]$ is the discharge concentration and S is the dilution factor under the mixing conditions characterizing the effluent discharge. As can be seen from Eq. (6.49) the permit limit may be substantially greater than the corresponding ambient water quality criterion based on the mixing that can be generated at an effluent discharge point. A mixing zone may be determined by computing a dilution factor or it may be delineated by a regulatory agency as a spatial area with fixed boundaries in which the required dilution must be achieved. In either case, it is an allocated region within receiving water in which the effluent is rapidly diluted due to momentum and buoyancy effects which include the ambient turbulence that may be present. From a regulatory point of view, water quality criteria may be exceeded within a mixing zone but must be met at its boundaries. USEPA water quality criteria guidance includes three components for each regulated pollutant (USEPA 1991): (i) Magnitude (the allowed concentration in ambient water); (ii) duration (the averaging period over which the ambient pollutant concentration is compared to the allowed value); and, (iii) frequency of occurrence, i.e. how often the criterion may be exceeded. As defined in the Code of Federal Regulations (CFR), a water quality standard, the boundaries of which are as described above, is a regulation promulgated by a state or the USEPA which designates the use or uses to be made of a water body and the criteria designed to protect them (USEPA 1994).

The surface water module of the ACTS software consists of three groups of models: (i) Near field mixing models; (ii) far field mixing models; and, (iii) sediment pathway models (Fig. 6.10). The near field mixing models are further subdivided into surface point, submerged point and submerged multi-port mixing models, which have additional subcategory models associated with ambient flow conditions that may exist at a site, such as strong or weak cross flows. In the far field mixing category, rivers models, estuary models, small lakes, reservoir models and oceans and great lakes models are considered. Each of these sub-groups has further subcategory models based on ambient flow and mixing conditions. The sediment pathway models are also divided into subgroups identified as rivers, estuaries, coastal waters and oceans and lakes. A summary of the surface water modeling system of the ACTS software is given in Fig. 6.10. Similar to the groundwater and air pathway module, the Monte Carlo analysis module of the ACTS software is linked to all surface water pathway models to provide uncertainty analysis associated with the variability in input parameters of the model under consideration.

6.3   Near Field Mixing Models

The region in the immediate vicinity of the effluent discharge point where mixing is dominated by buoyancy differences, the effluent jet momentum or both is characterized as the near field mixing zone (Fig. 6.11). The near field mixing zone is the zone that is most affected by outfall design. Thus, the "near field" mixing zone refers to the portion of the effluent plume between the diffuser outlet and the location where the discharged plume has effectively completed its initial mixing with the ambient water body, as determined by buoyancy and momentum differences. Accordingly, at the transition location between the "near" and "far" fields, the mixed effluent plume may be observed to be one of the entry conditions that is shown in Fig. 6.11, i.e. the plume has reached the surface, is attached to the bottom or is somewhere in between. As shown in Fig. 6.11, the pure plume condition refers to the case where buoyancy flux is the dominant mixing mechanism, the pure jet condition refers to the case where the momentum flux is the dominant mixing mechanism, the buoyant jet condition refers to the case where both fluxes are important in the mixing process. An effluent outfall has to be designed with a diffuser that would provide adequate initial mixing characteristics within the "near field" mixing zone as we have discussed in the NPDES permit limits.

The near field models estimate the behavior of the near field plume by considering the following flow domain parameters: (i) Effluent discharge and density; (ii)

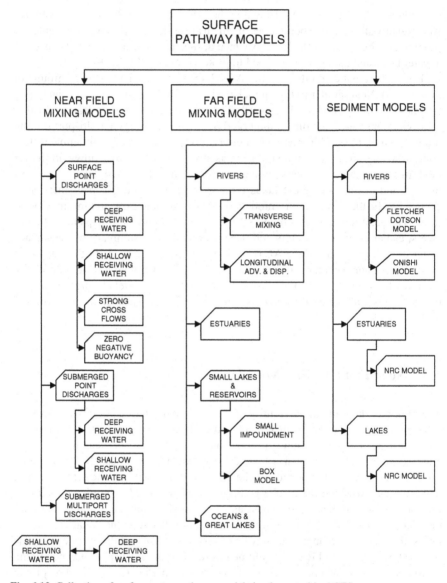

Fig. 6.10 Collection of surface water pathway models implemented in ACTS

ambient fluid density, depth and velocity at the discharge point; and, (iii) diffuser configuration, as it may influence the mixing conditions in the near field. Accordingly there are several models that may be used in the characterization of the near field mixing conditions, as will be discussed next.

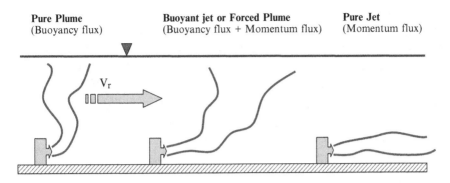

Fig. 6.11 Near field mixing processes

6.3.1 Surface-Point Discharges

Surface discharges consist of an outfall near the water surface, i.e. a discharge from a slightly submerged pipe. This configuration has historically been the most common discharge procedure for effluent disposal and thus has received substantial attention over the past decade. A particular application is the effluent discharge for waste heat disposal as buoyant surface jets. The properties of buoyant surface jets and buoyant surface discharges can be found in Tsanis and Valeo (1974), Jirka et al. (1981, 1983). A parameter that is used to describe the dynamic characteristics of the buoyant surface jet is the discharge densimetric Froude number, F_o which is given as,

$$F_o = \frac{U_o}{\sqrt{(\Delta\rho/\rho_o)g\ell_o}} \tag{6.50}$$

in which U_o is the mean effluent discharge velocity through the discharge port $[LT^{-1}]$, ρ_o is the ambient density $[ML^{-3}]$, $\Delta\rho$ is the density difference between the ambient fluid and the effluent ($\Delta\rho = \rho_o - \rho$), ρ is the effluent density, g is the gravitational acceleration $[LT^{-2}]$, and ℓ_o is a characteristic length scale of the discharge point such as the diameter of the port. In geometries other than circular ports the characteristic length scale is usually described in terms of the cross-sectional area A_o at the discharge point as,

$$\ell_o = \sqrt{\frac{A_o}{2}} \tag{6.51}$$

For example, a rectangular discharge port with discharge depth, h_o and half port width b_o would yield,

$$\ell_o = \sqrt{h_o b_o} \tag{6.52}$$

According to the definition of the densimetric Froude number F_o, the larger the effluent velocity, the lower the reduced acceleration due to density differences. Thus in this case it is more likely that the effluent discharge will resemble the pure jet condition. On the other extreme, a low densimetric Froude number would indicate a pure plume condition. Using these parameters the following discharge conditions can now be analyzed.

6.3.1.1   Stagnant and Weak Cross-Currents

Deep Receiving Water: A deep receiving water condition may be used when the vertical extent of the buoyant jet is sufficiently less than the water depth H at the discharge point. In this case, the effluent concentrations in the near field mixing zone can be evaluated by estimating the bulk dilution factor S defined in Eq. (6.49). For this case the transition distance x_t in which this dilution is expected to occur may also be computed as given below,

$$\left. \begin{array}{l} S = 1.4F_o \\ x_t = 15\ell_o F_o \end{array} \right\} \tag{6.53}$$

Based on the empirical equation (6.49) and Eq. (6.53), the contaminant concentrations and transition distance in the deep receiving water body may now be computed. The transition distance x_t can be used as a measure of the extent of the near-field zone. Another characteristic parameter which may be important is the maximum vertical penetration of the surface jet for this effluent discharge condition. This maximum penetration depth can be calculated using the empirical equation below,

$$\left. \begin{array}{l} h_{max} = 0.42\ell_o F_o \\ x_{max} = 5.5\ell_o F_o \end{array} \right\} \tag{6.54}$$

in which x_{max} is the estimated distance from the effluent discharge point where h_{max} is expected to occur.

In the case of stagnant ambient surface water conditions, the jet trajectory is almost a straight line and is vertically upward. If a weak cross-flow exists, the trajectory will be curved in the direction of the cross-flow. In this case the mixing mechanism will be affected by the influence of the cross-flow, (Adams et al. 1979, 1980). The empirical models given above can still be used for this case as conservative estimates. However, instances in which the receiving water bodies are too shallow or for water bodies which have strong cross currents within the confinement of lateral boundaries that affect the discharging effluent, need to be analyzed separately as discussed below.

Shallow Receiving Water: If the bottom of the water body affects the behavior of the discharged effluent jet, the receiving water can be identified as a shallow water body. Practically most river outfalls can be grouped under this category. As a criterion the shallow water conditions can be defined as follows which is based on experimental and field data (Jirka et al. 1981),

$$\frac{h_{max}}{H} > 0.75 \tag{6.55}$$

in which H is the water depth at the point of maximum plume depth, h_{max}. If this condition is observed at an effluent discharge point, an empirical correction factor, r_s can be applied to the deep-water equations to account for the dilution inhibiting effect of the shallow receiving water. Thus, bulk dilution under shallow water conditions is estimated by the dilution factor S',

$$S' = r_s S \tag{6.56}$$

The empirical correction factor r_s is given by,

$$r_s = \left(\frac{0.75H}{h_{max}}\right)^{0.75} \tag{6.57}$$

Strong Cross-Flow or Shoreline Attached Jets: For strong cross-flows, the effluent plume may be attached or pinned to the shoreline downstream of the discharge point. For this condition, the entrainment of uncontaminated water into the plume will be inhibited from one side of the plume. In shallow water where the plume is in contact with the bottom of the channel, the ambient cross-flow is also prevented from mixing under the jet. A relatively low cross-flow may cause shoreline attachment. The parameters that are used to characterize the shoreline attachment are the relative cross-flow velocity,$(R = U_a/U_o)$ in which U_a is the cross-flow velocity, and the shallowness factor that can be calculated as (h_{max}/H). On the basis of limited field and laboratory data, the criterion for shoreline attachment for a perpendicular discharge and a straight shoreline is described as, (Jirka et al. 1975, 1981, 1983),

$$R > 0.05\left(\frac{h_{max}}{H}\right)^{-3/2} \tag{6.58}$$

A simple predictive model is not available in the literature to estimate the near-field mixing of strongly deflected shoreline attached jets. Studies indicate that a re-circulation zone between the lee side of the jet and the shoreline re-entrains the already mixed water. Depending on the amount of blocking, based on R, it is estimated that the degree of re-entrainment may be up to 100% (Jirka et al. 1981). In that case the surface jet entrains ambient undiluted water from only one side. Hence, as a conservative estimate, the initial dilution of an attached shallow water surface jet may be defined as,

$$S_{attached} = \frac{1}{2}S' \tag{6.59}$$

in which S' is the bulk dilution estimated under shallow conditions as discussed above. The extent of the near-field zone may be estimated by the cross-flow deflection scale x_c, (Jirka et al. 1981).

$$x_c = 2\frac{\ell_o}{R} \tag{6.60}$$

or by x_t computed from Eq. (6.53), whichever is less.

Surface Discharge with Zero or Negative Buoyancy: All empirical models discussed above are only valid for buoyant discharges. Whenever the effluent discharge has some buoyancy, albeit small ($\Delta\rho \to 0$), and F_o is large, the results are still applicable and indicate that large dilutions are possible at considerable distances from the discharge point until the jet subsides. The fact that the ambient environment usually exhibits some variability in density should not be overlooked as a factor in the ultimate stabilization of practically all discharges. A truly non-buoyant jet is simply predicted by the classical analysis given in Albertson et al. (1950). In this case the dilution factor is defined as a function of longitudinal distance,

$$S(x) = 0.32\frac{x}{D_j} \tag{6.61}$$

in which D_j is the equivalent diameter of a round jet. This empirical equation indicates that there is continuous dilution with distance in the longitudinal direction. In practice, however, an ultimate transition is provided by an eventual stabilization of initial dilution or by the ambient turbulence levels beginning to dominate over the weakening jet turbulence.

Figure 6.12 shows the input window for the near-field mixing surface point discharge models. The menu options on this window are the same as the other models of the ACTS software. The functions of these menu options are described in Appendix 3.

6.3.2 Submerged-Point Discharges

When the effluent discharge point is located below the surface of the water body close to its bottom, the dilution effects are analyzed by means of empirical models developed for a submerged discharge. Numerous complications may arise when one works with the submerged discharges. First and foremost among these complications is the depth of the receiving water relative to the dynamic characteristics of the effluent discharge. As shown in Fig. 6.13, two fundamentally different conditions may exist:

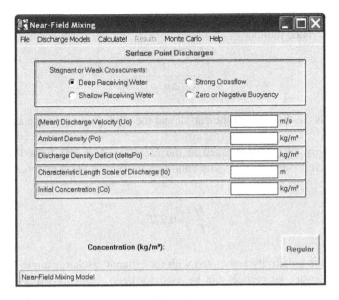

Fig. 6.12 Near-field mixing surface point discharge model input window

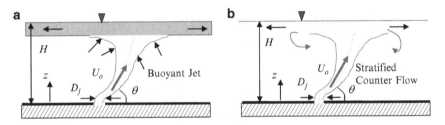

Fig. 6.13 Interaction of receiving water and effluent discharge characteristics. (**a**) Deep discharge with buoyant jet. (**b**) Shallow discharge with circulation

i. A deep receiving water condition, in which a distinct buoyant jet rises to the surface and dilution occurs because of turbulent jet entrainment up to the surface level. If the receiving water is sufficiently stratified, the jet trajectory can be shortened and the jet effects will no longer be effective when an equilibrium level is reached before the jet trajectory reaches the water surface.

ii. A shallow receiving water condition, in which the discharge momentum is sufficiently strong to cause a dynamic breakdown of the buoyant jet motion and creates a local re-circulation zone.

This dynamic distinction of deep and shallow water conditions is important, as entirely different empirical equations are used for the analysis of each case.

The discharge condition is characterized by the relative water depth H/D_j and the densimetric Froude number F_o,

$$F_o = \frac{U_o}{\sqrt{(\Delta\rho/\rho_o)gD_j}} \qquad (6.62)$$

in which D_j is the diameter of the outfall and the other parameters are as described earlier.

The stability analysis presented in the literature (Lee and Jirka 1974, 1981) indicates the following approximate condition to identify the deep receiving water conditions,

$$\frac{H}{D_j} > 0.22F_o \qquad (6.63)$$

including some sensitivity associated with the effluent discharge angle θ.

Simple buoyant jet models can be used for deep receiving water conditions. However, the mixing achieved in the local recirculation zone of a shallow discharge (Fig. 6.13b) must be analyzed on the basis of stratified counter flow models, (Jirka et al. 1983).

6.3.2.1 Stagnant or Weak Cross-Currents

Deep Receiving Water: Several submerged buoyant jet models may be found in the literature (Abraham 1963; Fan and Brooks 1969; Hirst 1972). For the vertically buoyant plume case, which is characteristic of deep discharges with reasonably small discharge Froude number, F_o, a centerline dilution S_c (i.e. minimum value in the plume) can be given as a function of normalized vertical distance (z/D_j), (Rouse et al. 1952).

$$S_c = 0.11\left(\frac{z}{D_j}\right)^{5/3} F_o^{-2/3} \qquad (6.64)$$

in which z is the distance above the nozzle and D_j is the effective diameter of the nozzle. This includes the effect of contraction expected for a sharp-edged orifice. When Eq. (6.64) is used for predictive purposes, it is always necessary to check if the deep-water condition (Eq. (6.63)) is satisfied.

The maximum vertical distance over which jet mixing takes place is given approximately by the total water depth H (Fig. 6.13). Since a mixing layer forms at the surface, based on experimental data, it is recommended to reduce this vertical distance by about 20%. For this case, the bulk dilution factor S for the entire near field mixing can be computed by,

$$S \approx 1.4S_c \qquad (6.65)$$

Shallow Receiving Water: The strong dynamic effect of the discharge within the shallow water column can create complicated flow patterns. For the vertical discharge, (Lee and Jirka 1981) provide a bulk dilution factor, which characterizes the local re-circulation cell,

$$S = 0.9 \left(\frac{H}{D_j} \right)^{5/3} F_o^{-2/3} \tag{6.66}$$

The equation above is applicable when the deep-water condition, given in Eq. (6.63), is not satisfied. If the discharge is horizontal, it is reasonable and also conservative to treat it simply as a surface discharge in shallow water, since the jet quickly rises to the surface and then behaves like a surface jet. In this case, Eq. (6.67) may be used with proper calculation of the variable ℓ_o.

$$\ell_o = \sqrt{\frac{1}{2} \left(\frac{\pi D_j^2}{4} \right)} \tag{6.67}$$

Figure 6.14 illustrates the input window for near-field mixing submerged point discharge models. The menu options on this window are the same as the other surface model input window menus. The functions of these menus are described in Appendix 3.

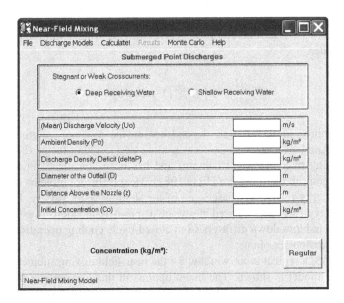

Fig. 6.14 Near-field mixing submerged point discharges model input window

6.3.3 Submerged Multiport Diffusers

A multiport diffuser is the most effective means for achieving a high degree of initial dilution. The diffuser is a linear structure consisting of many closely spaced ports, or nozzles, which inject high-velocity jets into the receiving water. The ports of the diffuser may be attached as risers to a buried pipe or may simply be openings in a pipe lying at the bottom of the receiving water.

As for a single port discharge, it is most important to realize that the dynamics of the discharge may form a stable deep water or an unstable shallow water discharge. These discharge types may be visualized by considering Fig. 6.13, replacing the round opening diameter D_j with a two-dimensional slot opening of width B. The equivalent slot width B for a diffuser with nozzles of diameter D_j and lateral spacing ℓ that ensures similar dynamic effects is given by,

$$B = \frac{\pi D_j^2}{4\ell} \tag{6.68}$$

The dynamic parameters for the discharge stability of a multiport diffuser are then associated with its equivalent flow densimetric Froude number and the relative water depth as given in Eq. (6.69),

$$F_s = \frac{U_o}{\sqrt{(\Delta\rho/\rho_o)gB}} \quad \text{and} \quad \frac{H}{B} \tag{6.69}$$

As reported in the literature, the stability analysis for multiport diffusers (Jirka and Harleman 1973) (see also Jirka 1982) gives the following definition for the deep receiving water,

$$\frac{H}{B} > 1.84 F_s^{4/3} \left(1 + \cos^2\theta\right)^2 \tag{6.70}$$

This definition indicates some dependence on the discharge angle θ with the horizontal. Ambient cross flow is often another destabilizing factor (i.e. it causes vertical mixing over the water column) and has been considered in a complete stability diagram in the literature (Jirka 1982). Most diffuser problems of practical interest in energy-related discharges are of the shallow water variety. Deep-water diffusers are typically encountered in sewage disposal applications in estuaries, and occasionally in blow-down diffusers from closed-cycle cooling operations encountered in nuclear power plants.

In Fig. 6.15, a typical input window for the near-field mixing submerged multiport diffuser model is shown. The menu options in this window are the same as those in the other surface model input window menus. The functions of these menus are described in Appendix 3.

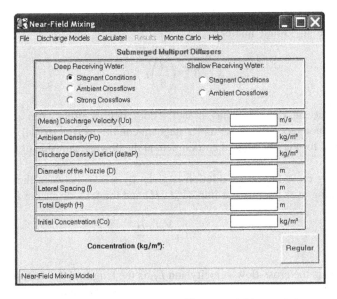

Fig. 6.15 Near-field mixing submerged multiport diffusers model input window

6.3.3.1 Deep Receiving Water

Stagnant Conditions: For this case, an estimate of bulk dilution can be obtained by considering the vertically buoyant plume model, $\theta_o = 90°$ and assuming that F_s is sufficiently small such that the deep receiving water condition given by Eq. (6.70) holds. Further, it should be noticed that as long as this condition is satisfied, all non-vertical discharges tend towards the rising buoyant plume. Also, a frequent design in the deep water condition is the alternating direction diffuser, in which adjacent nozzles point to opposite sides. In this case, $\theta = 90°$ is a reasonable approximation for the dynamic conditions created at the effluent discharge point.

For this case, empirical studies indicate that a bulk dilution coefficient S at a maximum vertical distance of $0.8H$ would be larger by a factor of $\sqrt{2}$, which accounts for the surface layer,

$$S = 0.44 \left(\frac{H}{B}\right) F_s^{-2/3} \tag{6.71}$$

Ambient Cross Flows: The direction of cross flow relative to the diffuser alignment, i.e., the axis of the main pipe, is an additional and critical parameter when ambient cross flows are present at the effluent discharge point. A perpendicular alignment is preferred, because this orientation intercepts the cross flow and maximizes mixing. An experimental study reported in the literature for a deep-water diffuser yielded the following dilution estimates for near-field bulk dilutions (Roberts 1977, 1979).

For weak cross flows, the dilution is still related to the buoyancy of the discharge, i.e. its Froude number,

$$S = 0.27\left(\frac{H}{B}\right)F_s^{-2/3} \tag{6.72}$$

In this case, the cross flow has, in fact, a conservative effect when compared to the case of deep receiving water under stagnant conditions (Eq. (6.71)). This effect compensates for the blocking of entrainment at the downstream side of the diffuser plume.

For strong cross flows, the dilution is given by a ratio between the total cross flow sweeping over the diffuser line, $(V_o L_D H)$ and the total discharge flow Q_o.

$$S = C_1 \frac{V_o L_D H}{Q_o} \tag{6.73}$$

in which V_o is the cross-flow velocity and L_D is the length of the diffuser. Ideally, the constant C_1 should be unity, but experimental studies indicate a smaller and hence a more conservative value of $C_1 = 0.58$ (Roberts 1977). Apparently, this more conservative value is due to incomplete mixing and buoyant restabilization. Based on these experimental studies, the value of C_1 is given as a function of the orientation and strength of cross flows and the buoyancy of the effluent (Roberts 1977). In this case the width of the plume at the surface is also characterized, so that the initial conditions to use in far-field models can be easily identified.

6.3.3.2 Shallow Receiving Water

Multiport diffusers in shallow water discharge conditions, which are frequently used for thermal discharges, can have a large number of possible flow configurations and mixing mechanisms. For these applications, highly site specific designs, i.e., different types of nozzle orientation and current alignments are possible and can be considered. Three major diffuser types that have been used in design are the following: (i) the unidirectional diffuser; (ii) the staged diffuser; and, (iii) the alternating diffuser.

Stagnant Receiving Water: The unidirectional and staged diffuser designs produce vertically mixed (uniform) diffuser plumes that sweep in the direction of the discharge nozzles. The bulk dilution factor for these diffusers is given by,

$$S = C_2\sqrt{\frac{H}{B}} \tag{6.74}$$

The factor C_2 is equal to $1/\sqrt{2}$ for unidirectional diffusers and 0.67 for staged diffusers.

The alternating diffuser with an unstable recirculation zone for shallow water is predicted by stratified flow theory to have a bulk dilution factor of,

$$S = C_3 \left(\frac{H}{B}\right) F_s^{-2/3}$$ (6.75)

in which the range of the constant is given as $(0.45 < C_3 < 0.55)$ depending on the friction effects in the counter flow.

Ambient Cross-Flows: For this case, depending on the diffuser type and alignment, a variety of interactions may be considered between the cross flows and the orientation of the diffuser (Jirka 1982). A diffuser type that is frequently employed in the design when the ambient current is steady and flows in only one direction is the co-flowing diffuser, i.e., a unidirectional design with perpendicular alignment. For this case the bulk dilution coefficient is given by the combined effect of cross flows and diffuser mixing:

$$S = \frac{1}{2}\left(\frac{V_o L_D H}{Q_o}\right)^2 + \frac{1}{2}\left(\left(\frac{V_o L_D H}{Q_o}\right)^2 + 2\left(\frac{H}{B}\right)\right)^{1/2}$$ (6.76)

This equation is particularly useful for diffuser applications in rivers as long as the diffuser length is sufficiently shorter that the river width. Equation (6.76) will be superseded by the proportional mixing ratio given below, if the diffuser covers the entire river,

$$S = \frac{Q_R}{Q_o}$$ (6.77)

since the diffuser induced mixing action cannot result in more mixing than is provided by the river flow Q_R.

6.4 Far Field Mixing Models

Far field mixing and models used to describe this process are based on the analysis of plume dispersion beyond the near field. In this case the concentrations that are diluted due to near field mixing conditions would be used as the source concentration. Mixing for these applications is usually associated with ambient lake and river fate and transport processes, and it tends to occur at a greatly reduced rate, when compared to the initial mixing that occurs in the near field. This is the reason why the NPDES regulations usually require that all chronic concentration standards must be met at the edge of the near field mixing zone.

After an effluent discharge passes through the relatively rapid initial mixing process in the near field, its further dilution is determined by the advective and diffusive transport processes in the ambient far field. Since these processes are much slower, much longer distances and time frames must be considered. These concerns result in two important consequences: (i) For these cases it may be important to include decay and other physical or chemical loss processes into the analysis; and, (ii) the net advective transport properties of the receiving water body must be considered.

The long term chemical accumulation in a coastal bay or in an inland reservoir is often controlled simply by the average through flow and flushing rate of the aquatic pathway and the internal mixing processes. Diffusion and circulation, may be largely irrelevant. Rivers are characterized with a well defined net advective transport mechanism. Estuaries are characterized with strongly oscillating tidal flow conditions but also with weak net transport. Small lakes or reservoirs are characterized with strong boundary limitations and weak transport, and the ocean and large lakes are characterized with practically unlimited dimensions. Accordingly, the dilution analysis for these environments is based on these distinct characteristics of the water bodies, which exhibit highly variable geometric or advective transport characteristics.

6.4.1 Mixing in Rivers

Rivers are typically wide and shallow water bodies with strong advective and turbulent flows. After the initial mixing process, the effluent that is mixed over the shallow depth is advected downstream by the river flow, and is diffused laterally across the river. After sufficient distance, the effluent may be assumed to be fully mixed across the entire width of the river. Hence, it is important to analyze the mixing conditions for two distinctly different stages: (i) Transverse mixing; and, (ii) longitudinal advection and dispersion.

6.4.1.1 Transverse Mixing in Rivers

Examples of transverse mixing in a shallow river with uniform depth, H and ambient velocity, V are illustrated in the definition sketch shown in Fig. 6.16, which shows point discharge and a rapidly mixed line source. If the plume width is much less than the total river width, then the boundary effects can be considered to be negligible. For this case the two dimensional analytical model which yields the transverse concentrations in the longitudinal direction can be given as,

$$C(x, y) = \frac{Q_o C_o}{H\sqrt{4\pi D_y V x}} \exp\left(-\frac{y^2 V}{4D_y x} - \frac{\lambda x}{V}\right) \qquad (6.78)$$

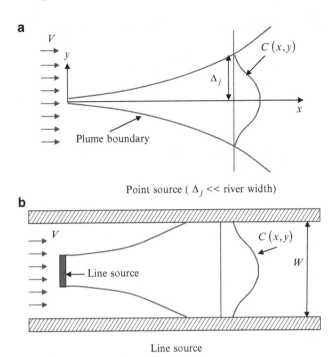

Fig. 6.16 Definition sketch of transverse mixing in a river with uniform depth and velocity

in which C_o is the initial concentration $[ML^{-3}]$, Q_o is the initial effluent flow rate $[L^3 T^{-1}]$, x is the longitudinal distance $[L]$, y is the transverse distance $[L]$, H is the depth of the river $[L]$, V is the ambient river velocity $[LT^{-1}]$, λ is the first-order decay coefficient $[T^{-1}]$, which is given as $(\lambda = \ln 2/(\text{half life}))$, and D_y is the transverse diffusion coefficient $[L^2 T^{-1}]$.

One should note that the parameters C_o and Q_o may represent the conditions after the initial mixing process, as determined by any of the other models in the previous section. The coefficient D_y, which represents the effect of transverse turbulent diffusion, is generally related to the energy dissipation characteristics of the channel as discussed before,

$$D_y = \beta_y u_* H \tag{6.79}$$

in which u_* is the shear velocity, defined as $u_* = \sqrt{gHS_o}$ and S_o is the channel slope. The coefficient β_y is typically of the order of 0.6 ± 0.3 for reasonably straight rivers. The standard deviation of the lateral Gaussian concentration distribution is given as,

$$\sigma_y = \sqrt{2D_y x / V} \tag{6.80}$$

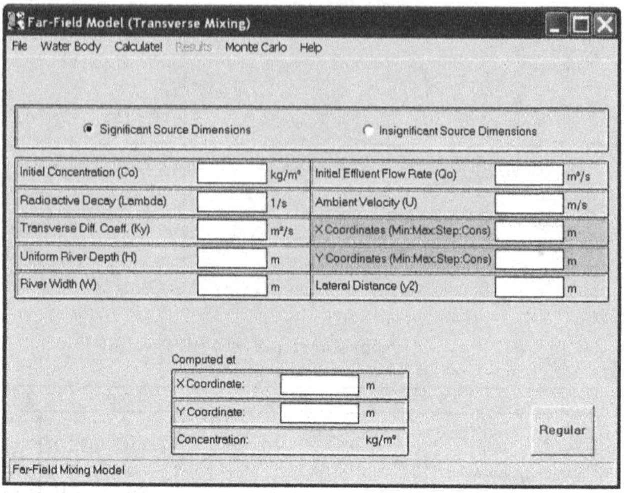

Fig. 6.17 Far-field transverse mixing river model input window

Equation (6.78) is applicable only as long as there is no significant interaction between the plume and the riverbank.

Figure 6.17 illustrates a typical input window for the far field river transverse mixing model. The menu options on this window are the same as the other surface model input window menus. The functions of these menu operations are described in Appendix 3.

When the initial source dimensions are significantly large and the plume interacts with the river banks, the concentration distribution is given by Eq. (6.81), in which the boundary effects of the banks on the transverse mixing is considered,

$$
C(x, y) = \frac{Q_o C_o}{Q_r} \exp\left(-\frac{\lambda x}{V}\right)
$$
$$
\bullet \left\{ 1 + 2 \sum_{n=1}^{\infty} \exp\left(\frac{n^2 \pi^2 D_y x}{Q_r^2}\right) \frac{2W}{n(y_2 - y_1)} \sin\left(n\pi \frac{(y_2 - y_1)}{W}\right) \right.
$$
$$
\left. \times \cos\left(n\pi \frac{(y_1 + y_2)}{W}\right) \cos\left(n\pi \frac{y}{W}\right) \right\}
\tag{6.81}
$$

Note that $(Q_r = VWH)$ is the river flow rate. In calculations, only two to three terms of the series in the above equation need to be included, as the plume approaches full mixing across the entire width rapidly. In the ACTS software the first 10 terms of this series are evaluated. The initial source width and location (y_1, y_2) may be given by the location of a diffuser or by the extent of the near field of a surface discharge, $(y_1 = 0)$. Similar to the case above, the variables C_o and Q_o are related to the discharge values that are evaluated based on the conditions of the near field model.

6.4.1.2   Longitudinal Advection and Dispersion

Once the effluent is laterally mixed, its transport under steady state conditions is affected mostly by simple advection in the longitudinal direction due to the river flow regime. If the effluent is rapidly decaying or if the flow regime is highly unsteady (e.g., in the case of an accidental release in a regulated river), it becomes important to also include the mechanism of longitudinal dispersion, that is, a combination of differential shear flow (non-uniform river velocity distribution) and cross-sectional turbulent mixing. The concentration profile in the longitudinal direction for a steady state release of effluents can be obtained from Eq. (6.82)

$$C(x) = \frac{C_o Q_o}{Q_r \sqrt{1 + 4 \frac{\lambda D_x}{V^2}}} \exp\left(\frac{xV}{2D_x} \left(1 - \sqrt{1 + 4 \frac{\lambda D_x}{V^2}} \right) \right) \qquad (6.82)$$

in which x is the downstream distance from the release point and D_x is the longitudinal dispersion coefficient $[L^2 T^{-1}]$. Generally, the shear flow is so large that the longitudinal dispersion coefficient D_x is several orders of magnitude larger than the transverse dispersion coefficient D_y. The empirical equation given below can be used to estimate the longitudinal dispersion coefficient (Fisher et al. 1979).

$$D_x = \frac{0.011 V^2 W^2}{H u_*} \qquad (6.83)$$

For an instantaneous accidental release of a chemical mass M_o, the time dependent concentration distribution in the longitudinal direction can be given as,

$$C(x,t) = \frac{M_o}{WH\sqrt{4\pi D_x t}} \exp\left(-\frac{(x - Vt)^2}{4D_x t} - \lambda t \right) \qquad (6.84)$$

In this equation, the downstream advection of a chemical spill at velocity V and its simultaneous longitudinal spread are modeled. Equation (6.84) is a useful first-order model for estimating exposure levels downstream of accidental releases. A measure of the longitudinal extent of the dispersing pulse can be evaluated by calculating its standard deviation at a downstream point given by,

$$\sigma_x = \sqrt{2D_x t} \qquad (6.85)$$

In other cases, such as long term releases to rivers, the transport of a continuous source in a downstream direction may become important. Two analytical solutions may be found in the literature for two different cases: (i) A continuous source of infinite duration; and, (ii) a continuous source of finite duration.

Continuous Source of Infinite Duration: In this case the one-dimensional advective–diffusive transport model can be given as,

$$\frac{\partial C}{\partial t} + V\frac{\partial C}{\partial x} = D_x\frac{\partial^2 C}{\partial x^2} - \lambda C \tag{6.86}$$

in which C is concentration $[ML^{-3}]$; t is time $[T]$; V is the mean river flow velocity $[LT^{-1}]$; x is the longitudinal distance $[L]$; D_x is the longitudinal dispersion coefficient $[L^2 T^{-1}]$; and λ is the first order decay rate coefficient. $[T^{-1}]$. Initial and boundary conditions for this case can be given by,

$$\begin{aligned}
&Initial\,Condition\ C(x,0) = 0 &&for\ x \geqslant 0\\
&Upstream\ B.C.\ C(0,t) = C_o &&for\ t \geqslant 0\\
&Downstream\ B.C.\ C(\infty,t) = 0 &&for\ t \geqslant 0
\end{aligned} \tag{6.87}$$

in which C_o is the concentration at the upstream boundary $[ML^{-3}]$. The analytical solution for this problem can be given as,

$$C(x,t) = \frac{C_o}{2}\left\{\exp\left(\frac{Vx}{2D_x}(1-\Gamma)\right)erfc\left(\frac{x-V\Gamma t}{2\sqrt{D_x t}}\right) + \exp\left(\frac{Vx}{2D_x}(1+\Gamma)\right)erfc\left(\frac{x+V\Gamma t}{2\sqrt{D_x t}}\right)\right\} \tag{6.88}$$

in which

$$\left.\begin{aligned}
\Gamma &= \sqrt{1+2H}\\
H &= 2\lambda D_x/V^2
\end{aligned}\right\} \tag{6.89}$$

Continuous Source of Finite Duration: Although the above solution is of interest, a far more useful problem in exposure analysis is that in which a source is present for a finite period of time. Let T $[T]$ represent the duration of the continuous source. The initial and boundary conditions of the problem described above can be given as,

$$\begin{aligned}
&Initial\,Condition\ C(x,0) = 0 &&for\ x \geqslant 0\\
&Upstream\ B.C.\ C(0,t) = C_o &&for\ \tau \geqslant t \geqslant 0\\
&Upstream\ B.C.\ C(0,t) = 0 &&for\ t \geqslant \tau\\
&Downstream\ B.C.\ C(\infty,t) = 0 &&for\ t \geqslant 0
\end{aligned} \tag{6.90}$$

Accordingly, the model given above describes a source of finite duration. An analytical solution for the conditions given in Eq. (6.90) can be given as shown in Eq. (6.88) for $t \leqslant \tau$. For $t \geqslant \tau$, the solution is,

$$C(x,t) = \frac{C_o}{2} \exp\left(\frac{-\lambda x}{V}\right) \left\{ erfc\left(\frac{x - Vt(1+H)}{2\sqrt{D_x t}}\right) + erfc\left(\frac{x - V(t-\tau)(1+H)}{2\sqrt{D_x(t-\tau)}}\right) \right\}$$

(6.91)

These two analytical solutions and the Monte Carlo applications for these solutions are included in the ACTS software as a component of the river transport module.

6.4.2 Estuaries

Transport and dispersion processes in estuaries are considerably more complicated than those in non-tidal rivers. The oscillatory tidal motion with cyclic variations in velocity and elevation causes complex hydrodynamic mixing conditions, which in turn affect the concentration distributions. The difference in density between the fresh water and saltwater superimposes additional vertical (baroclinic) circulations. Finally, wind-driven currents in wide, shallow (bay like) estuaries also play an important role. A detailed analysis of pollutant distributions in an estuary usually requires a thorough field investigation, including tracer studies, to determine their hydrodynamic and mixing patterns. The information and data thus obtained can be used in the selection and application of reasonably detailed estuary or coastal transport models.

It must be stressed that, depending on pollutant characteristics, higher dimensional models or very fine temporal resolution may be quite redundant and useless. For example, for steady-state releases of relatively conservative substances (small λ), the mean residence time, as dictated by the net freshwater flow through the estuary, determines the long term average concentrations. Since concentration gradients tend to be small, the details of the internal distribution process tend to be relatively unimportant.

The longitudinal distribution $C(x)$ of any pollutant that is released in a steady state fashion, at a distance L upstream of the estuary mouth, is given by (Stommel 1949, 1950),

$$C(x) = \frac{Q_o C_o}{Q_r \sqrt{1+\alpha}} \left\{ \frac{\exp\left(\frac{(x-L)V_f}{2D_T}\left(1 - \sqrt{1+\alpha}\right)\right) - \exp\left(\frac{(x-L)V_f}{2D_T}\left(1 + \sqrt{1+\alpha}\right)\right)}{\exp\left(\frac{LV_f}{2D_T}\left(1 - \sqrt{1+\alpha}\right)\right) - \exp\left(\frac{LV_f}{2D_T}\left(1 + \sqrt{1+\alpha}\right)\right)} \right\}$$

(6.92)

in which Q_r is the river discharge, $\alpha = 4\lambda\left(D_T/V_f^2\right)$, V_f is the mean fresh water river velocity determined by dividing the total fresh water inflow by the cross-sectional area, and D_T is the tidal dispersion coefficient. The origin of the x-axis is located at the release point, with the positive x-axis in the downstream direction.

6.4.3 Small Lakes and Reservoirs

Small natural or man-made impoundments, cooling ponds in particular, represent an extreme situation of geometric constraints and limited advective transport. The definition of "small" that is used here is based on the relative residence time, i.e. the through flow time relative to the decay time of the chemical. The half-life of many chemicals is considerably longer than impoundment residence time (typically on the order of a few days to weeks). Hence, except for a small initial mixing region, usually in the form of a buoyant surface jet for cooling ponds, the chemical concentration is essentially uniform within the entire impoundment, and simple bulk analysis may suffice for predictive purposes.

For these cases the contaminant concentration can be evaluated based on a box model concept. In this case the outcome represents a complete mixing in the system and the concentration will be a function of time only,

$$C = \frac{C_o}{(Q/Q_o) + \lambda V/Q_o} \left\{ 1 - \exp\left(-\left(\frac{Q}{V} + \lambda \right)t \right) \right\} \tag{6.93}$$

in which Q is the net through flow, either natural or in the form of an artificial blow down scheme, Q_o is the circulating water flow rate, such as the condenser flow, $(Q_o C_o)$ is the chemical release rate and V is the water body volume.

In the study of releases of chemicals with half-lives much longer than the impoundment replacement time (V/Q), it will be sufficient to consider the steady-state solution given below,

$$C = \frac{C_o}{Q/Q_o + \lambda V/Q_o} \tag{6.94}$$

In reservoirs where the assumption of horizontal homogeneity is not realistic, two- or three-dimensional numerical models should be used to estimate the reservoir effluent distribution in the water body.

Figure 6.18 illustrates a typical input window for a far-field small lake mixing model. The menu options on this window are the same as the other surface water pathway model input window menus. The functions of these menus are described in Appendix 3.

6.4.4 Oceans and Great Lakes

The main feature of pollutant dispersion in oceans or large lakes is the unlimited extent of their geometric configuration, which is without constraints on net advection or dispersion, except for possible shoreline boundary effects. The standard approach to pollution analysis for such environments is first to determine the

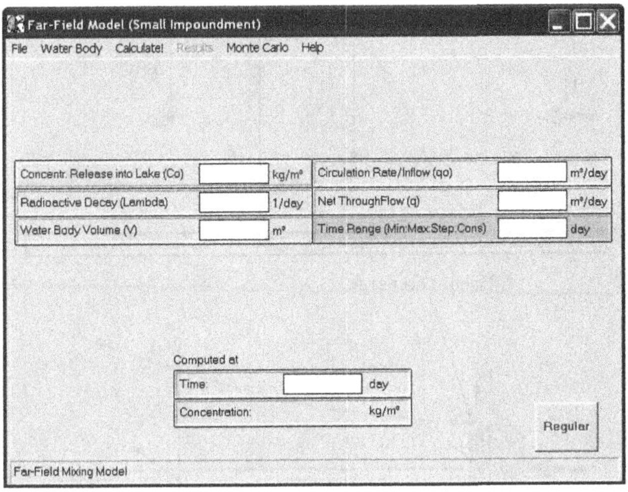

Fig. 6.18 Far-field mixing small lake model input window

velocity field and then to compute the dispersion of the release (instantaneous or continuous) which is affected by that velocity field. If small masses of chemicals with negligible buoyancy are involved, the dynamic coupling between these two phases of the analysis can always be neglected.

In these cases the concentration distribution can be computed using the Eulerian approach in the form of a solution of the advection-diffusion equation with a decay term. Analytic solutions of this equation exist for simple velocity fields. A useful analytical solution in this category is the steady-state solution for a uniform source of finite extent in steady uniform flow (Brooks 1960). Brooks solved the steady state advection–diffusion equation,

$$V\frac{\partial C}{\partial x} = \frac{\partial}{\partial y}\left(D_y\frac{\partial C}{\partial y}\right) - \lambda C \tag{6.95}$$

using several different assumptions for the definition of spatial variation of the eddy diffusivity D_y. In Eq. (6.95) λ is the first-order decay coefficient defined in terms of the half-life τ of the chemical as,

$$\lambda = \frac{\ln 2}{\tau} \tag{6.96}$$

These solutions are based on the assumption that at the interface of the near field and far field, the effluent is uniformly distributed over the width W and depth H and that beyond that point, one-dimensional advection and lateral diffusion are the primary transport mechanisms, Fig. 6.19.

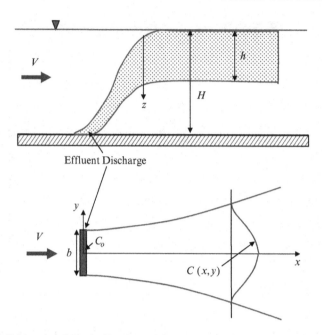

Fig. 6.19 Definition sketch for the Brooks model

The solution of the above equation for constant eddy diffusivity and the boundary conditions defined in Fig. 6.19 is,

$$C = \frac{C_o}{2} \exp\left(-\frac{\lambda x}{HV}\right)\left\{ erf\left(\frac{\left(y+\frac{W}{2}\right)}{2}\sqrt{\frac{V}{D_y x}}\right) - erf\left(\frac{\left(y-\frac{W}{2}\right)}{2}\sqrt{\frac{V}{D_y x}}\right)\right\} \quad (6.97)$$

in which H is the vertical extent of the water column and D_y is the Eddy viscosity.

Figure 6.20 illustrates a typical input window for the far field ocean and great lakes mixing model. The menu options for this window are the same as those for other surface water pathway model input window menus. The functions of these menus are described in Appendix 3.

6.5   Surface Water Sediments

The transport of chemicals in surface waters may cease permanently or slow down temporarily if the chemicals are adsorbed from the dissolved water phase onto sediments. Both suspended and bed sediments adsorb chemicals, but suspended sediment usually adsorbs more chemicals than bed sediment per unit weight of sediment. When adsorption occurs, the concentration of dissolved chemical in

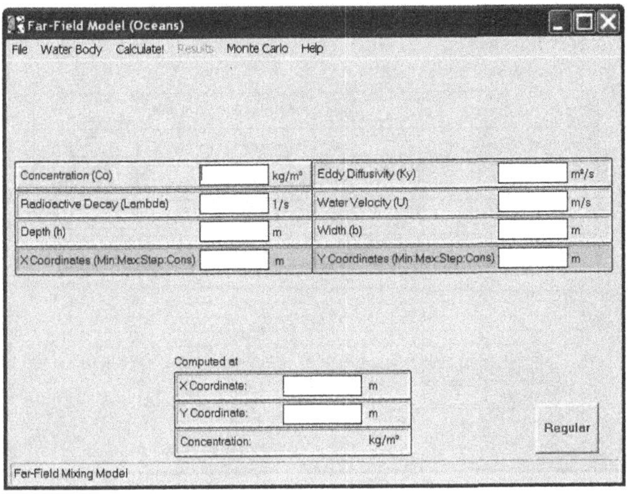

Fig. 6.20 Far-field mixing ocean model input window

the water body is lowered, and the chemicals may become less available to aquatic biota and man. This non-availability may be reversed, however, since it is possible for chemicals that have accumulated in bed sediments to be desorbed or become re-suspended, thus forming a long-term source of pollution. In the ACTS software, the concentration models for the following water bodies have been considered: (i) River bed sediments; (ii) estuary bed sediments; and, (iii) coastal water and ocean bed sediments.

6.5.1 River Bed Sediments

Fletcher and Dotson Model (1971): One of the first models developed to compute the chemical dose to man through liquid and gaseous pathways was the Fletcher and Dotson model, (Fletcher and Dotson 1971). This model uses an unsteady, one dimensional, liquid-pathway sub-model to calculate temporal and longitudinal distributions of dissolved effluent concentration, as well as the concentration of the chemicals that may be attached to suspended and bottom sediments of various sizes.

The dissolved chemical concentration at a given location is found by applying the mass conservation equation with decay as follows:

$$C(x,t) = \frac{1}{Q(x,t)} \left(Q(x - \Delta x, t - \Delta t) \exp(-\lambda \Delta t) + \sum_{1}^{n} Q_i C_i \right) \tag{6.98}$$

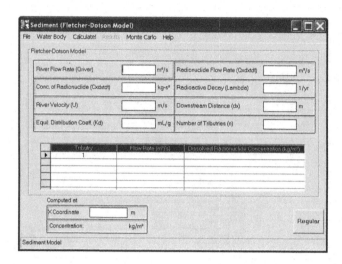

Fig. 6.21 River sediment Fletcher–Dotson model input window

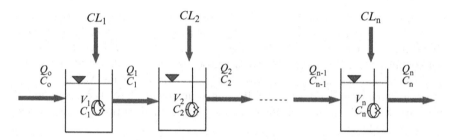

Fig. 6.22 Mixing tank model

in which $C(x,t)$ is the dissolved chemical concentration at location x and time t, C_i is the dissolved concentration in the tributary, $Q(x,t)$ is the flow rate at location x and time t, Q_i is the tributary flow rate, and λ is the decay coefficient.

Figure 6.21 illustrates a typical input window for the river sediment Fletcher–Dotson mixing model. The menu options on this window are the same as the other surface water pathway model menu options. The functions of these menu options are described in Appendix 3.

Mixing-Tank Model with Sediment Transport (Onishi 1981): A mixing tank transport model (Fig. 6.22) similar to one used for simulating pesticide transport in streams (Onishi et al. 1980), is described below as used in the ACTS software. In this model the following assumptions are made: (i) River reaches are divided into segments which are represented by a series of tanks. Within each segment (a tank) sediments and chemical concentrations are assumed to be completely mixed; (ii) chemicals and sediment contributions from point and non-point sources are treated as lateral influx that is uniformly distributed along the river reach for each segment; (iii) dissolved and particulate chemicals are linearly related by a distribution coefficient; (iv) dissolved and particulate chemicals reach their equilibrium conditions within one time step; and, (v) particulate chemical deposition to the riverbed and re-suspension from the riverbed does not occur.

The mass conservation of sediment in the nth tank leads to the following sediment transport equation:

$$\frac{\partial S_n}{\partial t} = -S_n\left(\frac{1}{V}\left(\frac{\partial V}{\partial t}+Q\right)\right)_n + \frac{Q_{n-1}S_{n-1}+SL_n}{V_n} \tag{6.99}$$

in which Q_n is the discharge from the nth tank, S_n is the sediment concentration in the nth tank, SL_n is the lateral influx of sediment, V_n is the water volume of the nth tank and t is the time.

The mass balance of the dissolved and particulate chemical in the nth tank is,

$$\frac{\partial C_n}{\partial t} = \frac{1}{V_n\left(1+S_nK_d\right)}\left(\begin{array}{l}\left(1+S_{n-1}K_d\right)Q_{n-1}C_{n-1}+\left(CL_n+C_pL_n\right)-\left(1+S_nK_d\right)Q_nC_n \\ -\lambda V_nC_n\left(1+S_nK_d\right)-C_n\dfrac{\partial}{\partial t}\left(V_n\left(1+S_nK_d\right)\right)\end{array}\right) \tag{6.100}$$

$$C_{Pn} = K_dC_n \tag{6.101}$$

in which C_n is the dissolved chemical concentration in the nth tank, C_{Pn} is the particulate chemical concentration in the nth tank, CL_n is the lateral influx of chemical, C_pL_n is the lateral influx of particulate chemical, and K_d is the distribution coefficient of the chemical.

By substituting Eq. (6.101) for (6.100), Eqs. (6.99) and (6.100) are then solved to obtain the sediment and dissolved chemical concentrations S_n, C_n in the nth tank. In general, Eqs. (6.99) and (6.100) must be solved numerically, as was done in Onishi et al. (1980). However, for the following simplified case an analytical solution, which is similar to that obtained in USNRC (1978) for a dissolved chemical case, can be given:

$$\begin{aligned} &C_o = 0 \\ &CL_n = 0 \text{ for all } n \\ &S_n = S_{n-1} = \ldots = S_i = \text{ constant for time and all } n. \\ &Q_n, V_n \text{ are not functions of time for all } n. \end{aligned} \tag{6.102}$$

The chemical release M_1 into the first segment during the time Δt is

$$M_1 = \left(CL_1 + C_pL_1\right)\Delta t \tag{6.103}$$

Hence, for an instantaneous release of M_1, the concentration of the dissolved phase chemicals in the nth river reach (or tank) is,

$$
C_n = \frac{\left(\dfrac{M_1}{1+S_n K_d}\right)\displaystyle\prod_{i=1}^{n-1} Q_i}{\displaystyle\prod_{i=1}^{n-1} \mathcal{V}_i} \left(\sum_{j=1}^{n} \frac{-R_j \exp\left(-R_j t\right)}{\displaystyle\prod_{\substack{k=1 \\ k \neq j}}^{n+1} \left(R_k - R_j\right)} \right) \tag{6.104}
$$

in which

$$
R_j = \frac{Q_j}{\mathcal{V}_j} + \lambda \quad ; \quad R_{n+1} = 0 \tag{6.105}
$$

A particulate chemical concentration is then obtained from Eq. (6.101).

6.5.2 Estuary Sediments

Estuarine water bodies are characterized by the distinct effect of: (i) Reversible tidal flow conditions; and, (ii) Salinity effects. Estuaries have substantially faster flowing water during the tidal cycle than their tidally averaged flow conditions would indicate, yet the net downstream transport during the tidal period is relatively small. This type of flow behavior allows for the re-suspension and subsequent re-deposition of some fine sediment during each tidal cycle. As such, the sediment and water are in close contact over a longer period in an estuary than in a reservoir or a lake. Salinity is also an important factor in this analysis, because salinity causes sediment flocculation at certain levels and also affects adsorption/desorption mechanisms. For these reasons it is difficult to identify a single partition coefficient value K_d for a study area (Wrenn et al. 1972; Onishi and Trent 1982; Schell and Sibley 1982).

None of the simple contaminant transport models can simulate reversible tidal flow and salinity impacts. However, if the tidal flow is averaged over several tidal cycles, than most of the models discussed for river bodies are applicable to estuaries. The following model accounts for sediment migration velocities and tidally averaged flow velocities.

NRC Estuaries Model with Sedimentation (USNRC 1978): The physical setting of the estuarine problem is given in Fig. 6.23. A water layer of thickness d_1 is in contact with a movable sediment layer of thickness d_2. The water layer is moving with a net tidally averaged velocity of V and the erodible bed is moving with a net velocity U_b. Diffusive transport occurring in tidal oscillations in water and sediment layers is assumed to be a constant defined in terms of the longitudinal dispersion coefficients, D_{dx} and D_{xb}, respectively. Sedimentation and burial occurs uniformly at a vertical velocity v_b.

Fig. 6.23 NRC estuarine model

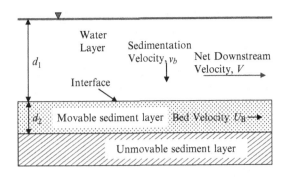

In this model, it is also assumed that dissolved and particulate chemicals are in equilibrium and are related by,

$$C_{Pn} = K_d C_n \qquad (6.106)$$

The differential equation describing the chemical concentration in the water phase becomes,

$$\frac{\partial C}{\partial t} + U' \frac{\partial C}{\partial x} = E'_L \frac{\partial^2 C}{\partial x^2} - \lambda' C \qquad (6.107)$$

in which

$$U' = \frac{fV + (1-f)U_B K_d}{f + (1-f)K_d} \qquad (6.108)$$

$$E'_L = \frac{fD_{dx} + (1-f)D_{xb}K_d}{f + (1-f)K_d} \qquad (6.109)$$

$$f = \frac{d_1}{d_1 + d_2}; \quad (1-f) = \frac{d_2}{d_1 + d_2} \qquad (6.110)$$

The solution to the model given above for an instantaneous release of $M_1 C_i$ at $x = 0$ is,

$$C = \frac{M_1}{aA\sqrt{4\pi E'_L t}} \exp\left(-\left(\frac{(x - U't)^2}{4E'_L t} + \lambda' t\right)\right) \qquad (6.111)$$

in which $a = f + (1-f)K_d$, A is the cross-sectional area of an estuary, and

$$\lambda' = \lambda + \frac{K_d \frac{v_b}{d_2}(1-f)}{f + (1-f)K_d} \qquad (6.112)$$

6.5.3   Coastal Waters and Ocean Sediments

In general, sediment effects on chemical transport in coastal waters and oceans are less important than in other surface waters, because both sediment concentrations and the distribution coefficients tend to be smaller. Since models must be at least two-dimensional to predict chemical distributions in coastal waters and oceans, the one-dimensional models discussed in earlier sections are not applicable. However, the two-dimensional models and the analytical solutions discussed above may be used to estimate dissolved and particulate chemicals if all particulate chemicals are in suspension.

6.5.4   Lake Sediments

Unique processes are responsible for the distribution and movement of chemicals in lakes. Basically, water flow is slower in lakes because they are relatively deep and confined. The major processes affecting chemical movement are: (i) Flow conditions; (ii) stratification and seasonal turnover; (iii) sediment interaction; and, (iv) biotic interaction. Because lakes have low flow velocities, sediments introduced into them tend to fall directly to the lake bottom. During this process, the sediment may adsorb chemicals and carry them to the lake bottom. In the absence of sediment movement, chemicals are either adsorbed or desorbed from the bed sediment.

NRC Lake Model (USNRC 1978): A two-layer lake model has been developed at the Nuclear Regulatory Commission (Codell et al. 1982). This unsteady state model divides a lake into water and bed sediment compartments through which chemicals are exchanged by direct adsorption mechanisms and sediment deposition (Fig. 6.24). The following assumptions were made for the model: (i) Water inflow and outflow are constant; (ii) sedimentation rate is constant; (iii) the thickness of the sediment layer remains constant. If sedimentation occurs, it is assumed that the

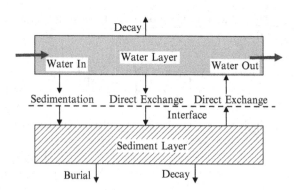

Fig. 6.24 NRC two-layer lake model

affected portion of the original bed layer becomes inactive, and is eliminated from the analysis; and, (iv) dissolved and particulate chemicals undergo decay.

In this model, mass balance equations for dissolved and particulate chemicals are

$$\frac{dC}{dt} = \frac{W(t)}{V} + C_p \lambda_1 - C \lambda_2 \tag{6.113}$$

$$\frac{dC_p}{dt} = C \lambda_3 - C_p \lambda_4 \tag{6.114}$$

in which V is the lake volume $[L^3]$, $W(t)$ is the input rate of the chemical spill (C_i/year), and

$$\lambda_1 = \frac{K_f}{d_1 K_d} \tag{6.115}$$

$$\lambda_2 = \frac{q}{V} + \lambda + \frac{v_b K_d}{d_1} + \frac{K_f}{d_1} \tag{6.116}$$

$$\lambda_3 = \frac{v_b K_d}{d_2} + \frac{K_f}{d_2} \tag{6.117}$$

$$\lambda_4 = \lambda + \frac{v}{d_2} + \frac{K_f}{d_2 K_d} \tag{6.118}$$

in which d_1 is the depth of the water layer, d_2 is the depth of the sediment layer, K_f is the coefficient of direct chemical transfer $[L/\text{year}]$, q is the freshwater flow rate $[L^3/\text{year}]$, v_b is the sedimentation velocity $[L/\text{year}]$, and λ is the chemical decay rate $[\text{year}^{-1}]$.

For an instantaneous release of effluent mass M_1, the water-phase concentration C_i can be obtained as,

$$C_i = \frac{M_1}{V(S_1 - s_2)} ((\lambda_4 + S_1) \exp(S_1 t) - (\lambda_4 + S_2) \exp(S_2 t)) \tag{6.119}$$

in which

$$S_{1,2} = \frac{-(\lambda_2 + \lambda_4) \pm \sqrt{(\lambda_2 + \lambda_4)^2 - 4(\lambda_2 \lambda_4 - \lambda_1 \lambda_3)}}{2} \tag{6.120}$$

Figure 6.25 illustrates a typical input window for the lake sediments model. The menu options on this window are the same as those for other surface water pathway model menus. The functions of these menu options are described in Appendix 3.

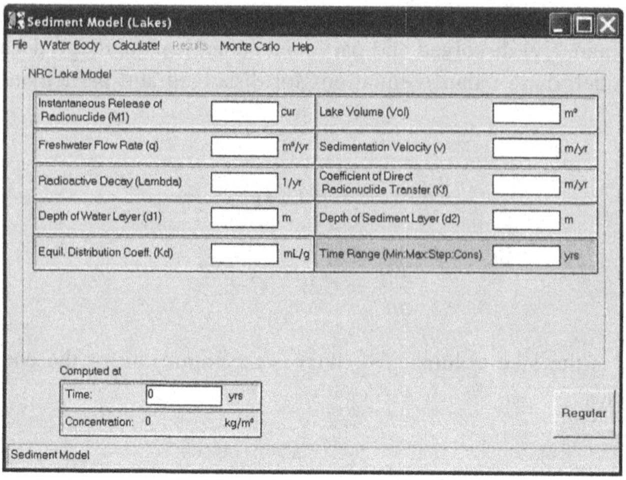

Fig. 6.25 Sediment NRC lake model input window

6.6 Applications

The aquatic pathway models discussed in this chapter cover a wide range of river, lake, estuarine and sediment fate and transport models. Providing applications for each of these cases would be an almost impossible task due to the multitude of cases that can be analyzed using these models. In this section several applications are selected and solved to demonstrate the use of the important features of the ACTS software. As the reader gets familiar with the ACTS software, he or she will recognize that the features and procedures discussed below are standardized for all other pathway applications of the ACTS software. These procedures can be repeated in other studies that involve these pathways to extend the analysis to a more sophisticated level. Thus, the purpose here is to introduce the reader to some applications in surface water pathway fate and transport analysis using the ACTS software, and in doing so familiarize the reader with the important features of the software.

Example 1: An effluent is discharging to a deep semi-infinite water body with a flow rate $Q_o = 0.6 \pm 0.02 \mathrm{m^3/s}$. The variability in the discharge rate is based on the variability of the operational conditions of the facility in handling the waste discharge. The outfall can be characterized as a rectangular cross section of 1 m width and 0.5 m height. The density of the effluent is given as 994 ± 3 kg/m^3 and the density of the water body as 998 kg/m^3. The variability in the effluent density is again attributed to the conditions in the facility generating the effluent. The concentration of the effluent is given as $C_o = 56 \pm 5$ μg/L with an MCL 5 μg/L. The variability is again associated with the variability of the waste stream concentration in the facility. Will these outfall conditions indicate a safe discharge design with respect to the MCL identified for this effluent?

Solution: The data given in this problem indicate that there is some variability in the conditions of the effluent and the effluent discharge. Thus, we first need to provide an answer for the problem using the "mean" values of the data given in the question. We will than provide an analysis of the effects of the variability on this answer using a Monte Carlo analysis of the parameters that are given as variables. The data given in Example 1 will yield the following data that will be entered into the deep receiving water near field surface water discharge model as shown in Fig. 6.26.

$$U_o = \frac{Q_o}{A} = \frac{0.6 \pm 0.02}{0.5} = 1.2 \pm 0.04 \text{ m/s}$$

$$\ell_o = \sqrt{0.5 \times 0.5} = 0.5 \text{ m} \quad (\text{Eq.}(6.49))$$

$$\rho_o = 998 \text{ kg/m}^3 \qquad\qquad\qquad\qquad (6.121)$$

$$\Delta\rho = 998 - (994 \pm 3) = 4 \pm 3 \text{ kg/m}^3$$

As shown in Fig. 6.26, the deterministic analysis of this problem would indicate that based on the MCL for this chemical, this outfall design indicates safe conditions in the near field for this effluent discharge $C = 4.67$ μg/L with MCL of 5 μg/L.

However, in this case there is some variability in the discharge conditions described in Example 1. To complete the analysis, the effects of this variability

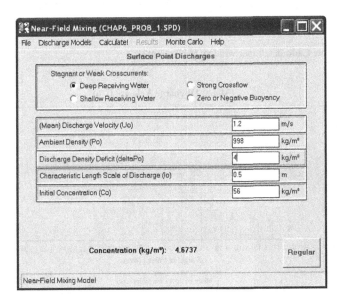

Fig. 6.26 Deterministic solution for Example 1

on the solutions will be analyzed next. For this analysis the Monte Carlo method can be utilized (Fig. 6.27). Selecting the mean discharge velocity, the density deficit and the initial concentration as the uncertain variables, we can enter the data to generate the probability density functions necessary to complete this analysis. As seen in Fig. 6.27, 5,000 random numbers will be generated, and all distributions are assumed to be normal distributions. The other input parameters shown in Fig. 6.27 are self explanatory given the data provided in Example 1. Calculating these probability density functions yields the distributions shown in Figs. 6.28–6.30 for the discharge velocity, the density deficit and the initial concentration respectively.

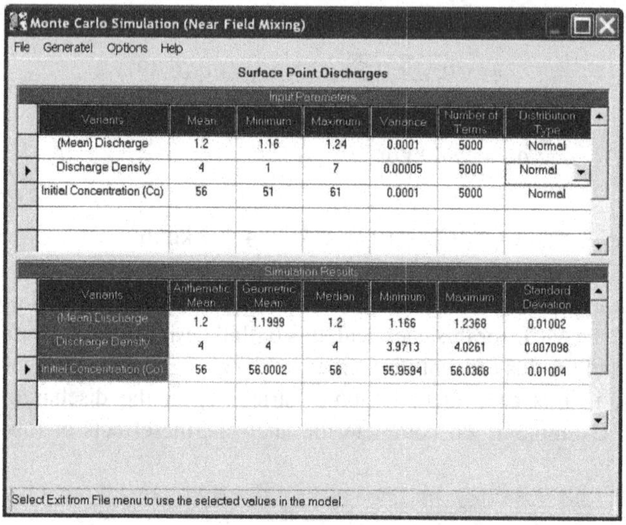

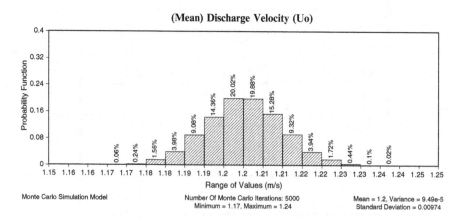

Fig. 6.27 Monte Carlo data input window for Example 1

(Mean) Discharge Velocity (Uo)

Monte Carlo Simulation Model   Number Of Monte Carlo Iterations: 5000   Mean = 1.2, Variance = 9.49e-5
                                     Minimum = 1.17, Maximum = 1.24   Standard Deviation = 0.00974

Fig. 6.28 Probability density function for discharge velocity for Example 1

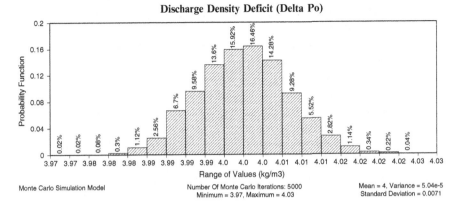

Fig. 6.29 Probability density function for density deficit for Example 1

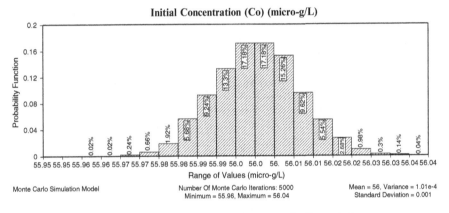

Fig. 6.30 Probability density function for initial concentration for Example 1

The user may now either perform a two stage Monte Carlo analysis by selecting all the data points or a one stage Monte Carlo analysis by selecting the mean or median for these distributions. As shown in Fig. 6.27 a two stage Monte Carlo analysis option is selected. Closing the Monte Carlo analysis window and returning to the data input window we see that the three uncertain input parameters are entered as "Monte Carlo" which indicates that there are now 5,000 parameter values in these data entry boxes. A two stage Monte Carlo analysis can now be completed for concentrations in the water body by clicking on the calculate button.

The results of this analysis can now be viewed in a graphics format, and the output files can be opened using any text editor available on the computer. Using the graphics package of the ACTS software the probability density function of the resulting concentrations can be obtained as shown in Fig. 6.32. As expected, the output in this case is also a normal distribution. The maximum concentrations expected at the site are higher than the deterministic solution provided earlier.

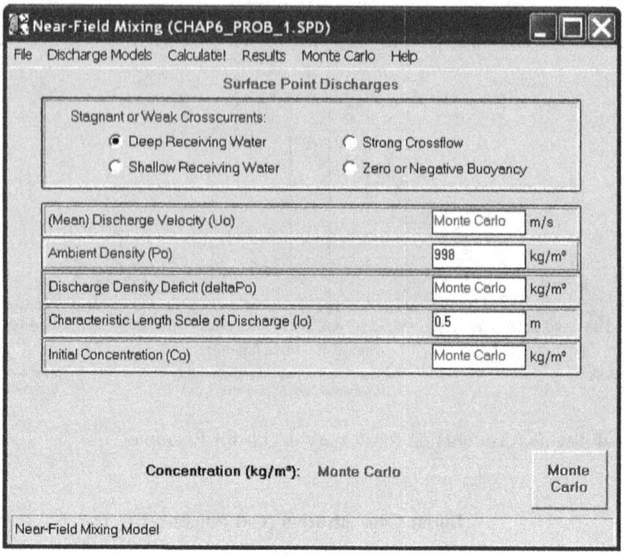

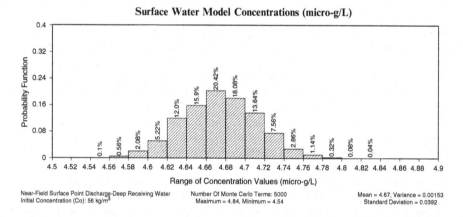

Fig. 6.31 Data entry window for Monte Carlo analysis of Example 1

Surface Water Model Concentrations (micro-g/L)

Near-Field Surface Point Discharge-Deep Receiving Water
Initial Concentration (Co): 56 kg/m³

Number Of Monte Carlo Terms: 5000
Maximum = 4.84, Minimum = 4.54

Mean = 4.67, Variance = 0.00153
Standard Deviation = 0.0392

Fig. 6.32 Probability density function for output concentrations for Example 1

However, the probability density function indicates that the maximum values will still be less than the MCLs of this effluent. This indicates that the effluent outfall design yields acceptable dilution for the problem at hand. If the concentrations exceeded the MCLs then there is a probability of exceeding the MCLs at the site. In that case this probability of exceedance can be computed using the complementary cumulative probability density function. Based on that analysis, the probability of exceedance can be calculated and the safety of the design can be evaluated based on the criteria of acceptance or rejection of the probability of exceedance level.

Example 2: An effluent is discharging to a shallow coastal water body under stagnant conditions at a depth of 6 m with a flow rate $Q_o = 10.0\text{m}^3/\text{s}$. The outfall can be characterized as an alternating diffuser with 100 nozzles of 0.15 m in diameter spaced at 0.5 m apart. The density of the effluent is given as 994 kg/m^3 and the density of the water body is 1,025 kg/m^3. The concentration of the effluent is given as $C_o = 100$ μg/L with an MCL of 10 μg/L. Will these outfall conditions indicate a safe discharge design with respect to the MCL identified for this effluent?

Solution: The data given in this problem is associated with a multiport diffuser which is discharging into a shallow coastline characterized as under stagnant sea water conditions. First we need to calculate the characteristic parameters of this problem, such as the density deficit and the discharge velocity. The calculation of these parameters yields the outcomes given in Eq. (6.122). Inserting these values into the ACTS model for near field dilution for multiport diffusers in shallow stagnant water discharge conditions and calculating the concentration gives the concentration of the effluent in the near field as $C = 15.52$ μg/L (Fig. 6.33).

$$A = \frac{\pi(0.15)^2}{4} = 0.0177 \text{ m}^2$$

$$U_o = \frac{Q_o}{A} = \frac{10.0}{100(0.0177)} = 5.65 \text{ m/s} \qquad (6.122)$$

$$\rho_o = 1025 \text{ kg/m}^3$$

$$\Delta\rho = 1025 - 994 = 31 \text{ kg/m}^3$$

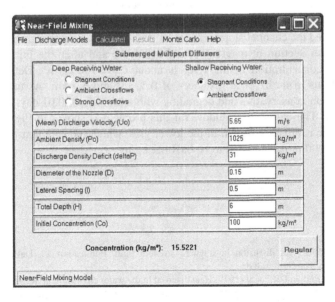

Fig. 6.33 Deterministic solution for Example 2

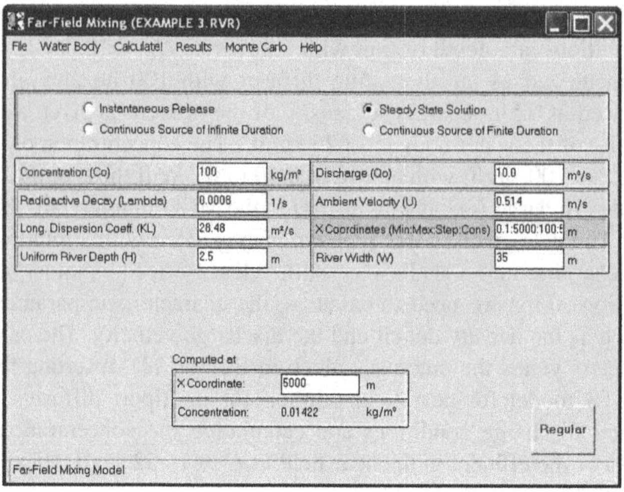

Fig. 6.34 Deterministic solution for Example 3

This outcome does not satisfy the MCL given for this effluent, which is 10 μg/L. Thus the design can be considered to be an unsatisfactory design. The uncertainty does not need to be evaluated, since the deterministic analysis already yields an unsatisfactory dilution for the conditions at the site. However, if such an uncertainty analysis is necessary, it can be constructed following the procedures described for Example 1.

Example 3: Consider a wide straight section of a river with discharge $Q = 45 \pm 5 \, \text{m}^3/\text{s}$. The width of the river is $W = 35 \, \text{m}$, the depth in the river section is $H = 2.5$ and the channel slope is $S_o = 10^{-4}$. A diffuser is discharging an effluent into this river section at a rate $Q_o = 10 \pm 3 \, \text{m}^3/\text{s}$, and after the initial dilution effects the concentration in the river is about $C_o = 100 \, \text{kg/m}^3$. Determine the concentrations in the river at distances of 0.5, 1.0 and 5.0 km. Assume that the effluent is non-conservative and degrading with a half life of 0.01 day.

Solution: The data entry window for this problem is shown in Fig. 6.34. The deterministic results are shown in the lower grid, which indicate that the concentration at 5,000 m is estimated to be 0.01422 kg/m^3.

References

Abraham G (1963) Jet diffusion in stagnant ambient fluid. Publication 29. Delft Hydraulics Laboratory, The Netherlands

Adams EE, Harleman DRF et al (1979) Heat disposal in the water environment. R. M. Parsons Lab for Water Resources and Hydrodynamics, Department of Civil Engineering, Massachusetts Institute of Technology, Cambridge, MA

Adams EE, Stolzenbach KD et al (1980) Near and far field analysis of buoyant surface discharges into large bodies of water. R.M. Parsons Lab for Water Resources and Hydrodynamics, Department of Civil Engineering, Massachusetts Institute of Technology, Cambridge, MA

Albertson ML, Dai YB et al (1950) Diffusion of Submerged Jets. ASCE Trans Am Soc Civil Eng 115:639–664

Bird RB, Stewart WE et al (2002) Transport phenomena. Wiley, New York

Brooks NH (1960) Diffusion of sewage effluent in an ocean current. In: 1st International conference on waste disposal in marine environment. Pergamon Press, New York

Chaudhry MH (2007) Open channelflow. Springer, New York

Chow VT (1959) Open-channel hydraulics. McGraw-Hill, New York

Codell RB, Key KT et al (1982) A collection of mathematical models for dispersion in surface water and groundwater. U.S. Nuclear Regulatory Commission, Washington, D.C

Elder JW (1959) The dispersion of marked fluid in turbulent shear flow. J Fluid Mech 5 (4):544–560

Fan LN, Brooks NH (1969) Numerical solution of turbulent buoyant jet problems. W. M. Keck Lab. Hydraulics and Water Resources, California Institute of Technology, Pasadena, CA, KH-R-18.

Fisher HB, List EJ et al (1979) Mixing in inland and coastal waters. Academic, New York

Fletcher HB, Dotson WL (1971) Hermes – a digital computer code for estimating regional radiological effects from the nuclear power industry. U. S. AEC, Hanford Engineering, Hanford, Washington, D.C, HEDL-TME-71-1968

Hirst EA (1972) Buoyant jets with three-dimensional trajectories. Proc Am Soc Civ Eng Hydraul J 98(HY11):1999–2014

Jirka GH (1982) Multiport diffuser for heat disposal: a summary. ASCE J Hydraul 108:1025–1068

Jirka GH, Abraham G et al (1975) Assessment of techniques for hydrothermal impact prediction. R.M. Parsons Lab for Water Resources and Hydrodynamics, Massachusetts Institute of Technology, Cambridge, MA

Jirka GH, Adams EE et al (1981) Properties of buoyant surface jets. ASCE J Hydraul 107:1467–1488

Jirka GH, Findikakis AN et al (1983) Transport of radionuclides in surface waters. Chapter 3 in Radiological assessment: a textbook on environmental dose analysis. Oak Ridge, TN

Jirka GH, Harleman DRF (1973) The mechanics of submerged multiport diffusers for buoyant discharges in shallow water. R.M. Parsons Lab for Water Resources and Hydrodynamics, Massachusetts Institute of Technology, Cambridge, MA

Lee JHW, Jirka GH (1974) Stability and mixing of a vertical round buoyant jet in shallow water. R.M. Parsons Lab for Water Resources and Hydrodynamics, Massachusetts Institute of Technology, Cambridge, MA

Lee JHW, Jirka GH (1981) Vertical round buoyant jet in shallow water. ASCE J Hydraul Eng 107 (HY12):1651–1675

NWLF (2009) Great Lakes Restoration. http://online.nwf.org/site/PageServer?pagename=glnrc_restoration_toxic_harbors. Accessed 17 June 2009

Onishi Y (1981) Sediment-contaminant transport model. J Hydraul Division – ASCE 107(9): 1089–1107

Onishi Y, Schreiber DL et al (1980) Mathematical simulation of sediment and radionuclide transport in the Clinch River, Tennessee. In: Baker RA (ed) Contaminants and sediments. Ann Arbor Science Publ. Inc., Ann Arbor, MI, 1–18, pp 393–406

Onishi Y, Trent DS (1982) Mathematical simulation of sediment and radionuclide transport in estuaries. Pacific Northwest Laboratory, Richland, Washington, D.C., NUREG/CR-2423

Roberts PJW (1977) Dispersion of buoyant wastewater discharged from outfall diffusers of finite length. W. M. Keck Laboratory, California Institute of Technology, Pasedena, CA, KH-R-35

Roberts PJW (1979) 2-Dimensional flow field of multiport diffuser. J Hydraul Division – ASCE 105(5):607–611

Rouse H, Yih CS et al (1952) Gravitational convection from a boundary source. Tellus 4:201–210

Schell WR, Sibley TH (1982) Distribution coefficient for radionuclides in aquatic environments – final summary report. College of Fisheries, University of Washington,Seattle, Washington, D.C., NUREG/CR-1869

Singh VP (1995) Watershed modeling. In: Computer models of watershed hydrology. Water Resources Publications, Colorado, pp 1–22

Stommel H (1949) Horizontal diffusion due to oceanic turbulence. J Mar Res 8(3):199–225

Stommel H (1950) Note on the deep circulation of the Atlantic Ocean. J Meteorol 7(3):245–246

Tsanis IK, Valeo C (1974) Mixing zone models for submerged discharges. Computational Mechanics Inc., Billerica, MA/Southhampton, UK

USEPA (1991) Technical support document for water quality based toxics control. Washington, D.C., EPA-505-2-90-001

USEPA (1994) Water quality standards handbook. Washington, D.C., EPA-823-B94-005a

USEPA (2009) Great lakes pollution and toxics reduction. http://www.epa.gov/glnpo/p2/pollsolu/. Accessed 17 June 2009

USNRC (1978) Liquid pathway generic study: impacts of accidental radioactivity releases to the hydrosphere from floating and land-based nuclear power plant. Nuclear Regulatory Commission, Washington, D.C

Wrenn ME, Lauer GJ et al (1972) Radioecological studies of the Hudson River. Progress Report. Consolidated Edison Company, New York

Chapter 7
Uncertainty and Variability Analysis

It is not certain that everything is uncertain.
Blaise Pascal

Statistical methods are an inseparable component of all modeling studies. Most environmental processes we have covered in this book do not lend themselves to a deterministic mode of analysis because of the inherent uncertainties involved in the parameter values that describe the physical system analyzed. These uncertainties may arise from the randomness of the natural processes, a lack of data to represent the parameters of the model or a lack of understanding of the processes that are used in the model that is built. Statistical methods may be used to account for these uncertainties. In most cases, for the completeness of the study, the deterministic analysis provided should always include a probability analysis of the occurrence or the likelihood of occurrence of the results presented. Thus, given the complex or simple nature of the models and the modeling tools developed in this book, it is important to provide an introduction to the statistical methods that are used in uncertainty and variability analysis that may be used in these models. The purpose of this chapter is to introduce the reader to these concepts.

The statistical analysis topics that will be emphasized in this review will be the frequency analysis and Monte Carlo techniques, as they were used extensively in the problems discussed in the previous chapters. However, before providing a working knowledge on frequency analysis and Monte Carlo methods it is important to review the uncertainty and variability concepts which were briefly discussed in earlier chapters of this book. The purpose of this is to provide a more structured perspective to these definitions and concepts.

A crucial step in building a mechanistic model is the development of the mathematical function which is composed of the primary unknowns of the problem and the parameters which characterize the physical system studied. There are many advantages to using mathematical models. They enable us to organize our theoretical understanding of the problem and our empirical observations about the system and they help us deduce logical implications about the system. This analysis may lead to

M.M. Aral, *Environmental Modeling and Health Risk Analysis (ACTS/RISK)*,
DOI 10.1007/978-90-481-8608-2_7, © Springer Science+Business Media B.V. 2010

improved understanding of the behavior of the system and provide a framework for testing the desirability of system modifications. In this approach there are at least three limitations which we must bear in mind while constructing and implementing a mathematical model and while evaluating the outcome of its solution. First, there is no guarantee that the time and effort devoted to modeling will return a useful result. Occasional failures are bound to occur when the level of scientific understanding of the problem solved is low. Second, utilization of the mathematical model to predict the system beyond the range of its applicability may produce unrealistic results. Finally, the most important limitation that the user should be aware of is the fact that real systems are, most of the time, too complex to be simulated by simple mathematical models, so the models constructed almost always include uncertainties which should be considered during implementation. Thus, the tendency of an investigator to treat his or her particular depiction of a problem as the best representation of reality without considering the uncertainties involved in the model may lead to unsatisfactory results.

An important issue in developing a proper model is also based on how one would evaluate the uncertainties imbedded in the model parameters. It is important to address this issue and distinguish between the relative contribution of true uncertainty in an event and uncertainty that is introduced into the predicted outcome of an event by inter-parameter variability effects (Bogen and Spear 1987). True uncertainty can be modeled using a random variable with a selected probability distribution that characterizes the variable. In contrast to true uncertainty, inter-parameter variability and its effect on a predicted outcome refers to quantities that are distributed empirically within a defined population, which may influence the outcome of an event based on a functional relation. Inter-parameter variability and true uncertainty have been referred to as, respectively, Type A uncertainty (inter-parameter variability), which is a function of the stochastic uncertainty due to input parameters of an event; and, Type B uncertainty (true uncertainty), which is due to a lack of knowledge or uncertainty about the event itself. When both Type A and Type B uncertainties are negligible we have a deterministic outcome, which is rare in environmental simulations and health risk assessment. However, there are situations in which true uncertainty (Type B) is negligible relative to inter-parameter variability (Type A) uncertainty. In these situations, the outcome of a variance propagation analysis can be represented in terms of the expected statistical variations for the parameters of the model. When neither inter-parameter variability nor true uncertainty is negligible, we have a situation in which there are multiple probability distributions representing variability, but correct distribution is unknown because of the presence of true uncertainty.

In the context of uncertainty analysis, one should also distinguish between inter-parameter uncertainty analysis and sensitivity analysis. Inter-parameter uncertainty analysis, as applied to mathematical models, involves the determination of the variation or imprecision in an output function based on the collective variation of model inputs. Sensitivity analysis, on the other hand involves the determination of changes in the model response as a result of the changes in the individual model parameters. Iman and Helton provide a discussion of three approaches which are useful for assessing uncertainty and sensitivity in mathematical models (Iman and Helton 1988). These are (i) differential analysis; (ii) response-surface replacement; and, (iii) Monte Carlo analysis. Among these techniques we will emphasize the Monte Carlo technique. However, all of these methods emphasize the use of the frequency analysis techniques. The frequency analysis is based on the definition of a certain variable or a set of variables according to a certain rule in terms of random numbers. Uncertainty and variability analysis is the technique of extracting probabilistic information from the behavior of a probabilistic definition in a mechanistic or stochastic model that represents the physical system.

In order to apply any of these methods, one can think of a model as producing an output Y, such as the contaminant distribution in an environmental pathway, which is a function of several input variables, X_k, and spatial and temporal coordinates.

$$Y = f(X_1, X_2, X_3, ..., X_k, x, y, z, t) \tag{7.1}$$

The variables X_k may represent various input parameters to the analytical model such as diffusion coefficients, velocity, source dimensions etc. in air, groundwater or surface water contaminant transport models. Most commonly, these input parameters are not known exactly due to measurement errors or other inherent spatial and temporal variability in data collection. In an unmodified Monte Carlo method, as illustrated in Fig. 7.1, each of the input parameters is represented by a probability-density function that defines both the range of values that the parameter can take and the likelihood that the parameter has a value in any subinterval of that range. In an unmodified Monte Carlo method, simple random sampling is used to select each member of the input parameter set. When a sufficient number of samples are used, the variance of the output Y reflects the combined impact of the variances in the parameters X_1, X_2 and X_k as the uncertainty is propagated through the model, given the relationship between these variables as described in the model.

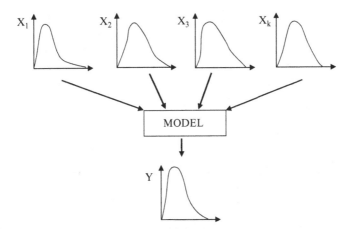

Fig. 7.1 Illustration of an unmodified Monte Carlo sampling method

Describing uncertainty in the output variable Y involves quantification of the range of Y that would be affected by the variability in the range of X_k. For this purpose the arithmetic mean value, the arithmetic or geometric standard deviation of Y and upper and lower quintile values of Y such as 5% lower bound and 95% upper bound, may be used. Convenient tools for presenting such information are the probability-density function (PDF) or the cumulative probability density function (CDF) for Y. However, the PDF or CDF of Y can often only be obtained when we have meaningful estimates of the probability distributions of the input variables X_k. If this information is missing or incomplete, one can still construct the PDF or CDF for Y based on simulated distributions of the parameters X_k. One should also be careful to characterize this variability as a screening distribution for parameter uncertainty, instead of characterizing it as a realistic representation of the uncertainty in Y.

The Monte Carlo method is particularly attractive when there are many input variables that are randomly distributed, because the computation efficiency is not a function of the dimensionality of the input vector. As discussed in the previous chapters, the ACTS software includes analytical solutions to a collection of models for air, groundwater and surface water pathway analysis. As demonstrated and discussed in the applications, these models use several input parameters, most of which may include uncertainty and variability. In order to provide the user with uncertainty and variability analysis for all models that are included in the ACTS software, a Monte Carlo analysis module is developed and incorporated for all models which are dynamically linked to all pathway modules of the software. Using this module, all or a selected set of parameters of the models used in the ACTS software can be treated as uncertain variables using statistical distributions included in the software.

The topic of statistical methods and the use of these methods in uncertainty and variability analysis is a mature field. For a more detailed description of these procedures and a review of the theory of uncertainty analysis the reader may refer to the following references: (Abdel-Magid et al. 1997; Ayyub and McCuen 1997; Benjamin and Cornell 1970; Bogen and Spear 1987; Churchill 1971; Conover 1999; Cothern and American Chemical Society. Meeting 1996; Crow et al. 1960; Fishman 1996; Günther 1961; Hoover and Perry 1989; Iman and Helton 1988; Kentel and Aral 2004, 2005; Law and Kelton 1991; Lemons 1996; Louvar and Louvar 1998; Mayer and Butler 1993; NRC 1983, 1994; Rosenbluth and Wiener 1945; Rubinstein 1981; Rykiel 1996; Saltelli et al. 2000; USEPA 1986, 1987, 1997; Zheng and Bennett 1995).

7.1 Probability Theory and Probability Distribution Functions

In this section we provide a review of the concepts in probability theory which form the basis of the Monte Carlo analysis that is discussed in the following section. The reader is referred to texts such as Hines and Montgomery (1990) and Benjamin and

Cornell (1970) and others for a more detailed review of the topics in probability and statistics in engineering applications.

Concepts in probability require an understanding of key terms such as random variable, sample, population and probability distribution. A random variable, by definition, is a numerical variable that cannot be precisely predicted. A set of observations of any random variable is called a sample. It is assumed that the sample is drawn from an infinite hypothetical population, which is defined as the complete assemblage of all the values that represent the random variable. A probability distribution is a mathematical expression that describes the probabilistic characteristics of a population. For example, a probability distribution may be used to characterize the chance that a random variable drawn from this population will fall in a specified range of numerical values of the population.

Definitions in Probability Theory: Let's suppose observations are made on a series of occasions, often termed "trials", and that on each occasion it is noted whether or not a certain event, which may be denoted by E, occurs. The "event" can be almost any observable phenomenon. For example, in the models of the ACTS software, the diffusion coefficient of an aquifer being greater than some value or the Darcy velocity magnitude being within a certain range, and so on, is going to be treated as an event of an observable phenomenon. For these events we will consider a series of "trials" which can be regarded as a part of a series. We denote the number of trials out of the first N in which E is observed by n_N. We assume that if there is a probability of occurrence E in this series, then we denote the probability of occurrence Pr[E] as follows,

$$\Pr[E] = \lim_{N \to \infty} (n_N/N) \tag{7.2}$$

Often, n_N is called the frequency and (n_N/N) the relative frequency of an event E in the first N trials. The definition given above equates probability to the limit of the relative frequency in a long series of trials. Since n_N and N are both non-negative numbers and n_N is not greater than N, the following should also hold true,

$$\left. \begin{array}{l} 0 \leqslant (n_N/N) \leqslant 1 \\ 0 \leqslant \Pr[E] \leqslant 1 \end{array} \right\} \tag{7.3}$$

If the event E occurs at every trial then $(n_N = N)$ and $(n_N/N = 1)$ for all N, and $(\Pr[E] = 1)$. Since it is possible that $(\lim_{N \to \infty}(n_N/N) = 1)$ without $(n_N = N)$ for all values of N, the converse of this statement need not hold. That is, a probability of 1 does not imply certainty. If the event E never occurs, then $(n_N = 0)$ and $(n_N/N = 0)$ for all N and thus $(\Pr[E] = 0)$. Again the converse of this statement is not necessarily true, because it is possible that $(\lim_{N \to \infty}(n_N/N) = 0)$ even though $(n_N > 0)$. Despite the above qualifications, it is useful to think of probability as being measured on a scale varying from (near) impossibility at 0 to (near) certainty at 1.

Random Variables: If we measure occurrence of a characteristic of an individual event we obtain a quantity x, which is usually a real number. We can define an event E as the event "x being less than or equal to X", in which X is a fixed real number on the probability scale. In a succession of "trials", that is, a measurement of each sequence of individuals, the event E will sometimes occur and sometimes will not. If the relative frequency of the event $E = (x \leqslant X)$ tends to a limit, whatever the value of X, or of $\Pr[x \leqslant X]$ exists for every real number X, then x is called a random variable.

The usefulness of these concepts should be very clear in the applications we discussed in earlier chapters. For example, in the Monte Carlo analysis based applications we introduced a selected set of the parameters of our application as a series of random variables, which had some statistical properties. In particular, we worked with discrete random variables and the probability distributions of random variables. A discrete random variable is a variable which takes a set of distinct ("discrete") values, and the probability of taking any one given value is greater than zero. Mathematically this can be represented as,

$$\Pr[x = X] = p_i; \ i = \dots, -1, 0, 1, \dots \tag{7.4}$$

If the number of possible values is large, it may be more convenient to represent the measurements by continuous random variables, which one can conceive of as having the possibility of taking any value over some interval.

Using the definition given above, it is possible to define the cumulative probability density function. The value of $\Pr[x \leq X]$ for a given random variable is a function of X. We denote this function in the form,

$$\Pr[x \leqslant X] = F(x)_{x=X} \tag{7.5}$$

where $F(x)$ is called the cumulative probability density function of the random variable x. Now suppose x to be a discrete variable with $\Pr[x = X_i] = p_i$, with $(\dots X_{-2} < X_{-1} < X_o < X_1 < X_2 <, \dots)$. Since the events $(x = X_i)$ are mutually exclusive, because x takes only one value at a time, then $\sum p_i = 1$. Based on these definitions, we can now evaluate the value of $\Pr[x \leqslant X]$, in which $X_{j-1} \leqslant X \leqslant X_j$. The event $(x \leqslant X)$ is then the logical sum of the mutually exclusive events $(x = X_m)$ for all $(m \leqslant j - 1)$. Thus we have,

$$F(x)_{x=X} = \sum_{m}^{j-1} p_l \qquad (X_{j-1} \leqslant X < X_j) \tag{7.6}$$

Similarly,

$$F(x)_{x=X} = \sum_{m}^{j-1} p_l + p_j \qquad (X_j \leqslant X < X_{j+1}) \tag{7.7}$$

Thus, $F(x)$ remains constant for $(X_{j-1} \leqslant x < X_j)$, and jumps by an amount $p_j (= \Pr[x = X_j])$ at the point $(x = X_j)$. $F(x)$ is therefore a step-function as seen in the bar graphs of the Monte Carlo analysis output in the ACTS software. This is typical of cumulative distributions for any discrete random variable.

We have devoted some attention to the definition of cumulative probability density functions. The properties of this function can be given as,

$$
\begin{aligned}
&(i) \quad 0 \leqslant F(x) \leqslant 1 \\
&(ii) \quad F(x) \text{ is a nondecreasing function of } x \\
&(iii) \quad \lim_{x \to \infty} F(x) = 1; \qquad \lim_{x \to -\infty} F(x) = 0
\end{aligned}
\tag{7.8}
$$

Most theoretical probability distributions are characterized by parameters such as the mean, standard deviation and skewness. These parameters cannot be determined precisely since we do not know all the values that may be included in the entire population. However, we can estimate these parameters based on the observations on the sample, from which come the definitions given below.

Expected Values: Consider a discrete random variable z,

$$
\Pr[z = Z_i] = p_j \; ; \; \sum_j p_j = 1
\tag{7.9}
$$

Let $(z_1, z_2, z_3, \ldots, z_n)$ be successive independent random variables, each distributed like z. They model repeated observations of some parameter "distributed as z." Let's denote the number of these z_i's which are equal to Z_j by n_j, so that $\sum n_j = N$. Then the arithmetic mean of the z_i's is,

$$
\bar{z} = \frac{1}{N} \sum_j n_j Z_j
\tag{7.10}
$$

Now let's consider what happens to this arithmetic mean as N increases; that is, as we consider more and more observations. We would have,

$$
\lim_{N \to \infty} \bar{z}_N = \sum_j Z_j \lim_{N \to \infty} \frac{n_j}{N} = \sum_j Z_j p_j
\tag{7.11}
$$

which is based on the original definition of the probability given earlier. We see, therefore, that the arithmetic mean of z_i's tend to be a fixed number $\sum Z_j p_j$. We can expect this quantity to be approximated by the average (arithmetic mean) of a large number of z_i's. This is also identified as the expected value of z. To indicate its relationship to the random variable z the expected value of z is written as $E(z)$, but it should be remembered that it is not a mathematical function of z. It is in fact a fixed number, and is a property of z just as the cumulative probability density function of z is a property of z. Note that $E(z)$ is a constant as given below,

$$E(z) = \sum_j Z_j \Pr[z = Z_j] \tag{7.12}$$

For completeness we introduce the definition of an expected value of a continuous variable z, with a probability density function $p(z)$ as,

$$E(z) = \int_{-\infty}^{\infty} zp(z)dz \tag{7.13}$$

For discrete variables, we define the expected value of any single-valued function $f(x)$ of x in the following manner (note that $f(x)$ is an ordinary mathematical function of x – unlike $F(x)$, $E(x)$ and $p(x)$). If $x = X_j$ then $f(x) = f(X_j)$ and thus,

$$E[f(x)] = \sum_j p_i f(X_j) \tag{7.14}$$

Moments: In the definitions given above, procedures to calculate the expected value of a function $f(x)$ of a random variable x are given. The expected values of certain functions are of special importance. These include the cases where $f(x) = (x - A)^r$, in which A and r are constants. For example $E[(x - A)^r]$ is called the rth moment of x about A. If A is equal to zero, this definition implies the rth moment about zero, often simply called the rth moment and denoted by $\mu'_r(x)$. If $A = E(x)$ the moment is called the rth central moment and is denoted by $\mu_r(x)$. If no confusion is likely to arise the term x is usually omitted from these expressions and the symbols μ'_r and μ_r are used for the rth moments about zero and the central moment, respectively.

$$\mu'_1 = E(x) \tag{7.15}$$

As we have already noted earlier and,

$$\mu_2 = E\left[(x - E(x))^2\right] \tag{7.16}$$

is termed the variance of x, often denoted as var(x). The larger the variation of x from its expected value $E(x)$, the larger the μ_2 will be. The variance measures the "variability" of x. If a measure of variability is required, the quantity σ which is called the *standard deviation* and has the same dimensions as the random variable x, can be used,

$$\sigma = \sqrt{\mu_2} \tag{7.17}$$

Standardization: Let $(y = a + bx)$ be a linear function of x, in which a and b are constants. Then the expected value of y is,

$$E(y) = a + bE(x) \tag{7.18}$$

The variance of y is,

$$\text{var}(y) = b^2 \text{var}(x) \tag{7.19}$$

In general the rth central moments of x and y are related by the equations,

$$\mu_r(y) = b^r \mu_r(x) \tag{7.20}$$

If we choose,

$$\left.\begin{array}{l} a = -E(x)/\sqrt{\text{var}(x)} \\ b = 1/\sqrt{\text{var}(x)} \end{array}\right\} \tag{7.21}$$

then,

$$y = \frac{(x - E(x))}{\sqrt{\text{var}(x)}} \tag{7.22}$$

has the expected value of zero and standard deviation 1. Such a variable is called the standardized variable corresponding to x. In particular, the standardized variable corresponding to x has the same shape factors as x. Standardizing affects the mean and standard deviation, but not the moment ratios.

7.2 Probability Density Functions

We now review the properties of the probability density functions (PDFs) that may be used in Monte Carlo analysis (Benjamin and Cornell 1970; Hines and Montgomery 1990; Snedecor and Cochran 1967). These PDFs can be used to generate input parameter distributions for all relevant parameters of a simulation model. Operational steps to implement this data generation process in the ACTS software are described in Appendix 3.

Normal Distribution: The normal distribution is the most commonly used distribution in applied statistics. It was studied extensively in the eighteenth century. Through historical error, it has been attributed to Gauss, whose first published reference to it appeared in 1809, and thus the term Gaussian distribution is also frequently associated with the normal distribution.

A random variable X is said to have a normal distribution with mean $\mu(-\infty < \mu < \infty)$ and variance $\sigma^2 > 0$ if it has the probability density function,

$$p(x) = \frac{1}{\sigma\sqrt{2\pi}} \exp\left(-\frac{1}{2}\left(\frac{x-\mu}{\sigma}\right)^2\right); \quad -\infty < x < \infty \quad (7.23)$$

The distribution is illustrated in Fig. 7.2. Note that the value of $p(x)$ approaches zero as x approaches $-\infty$ or $+\infty$. The distribution is symmetric about $x = \mu$ and $x = \mu \pm \sigma$ gives us the two points of the inflection of the distribution curve.

The normal distribution is used extensively so that the shorthand notation $X \approx N(\mu, \sigma^2)$ is often used to indicate that the random variable X is normally distributed with mean μ and variance σ^2. The normal distribution has several important properties as given below.

$$
\begin{aligned}
&i. \quad \int_{-\infty}^{\infty} p(x)dx = 1 \\
&ii. \quad p(x) \geqslant 0 \quad \text{for all } x \\
&iii. \quad \lim_{x \to \infty} p(x) = 0 \text{ and } \lim_{x \to -\infty} p(x) = 0 \\
&iv. \quad p[(x+\mu)] = p[-(x-\mu)]. \text{ The density is symmetric about } \mu. \\
&v. \quad \text{The maximum value of } p \text{ occurs at } x = \mu. \\
&vi. \quad \text{The points of inflection of } p \text{ are at } x = \mu \pm \sigma.
\end{aligned}
\quad (7.24)
$$

Using the definitions given earlier, the expected value $E(x)$ and variance var(x) can be given as,

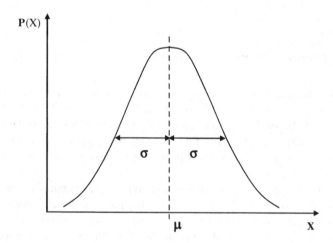

Fig. 7.2 The normal distribution

$$E(x) = \int_{-\infty}^{\infty} xp(x)dx$$

$$\text{var}(x) = \int_{-\infty}^{\infty} [x - E(x)]^2 p(x)dx \tag{7.25}$$

$$\text{var}(x) = E(x^2) - [E(x)]^2$$

From Eqs. (7.23) and (7.25) and the definitions given earlier, it can be shown that,

$$E(x) = \frac{1}{\sigma\sqrt{2\pi}} \int_{-\infty}^{\infty} x \exp\left(-\frac{(x-\mu)^2}{2\sigma^2}\right) dx = \mu$$

$$\text{var}(x) = \frac{1}{\sigma\sqrt{2\pi}} \int_{-\infty}^{\infty} x^2 \exp\left(-\frac{(x-\mu)^2}{2\sigma^2}\right) dx - \mu^2 = \sigma^2 \tag{7.26}$$

The normal distribution is then symmetrical with respect to the mean, median and the mode, which are all equal to μ. The standard deviation σ or the square root of the variance is a measure of the distance from μ to each of the two points of inflection.

The distribution given in Eq. (7.23) represents a family of curves which depend on the two parameters, μ and σ. Both of these parameters have the same unit as the parameter they represent, whether it is in inches, degrees centigrade, pounds per square inch, percentage concentrations, etc. In order to determine what proportion of the distribution is beyond $x = x_0$, it is necessary to integrate the distribution over the range $x > x_0$. This involves numerical integration and becomes quite tedious. To avoid this we make use of the standard normal distribution in which $\mu = 0$ and $\sigma = 1$. The area under this curve has been tabulated for a wide range of values of x. We can then solve simple problems relating to the proportion of this area by transforming to the case $\mu = 0$ and $\sigma = 1$, and by finding the appropriate area from the tables. This is done by standardizing x by the transformation,

$$u = \frac{x - \mu}{\sigma} \tag{7.27}$$

The mean and standard deviation of u are 0 and 1 respectively, and its probability density function is given as,

$$p(u) = \frac{1}{\sqrt{2\pi}} \exp\left(-\frac{u^2}{2}\right); \quad -\infty < u < \infty \tag{7.28}$$

and the cumulative probability density function is given as,

$$F(x) = \frac{1}{\sqrt{2\pi}} \int_{-\infty}^{x} \exp\left(-\frac{u^2}{2}\right) du \tag{7.29}$$

Now we can introduce a convenient notation. Let the symbol $N(\mu, \sigma)$ refer to the normal distribution with mean μ and the standard deviation σ. Using these definitions, the procedure in solving practical problems involving evaluation of the cumulative normal probabilities becomes very simple. For example, suppose $X \approx N(100, 4)$ and we wish to find the probability that X is less than or equal to 104; that is $P(X \leqslant 104) = F(104)$. Since the standard normal variable is,

$$Z = \frac{X - \mu}{\sigma} \tag{7.30}$$

We can standardize the point of interest $x = 104$ to obtain,

$$z = \frac{x - \mu}{\sigma} = \frac{104 - 100}{2} = 2 \tag{7.31}$$

Now the probability that the standard normal random variable Z is less than or equal to 2 is equal to the probability that the original normal random variable X is less than or equal to 104. Expressed mathematically,

$$\left. \begin{array}{l} F(x) = p\left(\frac{x - \mu}{\sigma}\right) = p(u) \\ \text{or} \\ F(104) = p(2) \end{array} \right\} \tag{7.32}$$

Cumulative standard normal probabilities for various values of u can be obtained from tables (Benjamin and Cornell 1970; Conover 1999; Hines and Montgomery 1990; Snedecor and Cochran 1967). For example, $p(2) = 0.9772$ which is the answer we were looking for.

Lognormal Distribution: In environmental modeling the parameters of the models we use cannot have negative values due to the physical definitions of these parameters. For these cases associating the parameter distributions with a lognormal distribution is a better choice, as it automatically eliminates the possibility of negative values. The lognormal distribution is the distribution of a random variable whose logarithm follows the normal distribution. It has been applied in a wide variety of fields from the physical sciences to the social sciences to engineering. Some practitioners in the environmental engineering area hold the idea that the lognormal distribution is as fundamental as the normal distribution. It arises from the combination of random terms of a multiplicative process.

We consider a random variable X with a range $(0 < x < \infty)$, in which $Y = \ln X$ is normally distributed with mean μ_Y and variance σ_Y^2. The probability density function of X is,

$$p(x) = \frac{1}{x\sigma_Y\sqrt{2\pi}} \exp\left(-(1/2)\left[\ln x - \mu_Y)/\sigma_Y\right]^2\right) \qquad 0 \leqslant x < \infty$$

(7.33)

$$p(x) = 0 \qquad\qquad\qquad otherwise$$

The definition sketch of a lognormal distribution is shown in Fig. 7.3. The mean and the variance of the lognormal distribution are,

$$E(x) = \mu_X = \exp\left(\mu_Y + \frac{1}{2}\sigma_Y^2\right)$$

$$\mathrm{var}(x) = \sigma_X^2 = \mu_X^2\left[\exp(\sigma_Y^2) - 1\right]$$

(7.34)

In some applications of the lognormal distribution, it is important to know the values of the median and the mode of the distribution. The median, which is the value $\tilde{x}$ such that $p(X \leqslant \tilde{x}) = 0.5$, is,

$$\tilde{x} = \exp(\mu_Y)$$

(7.35)

The mode is the value of x for which p(x) is maximum. For the lognormal distribution the mode is,

$$Mode = \exp(\mu_Y - \sigma_Y^2)$$

(7.36)

In Fig. 7.3 the relative location of the mean, median, and the mode for the lognormal distribution is indicated approximately.

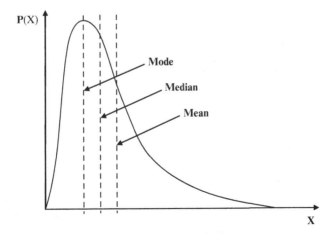

Fig. 7.3 Lognormal distribution

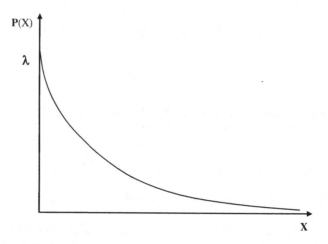

Fig. 7.4 Exponential distribution

Exponential Distribution: The exponential distribution is represented with a probability density function, which is decreasing as the value of the variant increases (Fig. 7.4). The probability density function is given as,

$$\left.\begin{array}{l} p(x) = \lambda \exp(-\lambda x); \ x \geqslant 0 \\ p(x) = 0 \quad \textit{otherwise} \end{array}\right\} \tag{7.37}$$

in which the parameter λ is a real, positive constant. The expected value and the variance of the exponential distribution can be given as,

$$\begin{aligned} E(x) &= \mu_x = \lambda^{-1} \\ \text{var}(x) &= \sigma_x^2 = \lambda^{-2} \end{aligned} \tag{7.38}$$

The cumulative distribution function for the exponential distribution can be given as,

$$\begin{aligned} F(x) &= \int_0^x \lambda \exp(-\lambda t) \ dt = 1 - \exp(-\lambda x); \ x \geqslant 0 \\ F(x) &= 0; \ x < 0 \end{aligned} \tag{7.39}$$

Gumbel Distribution: In the literature this distribution is also known as the extreme value Type I distribution. The theory of extreme values considers the distribution of the largest (or smallest) observations occurring in each group of repeated samples. The distribution of the n_1 extreme values taken from n_1 samples with each sample having n_2 observations depends on the distribution of the $n_1 n_2$

total observations. Gumbel was the first to employ extreme value theory in hydrologic analysis. Later on it was shown that the Gumbel distribution is essentially a lognormal distribution with constant skewness. The CDF of the density function for this distribution takes the form,

$$F(x) = \exp(-\exp[-\alpha(x - u)])$$ (7.40)

Parameters α and u are given in terms of the mean and the standard deviation.

Triangular Distribution: The triangular distribution is a relatively simple probability distribution defined by the minimum value, maximum value, and the most frequent value. Figure 7.5 shows an example of a triangle probability density function. The cumulative distribution for values of x less than the most frequent value x_m is given by,

$$F(x) = \frac{(x - x_1)^2}{(x_m - x_1)(x_2 - x_1)}$$ (7.41)

in which x_1, x_2 and x_m are the minimum, maximum and most frequent values respectively.

Uniform Distribution: The continuous uniform distribution is encountered in applied statistics mainly in situations where any of the values in a range of values is equally likely (Fig. 7.6). The probability density function for the uniform distribution is,

$$p(x) = \frac{1}{\beta - \alpha} \qquad \alpha \leqslant x \leqslant \beta$$
$$p(x) = 0 \qquad\qquad otherwise$$ (7.42)

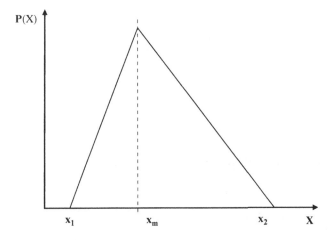

Fig. 7.5 Triangular distribution

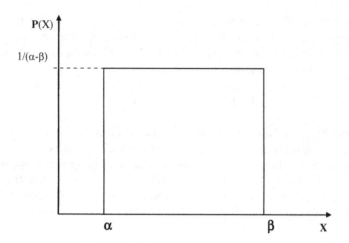

Fig. 7.6 Uniform distribution

where α and β are real constants. The mean and the variance of the uniform distribution are,

$$E(x) = \mu_x = \frac{\beta + \alpha}{2}$$
$$\mathrm{var}(x) = \sigma_x^2 = \frac{(\beta - \alpha)^2}{12}$$

(7.43)

and for a uniformly distributed random variable, the cumulative probability density function $F(x) = P(X \leqslant x)$ is given as,

$$F(x) = 0 \qquad\qquad\qquad x \leqslant \alpha$$

$$F(x) = \int_\alpha^x \frac{dx}{\beta - \alpha} = \frac{x - \alpha}{\beta - \alpha} \qquad \alpha \leqslant x \leqslant \beta$$

(7.44)

$$F(x) = 1 \qquad\qquad\qquad x \geqslant \beta$$

7.3 Monte Carlo Analysis

Monte Carlo simulation or analysis can be defined as a process that utilizes random numbers to define the parameters in a model in an effort to identify the stochastic behavior of the model with respect to the parameters considered. The concept relies on the observation that the statistical properties of the numerical distribution of the output can be interpreted as if the simulation results are a sample from the population of the output. Typically, as the number of samples and model replications is increased in a Monte Carlo analysis, the outcome is more likely to converge to a certain statistical distribution with stable parameter values. In this process, the number of simulations may be increased until a desired precision in the predicted output distribution is obtained. Alternatively, the central limit theorem can be used to estimate the number of runs required to obtain a desired degree of confidence in the expected value of the output.

The computational steps that are required for performing Monte Carlo simulations are: (i) definition of a system using a model; (ii) selection of parameters of the model that are uncertain; (iii) generation of random distributions for these variables; (iv) evaluation of the model using random variables that are generated, i.e. two stage Monte Carlo analysis, or using the mean or other representative values of the parameter distributions generated in the solution of the model, i.e. one-stage Monte Carlo analysis; (v) statistical analysis of the resulting behavior; and, (vi) study of the confidence levels of the outcome with respect to some compliance criteria.

As stated earlier, consider a model that is represented as $y = f(\mathbf{x})$, where $\mathbf{x} = \{x_1, x_2, x_3, \ldots, x_n\}$. According to the steps outlined above, the first step in the Monte Carlo analysis is to estimate the probability distribution function $g(\mathbf{x})$, that characterizes the joint uncertainty in the input parameters $\mathbf{x}$, If the inputs are assumed to be independent, then the joint distribution is characterized by the individual, marginal probability distribution functions for each $g_1(x_1)$, $g_2(x_2), g_3(x_3), \ldots, g_n(x_n)$. If the parameters are independent, each probability distribution can be defined by its mean and standard deviation or other parameters that would characterize the distribution. Thus, by defining the model input uncertainty distributions in this manner, the Monte Carlo analysis proceeds by sampling a random vector of inputs, $\mathbf{x}^j = \{x_1^j, x_2^j, x_3^j, \ldots, x_n^j\}$, which would identify the *j*th scenario in a model run. Using these scenarios the model is run m times yielding m outputs for the function y. This yields a sample of m outputs $\{y^1, y^2, y^3, \ldots, y^m\}$. This procedure represents a two stage Monte Carlo analysis, which is usually the preferred approach. This output is usually presented as histograms or cumulative distribution functions. The statistical behavior of the output, such as the mean, the standard deviation, or fractiles of the distribution can be estimated using standard statistical techniques. These estimates, when compared to a deterministic output or a compliance criteria, yield the statistical interpretation of the model predictions based on the uncertainty introduced in the model.

7.4 Interpretation of the Results of the Monte Carlo Analysis

As an example, using the randomly generated parameter values, the parent model may be used to estimate the values of concentrations at a point down-gradient from the waste facility or a contamination source. Using these values, some practical calculations can be made for the purposes of exposure analysis. Let C_r represent the normalized concentration at the exposure location calculated by the parent model based on an up-gradient source concentration, and let C_h represent the health based maximum concentration for the chemical at the receptor. Based on these values the maximum allowable concentration at the up-gradient source C_s can be calculated as,

$$C_s = \frac{C_h}{C_r} \qquad (7.45)$$

The maximum allowable source concentration defined by Eq. (7.45) is the concentration at the source for which the down-gradient receptor concentration does not exceed the health based maximum allowable concentration. This relationship may also be written as follows,

$$\frac{1}{C_r} = \frac{C_s}{C_h} \qquad (7.46)$$

Equation (7.46) indicates that the reciprocal of the computed normalized concentration at the receptor represents the maximum allowable ratio of the source concentration to the health based maximum allowable concentration. Thus, for a simulated normalized concentration of $C_r = 0.02$ mg/L at the receptor location, the up-gradient source concentration can be 50 times the health based maximum allowable concentration.

It is possible to repeat this calculation for a Monte Carlo analysis, in which case the results will be interpreted in a probabilistic manner. For each chemical, application of the Monte Carlo method results in an array of values for the normalized concentration at the receptor. Each of these predicted results represents a feasible solution for the scenario considered. These values can be statistically analyzed to derive the cumulative probability distribution function, which can be obtained from the ACTS software. The cumulative probability distribution, together with the health based maximum allowable concentration C_h and the calculation discussed above, provide the information necessary to calculate the maximum allowable source concentration. In particular, they give the value of source concentration C_s that leads to percentile realizations in compliance, i.e. the receptor concentration that is less than or equal to the health based maximum allowable concentration,

$$C_s = \frac{C_h}{C_p} \qquad (7.47)$$

in which C_p is the p-percentile concentration obtained from the cumulative distribution function of the downgradient receptor concentration.

As stated above, the Monte Carlo simulation provides an estimate of C_p, the p-percentile concentration obtained from a sample of n simulations. Since the sample size is finite, the estimates of C_p will be uncertain with the degree of uncertainty decreasing with increasing sample size.

A quantitative estimate of the uncertainty in the estimate of C_p can be obtained by computing a confidence interval around the estimate C_p. The upper and lower bounds of this confidence interval can be defined as,

$$Probability\ (C_L < C_p < C_U) = 1 - \alpha \tag{7.48}$$

where C_L is the lower bound of the confidence interval (mg/l), C_U is the upper bound of the confidence interval (mg/l) and α is the dimensionless measure of the significance level. The interval $\{C_L - C_U\}$ is usually referred to as the $100(1 - \alpha)$ confidence interval and implies that the true value of the estimate of the quantile C_p lies within the interval C_L to C_U with a probability of $100(1 - \alpha)$. The confidence interval is estimated using the binomial distribution as described below.

Define the Bernoulli random variable indicator, I, such that,

$$I = \begin{cases} 1 & if \quad C_r < C_p \\ 0 & if \quad C_r > C_p \end{cases} \tag{7.49}$$

Also, define the random variable K equal to the number of trials for which $I = 1$ from the Monte Carlo Simulations. The random variable K is then binomially distributed with a mean of $(n \times p)$ and a variance of $(n \times p)(1 - p)$, i.e.,

$$Probability\ \left\{ \sum_{i=1}^{n} I_i = k \right\} = \frac{n!}{k!(n-k)!} p^k (1-p)^{n-k} \tag{7.50}$$

in which n is the number of independent realizations of C_r computed by the Monte Carlo simulation corresponding to n independent realizations of I.

The probability that K is less than a given positive integer is also the probability that $C_K < C_p$, in which C_K is the Kth smallest simulated value of concentration. Thus a confidence interval on K, based on the binomial distribution, can be used to establish the confidence interval C_p. A procedure can now be defined for C_p that essentially involves looking up values in a table of cumulative probabilities of the binomial distribution to find the k values corresponding to probabilities of $(\alpha/2)$ and $(1 - \alpha/2)$ (Conover 1999; Iman and Conover 1980). For sample sizes greater than 20, an alternative procedure may also be defined based on the normal approximation to binomial probabilities for large n. This approximation requires the calculation of two values, o and s,

$$o = np + z_{\alpha/2}[np(1-p)]^{1/2} \tag{7.51}$$

$$s = np + z_{1-\alpha/2}[np(1-p)]^{1/2} \tag{7.52}$$

where $z_{\alpha/2}$ and $z_{1-\alpha/2}$ are quantiles of the standard normal (mean $= 0$, variance $= 1$) distribution. Note that $z_{\alpha/2} = -z_{1-\alpha/2}$. Rounding up the values of o and s to the next higher integers is recommended. The corresponding values of C_o and C_s are then estimated as the oth and s^{th} smallest values of C_r. The confidence interval is then of the form,

$$Probability \; \{C_{(o)} \leqslant C_p \leqslant C_{(s)}\} = (1 - \alpha)100\% \qquad (7.53)$$

Unfortunately, there is no way to calculate the number of Monte Carlo simulations required to establish a confidence interval on C_p a priori, with a given width without first having a very good estimate of the shape of $F_{C_r}(C_r)$ in the region of C_p. It is easier, however, to calculate the number of realizations required to bring the ranks o and s as close together as required. For realistically large n (typically in the hundreds or more), the normal approximation applies and n can be found by fixing $(s - o)$ to the width desired and solving for n using Eqs. (7.51) and (7.52). Thus,

$$n = \frac{(s - o)^2}{4p(1 - p)(z_{1-\alpha/2})^2} \qquad (7.54)$$

Notice that the smaller the specified range $(s - o)$, the smaller the number of realizations required. This should not be counterintuitive, because a fixed confidence interval on the C_r scale should naturally contain more simulated values, and thus a large value for $(s - o)$ if more simulations are performed. An alternative criterion for specifying n might be the fraction of the range of simulated ranks to be covered by the confidence interval. Thus,

$$\frac{s - o}{n} = f = \frac{2z_{1-\alpha/2}[np(1 - p)]^{1/2}}{n} \qquad (7.55)$$

or solving for n,

$$n = \frac{4p(1 - p)(z_{1-\alpha/2})^2}{f^2} \qquad (7.56)$$

These are typical alternatives for the interpretation of Monte Carlo analysis outcome. The reader is referred to Chapter 9 for a particular application in which this methodology was used.

7.5 When an Uncertainty Analysis Will be Useful and Necessary

In some cases, an uncertainty analysis may not be necessary. This decision must be based on the confidence one can place on the data used in a particular application and also the level confidence necessary for the outcome. If the decision to be made on the case is critical to the health and safety of populations, or ecological risk decisions are to be based on the outcome of the analysis, then the decision makers need to be informed on the uncertainty in the outcome or the confidence levels the outcome has in reference to the inherent uncertainty of the data. The following guidelines are suggested in the literature (NCRP 1996).

    i. If conservatively biased screening calculations indicate that the risk from possible exposure is clearly below regulatory risk levels of concern, a quantitative uncertainty analysis may not be necessary.

    ii. If the cost of an action required to reduce the exposure is low, a quantitative uncertainty analysis on the dose or risk may not be necessary.

    iii. If the data characterizing the nature and the extent of contamination at a site are inadequate to permit even a bounding estimate (an upper and lower estimate of the expected value), a meaningful quantitative analysis cannot be performed. Under these conditions, it may not be feasible to perform a probabilistic exposure or risk estimate, unless the assessment is restricted to a non-conservative screen that is designed not to overstate the true value. A non-conservative screen is obtained by eliminating all assumptions that are known to be biased towards an upper bound estimate. In this case, if the non-conservative screen suggests that potential exposures or risks will likely be above regulatory criteria, plans for remediation or exposure control can begin even before the full extent of the contamination is characterized or a full uncertainty analysis is conducted. Once the extent of contamination is more clearly characterized, a quantitative uncertainty analysis would be useful in decision making.

In contrast to the conditions identified above, a quantitative uncertainty analysis of exposure and risk based events will be necessary in the following cases (NCRP 1996):

    i. A quantitative uncertainty analysis should be performed when an erroneous result in exposure or a risk event may lead to unacceptable consequences. Such situations are likely to occur when the cost of regulatory or remedial action is high and the potential health risk associated with exposure is significant.

    ii. A quantitative uncertainty analysis is needed whenever a realistic rather than conservative estimate is needed. This is especially the case for epidemiologic investigations that attempt to evaluate the presence of a dose and exposure response. Failure to include uncertainties in these cases may lead to misclassification of the exposed individual and thus decrease the power of analysis.

    iii. A quantitative uncertainty analysis should also be used to set priorities for the assessment components for which additional information will likely lead to improved confidence in the estimate of the risk. Without an uncertainty analysis, inconsistencies in the application of conservative assumptions may obscure those assessment components that dominate the uncertainty in the estimation of the risk. With uncertainty analysis, the variables that contribute most to the overall uncertainty in the results are readily identified. These are the variables warranting the highest priority for further investigation and are those for which an increase in the base of knowledge will effectively reduce the uncertainty in the calculated results.

These criteria may be used as general guidelines in making a decision on the necessity of performing an uncertainty analysis in an application.

References

Abdel-Magid IS, Mohammed A-WH et al (1997) Modeling methods for environmental engineers. CRC Lewis, Boca Raton, FL, 518 pp

Ayyub BM, McCuen RH (1997) Probability, statistics, and reliability for engineers. CRC Press, Boca Raton, FL, 514 pp

Benjamin JR, Cornell CA (1970) Probability, statistics, and decision for civil engineers. McGraw-Hill, New York, 684 pp

Bogen KT, Spear RC (1987) Integrating uncertainty and interindividual variability in environmental risk assessments. Risk Anal 7:427–436

Churchill RV (1971) Operational mathematics. McGraw-Hill, New York, 481 pp

Conover WJ (1999) Practical nonparametric statistics. Wiley, New York, 584 pp

Cothern CR, American Chemical Society Meeting (1996) Handbook for environmental risk decision making: values, perceptions and ethics. Lewis, Boca Raton, FL, 408 pp

Crow EL, Davis FA et al (1960) Statistics manual. New York, Dover, 288 pp

Fishman GS (1996) Monte Carlo: concepts, algorithms, and applications. Springer, New York

Günther O (1961) Environmental information systems. Springer, Berlin

Hines WW, Montgomery DC (1990) Probability and statistics in engineering and management science. Wiley, New York

Hoover SV, Perry RF (1989) Simulation. Addison-Wesley, Reading, MA, 696 pp

Iman RL, Conover WJ (1980) Small sample sensitivity analysis techniques for computer-models, with an application to risk assessment. Commun Stat a-Theor 9(17):1749–1842

Iman RL, Helton JC (1988) An Investigation of uncertainty and sensitivity analysis techniques for computer models. Risk Anal 8:71–90

Kentel E, Aral MM (2004) Probabilistic-fuzzy health risk modeling. Stoch Env Res Risk A 18(5):324–338

Kentel E, Aral MM (2005) 2D Monte Carlo versus 2D fuzzy Monte Carlo health risk assessment. Stoch Env Res Risk A 19(1):86–96

Law AW, Kelton WD (1991) Simulation modeling and analysis. McGraw-Hill, New York

Lemons J (ed) (1996) Scientific uncertainty and environmental problem solving. Blackwell Science, Cambridge, MA

Louvar JF, Louvar BD (1998) Health and environmental risk analysis: fundamentals with applications. Prentice Hall PTR, Englewood Cliffs, NJ, 678 pp

Mayer DG, Butler DG (1993) Statistical validation. Ecol Model 68(1–2):21–32

NCRP (1996) A guide for uncertainty analysis in dose and risk assessment related to environmental contamination. National Council on Radiation Protection and Measurements, Bethesda, MD, NCRP-14, 54 p

NRC (1983) Risk assessment in the Federal Government: managing the process. N. R. Council. National Academy Press, Washington, D.C

NRC (1994) Science and Judgement in risk assessment. N. R. Council, National Academy Press, Washington, D.C

Rosenbluth A, Wiener N (1945) The role of models in science. Philos Sci XII(4):316–321

Rubinstein RY (1981) Simulation and the Monte Carlo method. Wiley, New York, 278 pp

Rykiel EJ (1996) Testing ecological models: the meaning of validation. Ecol Model 90(3):229–244

Saltelli A, Chan K et al (2000) Sensitivity analysis. Wiley, West Sussex, England, 475 pp

Snedecor GW, Cochran WG (1967) Statistical methods. The Iowa State University Press, Ames, IA

USEPA (1986) Guidelines for the health risk assessment of chemical mixtures. Office of Health and Environmental Assessment, Washington, D.C

USEPA (1987) The risk assessment guidelines of 1986. Office of Health and Environmental Assessment, Washington, D.C

USEPA (1997) Guiding principles for Monte Carlo analysis. Washington D.C., Office of Health and Environmental Assessment. EPA/630/R-97/001

Zheng C, Bennett GD (1995) Applied contaminant transport modeling: theory and practice. Van Nostrand Reinhold York, New York, 440 pp

Chapter 8
Health Risk Analysis

The most important pathological effects of pollution are extremely delayed and indirect.
Rene Dubos

The Centers for Disease Control and Prevention defines public health assessment as: "A systematic approach to collecting information from individuals that identifies risk factors, provides individualized feedback, and links the person with at least one intervention to promote health, sustain function and/or prevent disease" (CDC 2010). Within this general framework, a health risk assessment is defined as "An analysis that uses information about toxic substances at a site to estimate a theoretical level of risk for people who might be exposed to these substances". Similarly, the National Academy of Sciences defines health risk assessment as a process in which information is analyzed to determine if an environmental hazard might cause harm to exposed persons (NRC 1983). In essence a health risk assessment study provides a comprehensive scientific estimate of risk to persons who could be exposed to hazardous materials that are present at a contaminated site. The environmental information that is necessary to conduct a health risk study is extensive and is obtained from scientific modeling studies and also from data from the site.

The health risk assessment helps answer the following questions for populations or people who might be exposed to hazardous substances at contaminated sites:

   i. The potential condition and route of exposure to hazardous substances.
  ii. The potential of exposure to hazardous substances at levels higher than those that are determined to be safe.
 iii. If the levels of hazardous substances are higher than regulatory standards, how low do the levels have to be for the exposure risk to be within regulatory standards?

Exposures to environmental contaminants are significant risk factors in human health and disease. To understand and manage these risk factors, environmental and public health managers must have knowledge of the source of the exposure,

M.M. Aral, *Environmental Modeling and Health Risk Analysis (ACTS/RISK)*,
DOI 10.1007/978-90-481-8608-2_8, © Springer Science+Business Media B.V. 2010

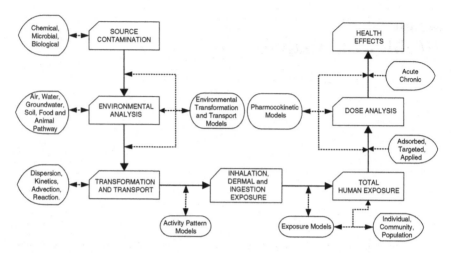

Fig. 8.1 Exposure assessment and environmental modeling continuum

the transformation and transport of contaminants in several environmental pathways, the exposed population, exposure levels, and the routes of the exposure as contaminants come in contact with the human body. Thus, a description of the relationship between source concentration, exposure, dose, and risk of disease must be understood and quantified through a sequence of studies which involve multidisciplinary teams. This relationship and the sequence of studies that are the components of this effort are shown in Fig. 8.1, which has been adapted from several publications that have described the source-to-dose human exposure continuum (Lioy 1990; NRC 1991a, b; Johnson and Jones 1992; Piver et al. 1997; Maslia and Aral 2004). Exposure to contaminants can be determined by direct or indirect methods (Johnson and Jones 1992), as such, models play an important role in this spectrum of analysis by providing insight and information when data obtained from direct measurements are missing, insufficient, or unavailable (Fig. 8.1). Because of the quantity, complexity, and choice of models that is available within the human exposure paradigm, previous chapters of this book were allocated to the discussion and introduction of those topics. In this chapter we will focus on specific exposure models that are recommended by USEPA and are currently in use.

8.1 USEPA Guidelines on Baseline Health Risk Assessment

As introduced above, we can define the goal of human health risk assessment as "to estimate the severity and likelihood of harm to human health from exposure to a potentially harmful substance or activity". Since exposure is the key element that leads to health or ecological risk, it is important to provide the definition of the term exposure. Exposure is defined as: "The contact of an organism (humans in the health risk assessment process) with a chemical or physical agent for a duration of time" (USEPA 1988). The magnitude of exposure can be determined by measuring the amount of chemical present in the contact media if the exposure analysis is conducted for current conditions or by estimating the amount of contaminants that may be present in the contact media in the past or in the future using modeling techniques. Exposure assessment is the determination or estimation of the magnitude, frequency, duration, and route of exposure. The other definition which is linked to exposure analysis is "dose" which is defined as: "The amount of a substance available for interaction with the metabolic processes or biologically significant receptors after crossing the outer boundary of an organism", i.e., penetrates a barrier such as the skin, gastrointestinal tract or lung tissue. Levels of internal dose may be measured in some body compartments through biologic sampling, e.g., medical testing for biologic markers of exposure in blood or urine. Thus, exposure is a measure of an external contact to a human body or organism and dose is an internal process.

Quantification of exposure can be achieved through the following:

i. Measure at the point of contact while exposure is occurring. In this case one has to measure both exposure to contaminant concentrations and time of contact and integrate them to arrive at total exposure.

ii. Estimating exposure concentrations and duration of exposure through environmental models, and then combining the information to arrive at total exposure.

iii. Estimating exposure from dose, which in turn can be reconstructed through internal indicators (biomarkers, body burden, excretion levels, etc.) after the exposure process has taken place. This process identified as inverse analysis.

Accordingly, exposure is quantified based on the following equation,

$$E = \int_{T_o}^{T_1} C(t)dt \tag{8.1}$$

in which $E[MW^{-1}T^{-1}]$ is exposure quantity, $C(t)$ is concentration which can be determined by direct measurement or through the utilization of the environmental models discussed in the previous chapters, and the interval (T_o, T_1) is the exposure duration. Similarly, potential dose that is linked to this exposure pattern is quantified based on the equation below,

$$D_p = \int_{T_o}^{T_1} C(t)IR(t)dt \tag{8.2}$$

in which D_p is the potential dose and $IR(t)$ is the intake rate which is the amount of a contaminated medium to which a person is exposed during a specified period of time. The amount of water, soil, and food ingested on a daily basis, the amount of air inhaled, or the amount of water or soil that a person may come into contact with through dermal exposures are typical examples of intake rates.

Individuals may be exposed to contaminants in environmental media in one or more of the following ways:

i. Ingestion of contaminants in groundwater, surface water, soil, and food;
ii. Inhalation of contaminants in air (dust, vapor, gases), including those volatilized or otherwise emitted from groundwater, surface water, and soil; and,
iii. Dermal contact with contaminants in water, soil, air, food, and other media, such as exposed wastes or other contaminated material.

The exposure assessment proceeds along the following steps:

i. The Characterization of Exposure Environment: In this step the exposure setting is characterized with respect to the physical characteristics of the site and the characteristics of the population at and around the site. At this step, characteristics of the current population or the population at the time of exposure (future or past) will be considered.
ii. Identification of Exposure Pathways: In this step, exposure pathways are identified. For total exposure characterization of all potential pathways of exposure need to be considered.
iii. Quantification of Exposure: In this step the magnitude, frequency and duration of exposure are quantified using either field data collection or modeling techniques. Eventually, the exposure estimates are expressed in terms of the mass of the substance in contact with the body per unit weight per unit time. These estimates may be identified as "intakes."

Typically, exposure assessments at contaminated sites are based on an estimate of the reasonable maximum exposure (RME) that is expected to occur under both past, current, and future conditions of the site. The reasonable maximum exposure implies the highest exposure that is reasonably expected to occur at the site. If a population at a site is exposed to a contaminant or to a multitude of contaminants through multiple pathways, the combination of exposures from all pathways will be included into the RME for total exposure analysis. Quantification of the multiple exposure pathway analysis can be conducted using the environmental pathways discussed in the previous chapters of this book.

After the site specific environmental pathway concentrations are identified, the next stage is the determination of pathway-specific intakes. Generic equations for calculating chemical intakes, exposure-dose and exposure factor are given by the following equations (USEPA 1987, 1991):

$$I = \frac{C \times CR \times EFD}{BW \times AT}$$

$$D = C \times IR \times AF \times EF/BW \qquad (8.3)$$

$$EF = (F \times ED)/AT$$

in which I is the pathway specific intake, or the amount of chemical at the exchange boundary represented as mg/kg body weight-day; C is the chemical concentration, or the average concentration contacted over the exposure period as mass per unit volume at the exposure point; CR is the contact rate, or the amount of contaminated medium contacted per unit time or event expressed as volume per day; EFD is the exposure frequency and duration which describes how long and how often the exposure occurs. This is calculated using two terms: EF, which is the exposure frequency expressed as days/year, and ED, which is the exposure duration expressed in years. BW is the body weight, or the average body weight over the exposed period expressed in kg, and AT is the averaging time, or the period over which exposure is averaged expressed in days. D is the exposure–dose and IR is the intake rate of the contaminated medium, AF is the bioavailability factor, F is the frequency factor expressed as days/year. Values of the variables used in Eq. (8.3) for a given pathway are selected such that the resulting intake value is an estimate of the reasonable maximum exposure (RME) for that pathway. Determination of RME is based on quantitative information and professional judgment. Based on this generic equation the collection of exposure models that is included in the RISK software is shown in Fig. 8.2.

In this context, there are several other health risk criterion that need to be defined which may be used as guidelines for screening level analysis.

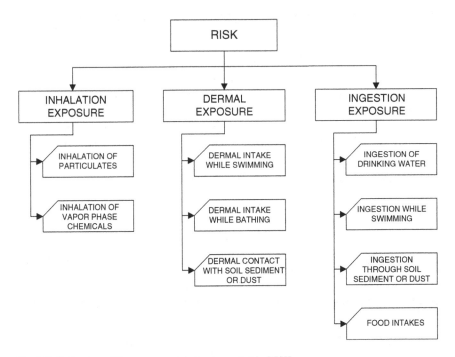

Fig. 8.2 Collection of Exposure models implemented in RISK

Minimal Risk Levels (MRLs): CDC/ATSDR in cooperation with USEPA has developed a priority list of hazardous substances that are most commonly found at hazardous waste sites. For these substances, toxicological profiles are developed which are used in the derivation of MRLs. Accordingly, MRLs are an estimate of the daily human exposure to a substance that would not lead to appreciable risk of adverse health effects during a specified duration of exposure. MRLs are based only on noncarcinogenic effects. MRLs are screening values only and are not indicators of health effects. Exposures to substances at doses above MRLs will not necessarily cause adverse health effects. MRLs are set below levels that might cause adverse health effects in most people, including sensitive populations. MRLs are derived for acute (1–14 days), intermediate (15–365 days), and chronic (365 days or longer) durations for the oral and inhalation routes of exposure. Currently, there are no MRLs for dermal exposure. MRLs are derived for substances by factoring the most relevant documented no-observed-adverse effects level (NOAEL) or lowest-observed-adverse-effects level (LOAEL) and an uncertainty factor as shown in Eq. (8.4) for oral exposure.

$$MRL = NOAEL/UF \tag{8.4}$$

in which MRL is the minimum risk level expressed as mg/kg/day; $NOAEL$ is the no-observed adverse effect level expressed as mg/kg/day and UF is the dimensionless uncertainty factor. An uncertainty factor between 1 and 10 may be applied for extrapolation from animal doses to human doses and/or a factor between 1 and 10 may be applied to account for sensitive individuals. When more than one uncertainty factor is applied, the uncertainty factors are multiplied.

Subchronic and Chronic Reference Doses (RfDs) and Reference Concentrations (RfCs): The subchronic RfD or RfC is an estimate of an exposure level that would not be expected to cause adverse effects when exposure occurs during a limited time interval. Subchronic values are determined from animal studies with exposure durations of 30–90 days. Subchronic human exposure information is usually derived from occupational exposures and accidental acute exposures. For example for oral exposure the subchronic RfD is determined by Eq. (8.5),

$$RfD = NOAEL/(UF \times MF) \tag{8.5}$$

where RfD is the reference dose expressed as mg/kg/day, MF is the dimensionless modifying factor which is based on a professional judgment of the entire database of the chemical and the other parameters are as defined earlier.

Cancer Slope Factor (CSF) and Inhalation Unit Risk (IUR): For known or possible carcinogens, CSFs and IURs are used as a quantitative indication of the carcinogenicity of a substance. A CSF is an estimate of possible increases in cancer cases in a population. A CSF is expressed in dose units $(mg/kg/day)^{-1}$. CSFs and IURs are usually derived from animal experiments that involve exposures to a single substance by a single route of exposure (i.e., ingestion or inhalation). USEPA extrapolates CSFs and IURs from experimental data of increased tumor incidences at high doses to estimate theoretical cancer rate increases at low doses. The experimental data often represent exposures to chemicals at concentrations which are orders of magnitude higher than concentrations found in the environment. Accordingly, the population cancer estimate can be calculated using Eq. (8.6),

$$ER = CSF(or\ IUR) \times dose(or\ air\ concentration) \tag{8.6}$$

where ER is the estimated theoretical risk which is dimensionless and CSF is the cancer slope factor expressed as $(mg/kg/day)^{-1}$.

Environmental Media Evaluation Guides (EMEGs): EMEGs represent concentrations of substances in water, soil, and air to which humans may be exposed during a specified period of time (acute, intermediate or chronic) without experiencing adverse health effects. EMEGs are based on MRLs and conservative assumptions about exposure, such as intake rate, exposure frequency and duration, and body weight. Acute exposures are defined as those of 14 days or less, intermediate exposures are those lasting 15 days to 1 year, and chronic exposures are those lasting longer than 1 year. For example, EMEG for drinking water can be calculated as,

$$EMEG_w = (MRL \times BW)/IR \tag{8.7}$$

in which $EMEG_w$ is the drinking water evaluation guide which is expressed as mg/L; BW is the body weight expressed as kg; IR is the ingestion rate expressed as L/day. Similarly, $EMEG_s$ for soil ingestion is given by,

$$EMEG_s = (MRL \times BW)/(IR \times CF) \tag{8.8}$$

in which CF is the conversion factor of 10^{-6} for (kg/mg) conversion.

Reference Dose Media Evaluation Guides (RMEGs): RMEGs are derived from USEPA's oral reference doses, which are developed based on USEPA evaluations. RMEGs represent the concentration in water or soil at which daily human exposure is unlikely to result in adverse noncarcinogenic effects.

Cancer Risk Evaluation Guides (CREGs): The *CREGs* are media-specific comparisons that are used to identify concentrations of cancer-causing substances that are unlikely to result in an increase of cancer rates in an exposed population. CREGs are calculated from USEPA's cancer slope factors (CSFs) for oral exposures or unit risk values for inhalation exposures. These values are based on USEPA evaluations and assumptions about hypothetical cancer risks at low levels of exposure. *CREGs* for drinking water or soil ingestion are calculated by,

$$CREG_{w/s} = (TR \times BW)/(IR \times CSF) \tag{8.9}$$

in which $CREG_{w/s}$ is the cancer risk evaluation guide expressed as mg/L for water or mg/kg for soil, TR is the target risk level 10^{-6}, and IR is the ingestion rate expressed as L/day for water or mg/day for soil. To calculate the *CREGs* a conversion factor of 10^{-6} needs to be included in the denominator of Eq. (8.9). $CREG_I$ for inhalation can be calculated from,

$$CREG_I = TR/IUR \qquad (8.10)$$

in which $CREG_I$ is the inhalation cancer evaluation guide expressed as $\mu g/m^3$, TR is the target risk level 10^{-6}, IUR is the inhalation unit risk $(\mu g/m^3)^{-1}$.

USEPA Maximum Contaminant Levels (MCLs): MCL is the maximum permissible level of a contaminant in water that is delivered to the free-flowing outlet of the ultimate user of a public water system. Contaminants added to the water by the user, except those resulting from corrosion of piping and plumbing caused by water quality, are exempt from meeting MCLs. In setting MCLs, USEPA considers health implications from possible exposures, as well as available technology, treatment techniques, and other means to reduce contaminant concentrations. In this analysis the cost of implementing technologies is also considered. MCLs are deemed protective of public health during a lifetime (70 years) at an exposure rate of 2 L/day. MCLs are dynamic values, subject to change as water treatment technologies and economics evolve and/or as new toxicologic information becomes available.

USEPA Maximum Contaminant Level Goals (MCLGs), Drinking Water Equivalent Levels (DWELs), and Health Advisories (HAs): The USEPA establishes several guidelines for permissible levels of a substance in a drinking water supply, including maximum contaminant level goals (MCLGs), drinking water equivalent levels (DWELs), and health advisories (HAs). MCLGs, formerly known as recommended maximum contaminant levels, are drinking water health goals. MCLGs are set at a level at which USEPA has found that "no known or anticipated adverse effects on human health occur and which allows an adequate margin of safety". USEPA considers the possible impact of synergistic effects, long-term and multistage exposures, and the existence of more susceptible groups in the population when determining MCLGs. For carcinogens, the MCLG is set at zero, unless data indicate otherwise, based on the assumption that there is no threshold for possible carcinogenic effects. The DWEL is a lifetime exposure level specific for drinking water (assuming that all exposure is from drinking water) at which adverse, noncarcinogenic health effects would not be expected. USEPA developed HAs as substance concentrations in drinking water at which adverse noncarcinogenic health effects would not be anticipated with a margin of safety. Drinking water concentrations are developed to establish acceptable 1- and 10-day exposure levels for both adults and children when toxicologic data (NOAEL or LOAEL) exist from animal or human studies.

8.2 Exposure Intake Models

The quantitative evaluation of human exposure through water ingestion, dermal contact, and inhalation; soil ingestion, dermal contact, and dust inhalation; air inhalation and dermal contact; and food ingestion can be performed using the models given below. Note that estimating an exposure or administered dose as described in the sections below does not take into account the relatively complex physiological and chemical processes that occur once a substance enters the body. Depending on the exposure situation being studied, one may need to consider additional factors to consider appropriately the exposure. This additional evaluation is particularly appropriate when determining the public health significance of an estimated exposure dose that exceeds an existing health guideline (USEPA 1991, 1992). The in-depth analysis will allow the health scientist to gain a better understanding of what is known and not known about the likelihood that a particular exposure will result in a harmful effect.

There are several terms in these models that are common to all cases with common default values for the parameter considered. Thus it is appropriate to give the definitions of these terms first in alphabetical order.

AB_f[dimensionless] = The absorption factor; (1×10^{-3}) for arsenic, beryllium and lead, (1×10^{-1}) for chlorobenzene, napththalene and trichlorophenol

BW[kg] = Body weight; 70 kg for adult approximate average, 16 kg for children 1–16 years old, 10 kg infant 6–11 months old

C_a[mg/m^3] = Contaminant concentration in air

C_f[mg/kg] = Contaminant concentration in air

C_{med}[mg/kg] = Contaminant concentration in meat egg and dairy products

C_s[mg/kg] = Contaminant concentration in fish

C_{vg}[mg/kg] = Contaminant concentration in vegetable and produce

C_w[mg/L] = Contaminant concentration in water

CF[10^{-6}kg/mg] = Conversion factor

ED[years] = Exposure duration; 70 years lifetime by convention, 30 years national upper-bound time (90th percentile) at one residence, 9 years national median time (50th percentile) at one residence, 6 years children 1–6 years old

EF[dimensionless] = Exposure factor

f_E[day/year] = Exposure frequency

f_f[dimensionless] = Fraction Ingested from the contaminated source which depends on local patterns

f_I[dimensionless] = Fraction ingested from the contaminated source; estimates are based on contamination pattern and population activity pattern

f_{med}[dimensionless] = Fraction ingested from the contaminated source; 0.44 average for beef, 0.4 average for dairy products

f_{vg}[dimensionless] = Fraction ingested from the contaminated source; 0.2 is the average

I_a[mg/kg/day] = Inhalation intake

I_d[mg/kg/day] = Dermal absorption

I_f[mg/kg/day] = Ingestion with fish

I_{da}[mg/kg/day] = Dermal intake while swimming or bathing

I_{med}[mg/kg/day] = Ingestion from meat egg and dairy products

I_w[mg/kg/day] = Ingestion of drinking water

I_{ssd}[mg/kg/day] = Ingestion of soil, sediment or dust

I_{sw}[mg/kg/day] = Ingestion while swimming;

I_{vg}[mg/kg/day] $=$ Ingestion from vegetable and produce

IR[L/day] $=$ Intake Rate of contaminated medium

IR_a[m³/h] $=$ Inhalation rate; 30 m³/day for adult upper bound value, 20 m³/day adult average

IR_f[kg/meal] $=$ Ingestion rate of fish; 0.284 kg/meal 95th percentile, 0.113 kg/meal 50th percentile, 132 g/day 95th percentile, 6.5 g/day daily average over year

IR_{med}[kg/meal] $=$ Ingestion rate of meat egg and dairy products; 0.3 kg/day for milk, 0.1 kg/day for meat, 0.28 kg/meal beef 95th percentile, 0.15 kg/meal for eggs 95th percentile

IR_s[mg/day] $=$ Ingestion rate of soil; 200 mg/day, children from 1–6 years of age, 100 mg/day anyone older than 6 years

IR_{vg}[kg/meal] $=$ Ingestion rate vegetable and produce; 0.05 kg/day for root crops, 0.25 kg/day for vine crops, 0.01 kg/day for leafy reports

RD[cm/h] $=$ Dermal permeability constant

RA[mg/cm²] $=$ Soil to skin adherence factor; 1.45 mg/cm² for commercial potting soil, 2.77 mg/cm² for kaolin clay

SA[cm²] $=$ Skin surface area available for contact

t_E[h/event] $=$ Exposure time pathway specific

t_{Ea}[h/day] $=$ Depends on duration of exposure; 12 min showering 90th percentile, 7 min showering 50th percentile

T_{ave}[day] $=$ Average time period of exposure; pathway specific for noncarcinogens ($ED \times 365$ day/year), 70 year for carcinogens (70 years $\times$ 365 day/year)

8.2.1 Intake Model for Ingestion of Drinking Water

Exposure to chemicals in drinking water through ingestion of drinking water is often the most significant source of hazardous substances. The ingestion exposure for this case may be estimated using the equation below. In this case the concentration of the chemical in the drinking water may be estimated using various pathway models described earlier or may be determined through direct measurements of the tap water concentration.

$$I_w = \frac{C_w \times IR \times f_E \times ED}{BW \times T_{ave}} \quad (8.11)$$

8.2.2 Intake Model for Ingestion while Swimming

Exposure to chemicals in water through ingestion while swimming may be estimated using the equation below.

$$I_{sw} = \frac{C_W \times IR \times t_E \times f_E \times ED}{BW \times T_{ave}} \quad (8.12)$$

8.2.3 Intake Model for Dermal Intake

Dermal absorption of contaminants from water is a potential pathway for human exposure to contaminants in the environment. Dermal absorption depends on several factors including the area of exposed skin, the anatomical location of the exposed skin, the duration of contact, the concentration of the chemical on skin, chemical specific permeability of the skin, the medium in which the chemical is applied and the skin condition and integrity. Dermal absorption of contaminants in water may occur during bathing, showering or swimming, all of which may be a significant exposure routes. Worker exposure for this pathway will depend on the type of the work performed, the protective clothing worn and the extent and duration of water contact. Chemical specific permeability constants should be used to estimate the dermal absorption of a chemical from water, which may not be readily available or may vary over a large range. For such cases uncertainty analysis of the case studied may be warranted. Dermal absorption of a chemical from water can be estimated using the equation given below, in which the parameters used in this model are given below (Tables 8.1 and 8.2).

Table 8.1 Standard values for total body surface area

Age (years)	50th Percentile (cm^2)	
	Male	Female
3 < 6	7,280	7,110
6 < 9	9,310	9,190
9 < 12	11,600	11,600
12 < 15	14,900	14,800
15 < 18	17,500	16,000
18 < 70	19,400	16,900

Table 8.2 Specific standard values for body part specific surface area

Age (years)	50th Percentile (cm^2)		
	Arms	Hands	Legs
3 < 4	960	400	1,800
6 < 7	1,100	410	2,400
9 < 10	1,300	570	3,100
18 < 70	2,300	820	5,500

$$I_d = \frac{C_w \times SA \times RD \times t_E \times f_E \times ED}{BW \times T_{ave}} \qquad (8.13)$$

8.2.4 Intake Model for Intake Via Soil, Sediment or Dust

If the intake of chemicals from soil, sediment or dust occurs by accidental ingestion then the equation below may be used to estimate the exposure.

$$I_{ssd} = \frac{C_s \times IR_s \times CF \times f_I \times f_E \times ED}{BW \times T_{ave}} \qquad (8.14)$$

8.2.5 Intake Model for Dermal Absorption of Soil, Sediment and Dust

If the intake of chemicals from soil, sediment or dust occurs by dermal exposure then the equation below may be used to estimate the exposure. Dermal adsorption will again be a function of the exposed surface area of the skin. In this case the skin surface area estimates will be based on the skin surface area on which the soil, sediment or dust is adhered to. If this data is not available, the percent area exposed may be estimated and the body surface area values given in Tables 8.1 and 8.2 may be used.

$$I_{da} = \frac{C_s \times CF \times SA \times RA \times AB_f \times f_E \times ED}{BW \times T_{ave}} \qquad (8.15)$$

8.2.6 Intake Model for Air Intakes

Inhalation is an important pathway for human exposure to contaminants. It may occur by direct inhalation of gases or by the inhalation of chemicals adsorbed to airborne particles or fibers. In order to estimate an inhalation exposure, the air concentration must be accurately estimated using the models described earlier, given a particular scenario of air pollution whether it occurs indoors or outdoors. Once that determination is made, the model given below may be used to estimate the exposure due to inhalation.

$$I_a = \frac{C_a \times IR_a \times t_{Ea} \times f_E \times ED}{BW \times T_{ave}} \qquad (8.16)$$

8.2.7 Intake Model for Ingestion of Fish and Shellfish

Exposure to chemicals from ingestion of fish and shellfish can be estimated using the equation below.

$$I_f = \frac{C_f \times IR_f \times f_f \times f_E \times ED}{BW \times T_{ave}}$$

(8.17)

8.2.8 Intake Model for Ingestion of Vegetables and other Produce

Exposure to chemicals from ingestion of vegetables and other produce can be estimated using the equation below.

$$I_{vg} = \frac{C_{vg} \times IR_{vg} \times f_{vg} \times f_E \times ED}{BW \times T_{ave}}$$

(8.18)

8.2.9 Intake Model for Ingestion of Meat, Eggs and Dairy Products

Exposure to chemicals from ingestion of meat, eggs and other dairy products can be estimated using the equation below.

$$I_{med} = \frac{C_{med} \times IR_{med} \times f_{med} \times f_E \times ED}{BW \times T_{ave}}$$

(8.19)

8.3 Applications

The RISK application starts with the selection of either the "New Application" or the "Open Application" options from the pull down menu "File". When a new application option is selected the window shown in Fig. 8.3 will appear.

In the upper grid of this window, the options to link the RISK software input data to the environmental computations performed by the ACTS software is available to the user. In the lower grid, the option available is the manual input option. When the "Manual Input" option is selected one enters the window shown in Fig. 8.4. This is the general RISK exposure data entry and calculation window for all exposure models discussed above. In Fig. 8.4 the third folder is shown, which is the "Intake Parameters" folder.

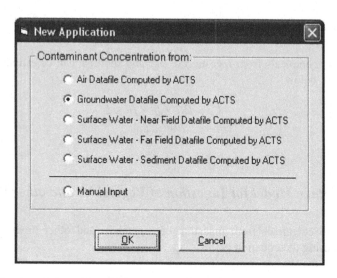

Fig. 8.3 Opening window of the RISK software

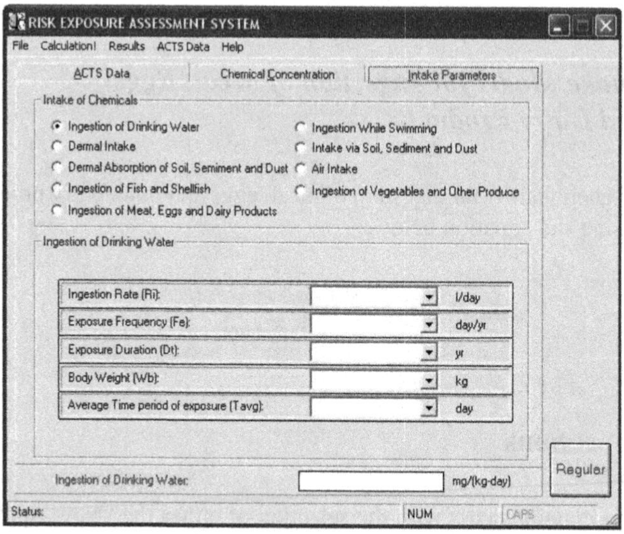

Fig. 8.4 Exposure models window of the RISK software

In this window, in the upper grid, the options available to the user are the list of exposure models that can be used for exposure analysis. A discussion of these exposure models can be found in Section 8.2. When an exposure model is selected, the data entry grid changes to reflect the input data necessary for the model selected. This data needs to be entered by the user. The other two folders referred to in this window are the "ACTS Data" folder and the "Chemical Concentration" folder.

Through the "ACTS Data" folder it is possible to select different ACTS output files for use in the RISK model. The user should recognize that the concentration value that will be transferred to the RISK model will be at the spatial and temporal constants identified in the ACTS data input folder. The user is referred to Appendix 3 which explains the purpose of the fourth data entry value in the spatial and temporal data entry grid in the ACTS software. The second folder in this window (Fig. 8.4) is the "Chemical Concentration" folder, which needs to be filled by the user if the data is not transferred from the ACTS software. Once all necessary input data is entered, one may click on the calculate button to complete the deterministic analysis of the problem, which will appear in the lowest grid of the window shown in Fig. 8.4. The other options that are available to the user can be seen on the main menu options grid. By selecting the results option, one may look at the text files of output or plot the results for the case of Monte Carlo analysis. It should be clear that in the deterministic analysis mode the graphical plotting of the results is not possible since the output is a single value. Graphical output in this case is only possible when a Monte Carlo analysis of the exposure model selected is performed. The options available to the user for this case are the same as those in the use of the graphics package that is described for the ACTS software. For more details on graphing, the reader is referred to the applications sections of the previous chapters and also to Appendix 3. The Monte Carlo analysis mode of the RISK software can be entered by clicking on the "Regular" button seen on the lower right hand corner of the window shown in Fig. 8.4. As can be seen, the RISK software is a relatively simple application platform when compared to the ACTS software. However, the introduction of the Monte Carlo analysis to this simple computational platform leads to uncertainty and sensitivity analysis of all exposure models, which is a unique feature of the RISK software. This option is extremely useful and valuable when some of the input parameters of the models used are not known precisely. Using the Monte Carlo analysis mode, uncertainty and sensitivity analysis of the exposure models can be directly made under the same computational platform. Other applications of the Monte Carlo platform can be found in Maslia and Aral (2004).

Example 1: It is estimated that the benzene concentration in a rural well is $C_w = 10 \times 10^{-3}$ mg/L. The concern is the potential ingestion exposure of the inhabitants of the household to benzene, who use this well water as their drinking water supply. Calculate the ingestion exposure of an adult who lives in this household.

Solution: The potential exposure due to drinking water is determined using Eq. (8.11). The parameters that are needed to implement this model are given as:

$$\left. \begin{array}{l} \text{Ingestion rate} = IR = 2 \, L/\text{day} \\ \text{Exposure frequency} = f_E = 365 \, \text{day/year} \\ \text{Exposure duration} = ED = 70 \, \text{years} \\ \text{Body weight} = BW = 70 \, \text{kg} \\ \text{Average time period of exposure} = T_{ave} = 70 \times 365 \, \text{day} \end{array} \right\} \qquad (8.20)$$

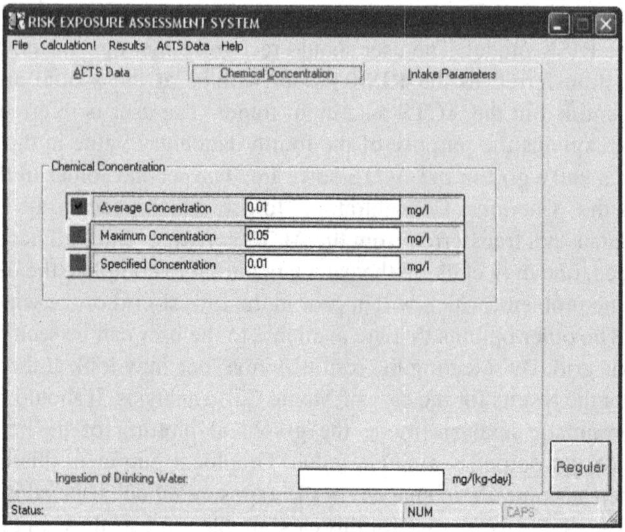

Fig. 8.5 Concentration data entry window for Example 1

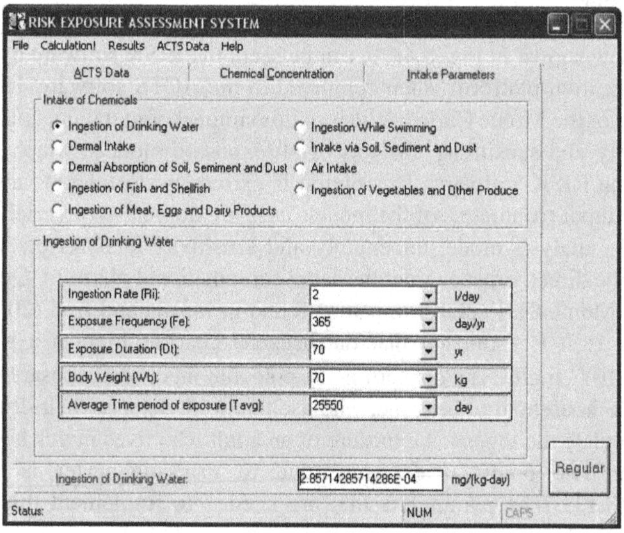

Fig. 8.6 Parameter data entry window for Example 1

The data entry window for this example is shown in Figs. 8.5 and 8.6. The reader should note that in the concentration data entry window three concentration values can be entered (Fig. 8.5). These are the "Average Concentration", the "Maximum Concentration" and the "Specified Concentration". These three options are available to the user in the deterministic mode of analysis. Once these data are entered, the user may select any one of these values by clicking on the green box to the left of the value entered. In this example, the "Average Concentration" value is selected as indicated by the check mark on the green box. The parameter value entry window for this problem is shown in Fig. 8.6. As can be seen this window now also shows the answer at the bottom grid, which is $I_w = 2.85 \times 10^{-4}$ mg/kg/day, since this image is taken after the calculate button is clicked.

Example 2: There are several uncertainties in Example 1. First the concentration of benzene in the well, $C_w = 10 \times 10^{-3}$ mg/L, is an estimate. The concentration range, based on several observations at the well, is determined to be $C_w = 35 \times 10^{-3}$ to 1×10^{-3} mg/L. The concern is the variability in potential ingestion exposure of the inhabitants of this household to benzene under this uncertainty.

Solution: The solution to this problem can be obtained through the Monte Carlo analysis. The user may enter the Monte Carlo analysis mode by clicking on the "Regular" button on the lower left hand side of the window shown in Fig. 8.6. This operation starts the window shown in Fig. 8.7, which is the same Monte Carlo application window that was used in the ACTS software. By now, the user should be familiar with the functions of this window. For more details the user may refer to Appendix 3.

In this window, under the "Variants" column, several parameters of this model can be selected to be uncertain. In this example, to keep the analysis simple, only concentration is selected to be uncertain (Fig. 8.7). Normal distribution, minimum and maximum benzene concentrations, the variance and the random number data point selections are entered into the Monte Carlo window to start the analysis. When the generate button is clicked, the output data calculated will be displayed in the lower grid of this window (Fig. 8.7). The probability density function obtained for the concentration in the well may also be plotted as shown in Fig. 8.8. This check of

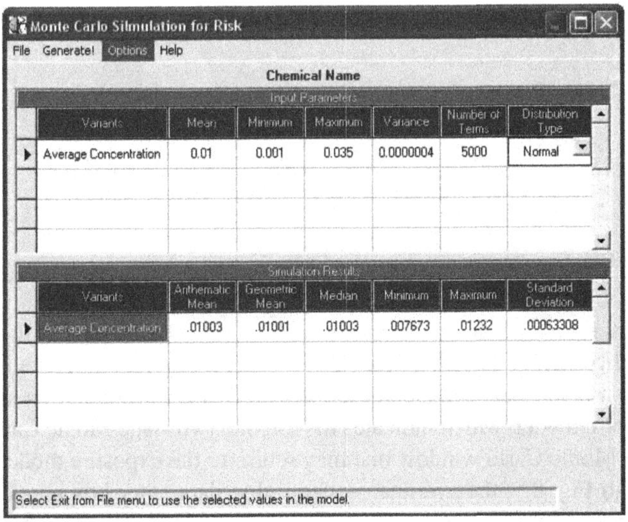

Fig. 8.7 Monte Carlo data entry window for Example 2

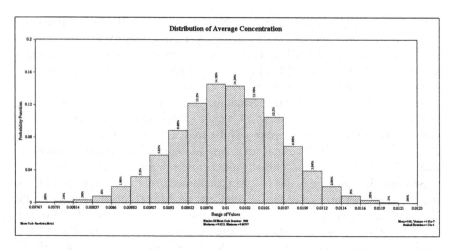

Fig. 8.8 Probability density function for benzene concentration for Example 2

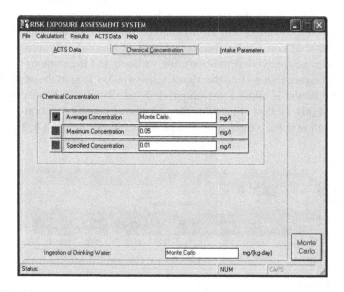

Fig. 8.9 Concentration data entry window for Example 2

the output is a good choice before moving forward to confirm that the output for the probability density function has the desired characteristics. Double clicking on the variant name selects the complete probability density function as data input to the exposure model, which indicates the use of a two stage Monte Carlo analysis. Closing the Monte Carlo window one may return to the exposure model window.

As seen in Fig. 8.9, the average concentration data entry box now contains the words "Monte Carlo", which indicates that the probability density function is properly transferred to the exposure model.

At this stage one may click on the calculate button to perform the Monte Carlo analysis for this problem. As seen in Fig. 8.9, this operation results in a probability density function output for the ingestion exposure, as indicated by the appearance of the words "Monte Carlo" in the output grid. These results may now be analyzed using standard statistical techniques for the probability of exceedance analysis when compared to a criterion of concern using the resulting probability density function, and the cumulative and complementary cumulative probability density function properties. These three graphs, as obtained from the graphics module of the RISK software are shown in Figs. 8.10–8.12. It can be concluded from Fig. 8.12

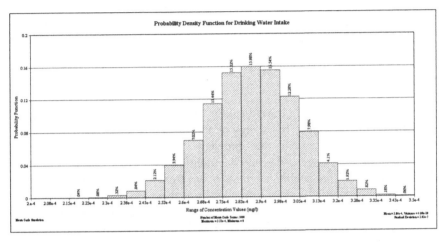

Fig. 8.10 Probability density function output for Ingestion Exposure for Example 2

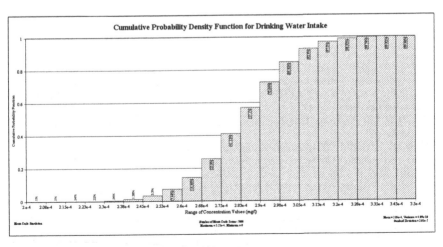

Fig. 8.11 Cumulative probability density function output for ingestion exposure for Example 2

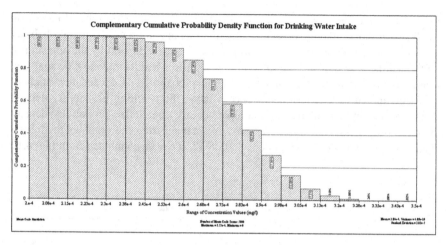

Fig. 8.12 Complementary cumulative probability density function output for ingestion exposure for Example 2

that there is about 7% probability for the ingestion exposure to exceed $I_w = 3.0 \times 10^{-4}$ mg/kg/day based only on the variability of the concentrations in the range as indicated in this problem. If the criteria of concern is $I_w = 3.0 \times 10^{-4}$ mg/kg/day, then the decision that needs to be made is whether the 7% exceedance probability is critical or not from a health effects perspective. One should remember that in the deterministic analysis, the outcome obtained was $I_w = 2.85 \times 10^{-4}$ mg/kg/day, which indicates a safe condition based on the same criteria of concern. Further details of this type of analysis can be found in Maslia and Aral (2004), and also Chapter 9.

References

CDC (2010) Citizens Guide to Health Risk Asserssment and Public Health. http://www.atsdr.cdc. gov/publications/CitizensGuidetoRiskAssessments.html. Accessed 21 Jan 2010

Johnson BL, Jones DE (1992) ATSDR's activities and views on exposure assessment. J Expo Anal Environ Epidemiol 1:1–17

Lioy PJ (1990) Assessing total human exposure to contaminants. Environ Sci Technol 24 (7):938–945

Maslia ML, Aral MM (2004) Analytical contaminant transport analysis system (ACTS), multimedia environmental fate and transport. Pract Period Hazard Toxic Radioact Waste Manage – ASCE 8(3):181–198

NRC (1983) Risk assessment in the Federal Government: managing the process. N. R. Council. National Academy Press, Washington, D.C

NRC (1991a) Environmental epidemiology - public health and hazardous wastes. N. R. Council. National Academy Press, Washington, D.C

NRC (1991) Human exposure assessment for airborne pollutants. Committee on Advances in Assessing Human Exposure to Airborne Pollutants, Washington, D.C.

Piver WT, Jacobs TL et al (1997) Evaluation of health risks for contaminated aquifers. Environ Health Perspect 105(1):127–143

USEPA (1987) The risk assessment guidelines of 1986. Office of Health and Environmental Assessment, Washington D.C.

USEPA (1988) Superfund exposure assessment mannual. Office of Health and Environmental Assessment, Washington, D.C.

USEPA (1991) Superfund exposure assessment manual. Office of Health and Environmental Assessment, Washington, D.C.

USEPA (1992) Guidelines for exposure assessment. Office of Health and Environmental Assessment, Washington, D.C.

Chapter 9
Application: Pesticide Transport in Shallow Groundwater and Environmental Risk Assessment

Knowledge becomes wisdom only after it has been put to practical use.
Albert Einstein

B.A. Anderson, M. L. Maslia, D. Ausdemore, ATSDR/CDC, Atlanta, GA, USA
J. L. Caparoso, US Army, Health Promotion and Preventative Medicine, Germany.
M. M. Aral, Georgia Institute of Technology, Atlanta, GA, USA
Water Quality Exposure and Health Journal, Volume 2, No. 2, 2010.

An analytical model dynamically linked with Monte Carlo simulation software such as ACTS can provide a relatively rapid, cost-effective way to conduct probabilistic analysis of contaminant fate and transport in simplified groundwater systems. This approach was used to evaluate migration of an existing organochlorine pesticide plume in a shallow, unconfined aquifer underlying a barrier island in coastal Georgia, USA (Anderson et al. 2007) Probabilistic analysis provided an estimate of the likelihood that the pesticide plume would reach coastal wetlands 244 m downgradient of the source area. The contaminant plume consists of four isomers of benzene hexachloride (BHC), also known as hexachlorocyclohexane (HCH). To analyze this problem, the Analytical Contaminant Transport Analysis System (ACTS) was used to simulate two-dimensional, saturated zone contaminant fate and transport. Deterministic simulations using calibrated, single-value input parameters indicate that the contaminant plume in the barrier island shallow aquifer will not reach the wetlands. Probabilistic analyses consisting of two-stage Monte Carlo simulations using 10,000 realizations and varying eight input parameters indicate that the probability of exceeding the detection limit (0.044 μg/L) of BHC in groundwater at the wetlands boundary increases from 1% to a maximum of 13% during the period from 2005 to 2065. This represents an 87% or greater confidence level that the pesticide plume will not reach downgradient wetlands. Based on this outcome environmental decisions for the contamination at the site can be made more reliably by managers.

M.M. Aral, *Environmental Modeling and Health Risk Analysis (ACTS/RISK)*,
DOI 10.1007/978-90-481-8608-2_9, © Springer Science+Business Media B.V. 2010

9.1   Problem Description

Contamination of water resources has long been recognized as having dual impacts –
environmental consequences leading to potential health effects (when contaminated water is used as a source of drinking water) and ecological consequences
leading to adverse biodiversity effects. Thus, during the past 50 years, computational methods have been developed that provide engineers, scientists, environmental managers and public health officials with tools with which they can assess,
analyze, and formulate technical, regulatory, and policy decisions regarding the
impacts of contaminated water resources. These computational methods, referred to
in a general sense as models, are typically grouped into three categories, depending
on the purpose for which they are to be used and applied. These categories are: (i)
Methods and models used for screening-level analyses; (ii) methods and models
that are used for research; and, (iii) models that are used in assessment and decision
making (Cullen and Frey 1999). Screening-level analyses are based upon the premise
of using simplified models to identify key system parameters of interest for further
focus in data gathering, research, or decision making. Typically, analytical models
fall into this category. Research models, on the other hand, are developed and applied
so that the functioning of real-world systems can be better understood by investigat-
ing functional relationships among multiple system parameters. Quasi-numerical and
numerical models fall into this category. Assessment and decision making models are
developed to assist with regulatory and rule-making decisions when issues related to
regulatory compliance need to be addressed. For example, certain air-dispersion,
surface water and groundwater models that are discussed in this book are considered
regulatory models because they are used to assess if plant stack emissions exceed
specified regulatory limits or to evaluate near field and far field pollution levels based
on effluent discharge conditions.

The models described above, when developed for the appropriate purpose, can
be applied using three modes of analyses – deterministic, sensitivity, or probabilistic (Cullen and Frey 1999). A deterministic analysis uses single-point estimates for
model input parameters, and results are obtained in terms of single-value output
(e.g., concentration of a contaminant at a specified location or time of interest).
A sensitivity analysis is a method used to ascertain the dependence of a given model
output (e.g., exposure concentration) on the variation of one or more model input
parameters (e.g., source concentration or stack height). A probabilistic analysis is
used to evaluate the effect of uncertainties in model inputs on the model outputs.
Thus, results of a probabilistic analysis are obtained in terms of distributed-value
output that provides quantitative information about the range and relative likelihood of model output values (e.g., the range of benzene concentrations in drinking
water is 5–15 µg/L, assuming a 95% confidence interval).

The choice and complexity of a model and the type of analysis that is undertaken
depends on site conditions, available information and data, the questions to be
answered, and decision-making applications desired from the model and analyses.
A detailed discussion of extremely complex analyses utilizing several research-type

(numerical) models in a probabilistic analysis can be found in the literature (Maslia et al. 2009). These models and analyses were chosen because environmental health scientists and public health officials need to determine if associations exist between historical exposures to contaminated drinking water, determined on a monthly basis, and specific birth defects and childhood cancers.

For the study described in this chapter, analytical modeling was selected as an efficient and computationally economical screening-level tool to assess contaminant fate and transport in a shallow, unconfined aquifer underlying a barrier island in coastal Georgia, USA. The analytical modeling was not intended for assessment and implementation of remediation design. To evaluate the potential adverse effects of contamination on the biodiversity at the site, probabilistic analysis using Monte Carlo simulation techniques was combined with the analytical model by varying selected input parameters across a wide range of values that represented the uncertainty and variability inherent in subsurface contaminant transport.

9.2 Background on Site and Contamination Conditions

The site in this study is located on Oatland Island, immediately east of Savannah, Georgia, USA (Fig. 9.1). During 1943–1973, the Communicable Disease Center (now known as the Centers for Disease Control and Prevention, CDC) and its predecessor agency, the Office of Malaria Control on War Areas, operated a research laboratory on the site. In 1998, the discovery of an historical map showing two onsite disposal areas prompted CDC to return to the site, conduct environmental investigations, and begin remediation activities. Source removal during 2000 included excavation of buried containers and contaminated soil to the depth of the water table (Montgomery Watson Harza [MWH] 2001). The footprint of the excavated area was about 1,254 m^2. As part of voluntary remediation activities, CDC tested an in-situ groundwater remediation technique recommended by consultants. Three full-scale in-situ treatment events utilizing a modified Fenton's Reagent (hydrogen peroxide and iron catalysts) were conducted during December 2003–March 2004 (MWH 2005).

9.2.1 Health and Regulatory Considerations

No current public health issues are associated with the pesticide-contaminated groundwater at the site, primarily because shallow groundwater in the area is not used for drinking water (ATSDR 2005a). However, future southward migration of the contaminant plume into the downgradient coastal wetlands is a potential ecological concern. Additionally, the pesticide concentrations in groundwater at the site exceed applicable state environmental regulatory levels (Georgia Environmental Protection Division 1993). Although aldrin, dieldrin, endosulfan, dichlorodiphenyltrichloroethane (DDT)-related compounds, and benzene hexachloride (BHC) were detected

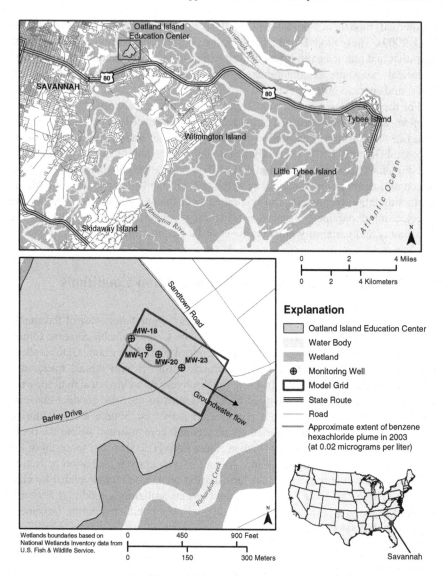

Fig. 9.1 Study area and ACTS model domain, Oatland Island Education Center, Oatland Island, Georgia, USA

in groundwater within the source area, the dissolved-phase contaminant plume extending downgradient from the source area consists of only four BHC isomers: alpha- $(\alpha-)$, beta- $(\beta-)$, delta- $(\delta-)$, and gamma- $(\gamma-)$ BHC. The BHC groundwater contamination is limited to the unconfined zone of the surficial groundwater aquifer that is approximately 4.3 m thick, with a clay confining layer at its base (Fig. 9.2). The maximum concentration of total BHC in groundwater during 2000–2005 monitoring events was 5.4 µg/L at monitoring well MW-17 during

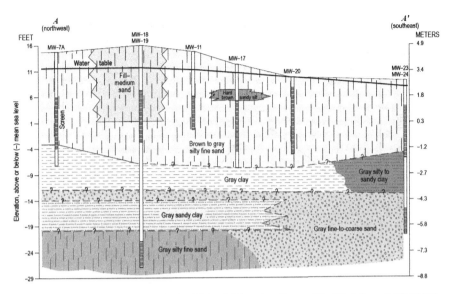

Fig. 9.2 Geologic cross-section, Oatland Island Education Center (Modified from MWH 2005; 1 ft equals 0.3048 m)

November 2001 (Figs. 9.1 and 9.3). Corresponding individual isomer concentrations ranged from less than 0.5 µg/L for γ-BHC to 3.6 µg/L for γ-BHC (MWH 2001, 2005). During annual groundwater monitoring in 2005 (after source removal and in-situ treatment), the maximum concentration of total BHC detected in the surficial aquifer was 2.5 µg/L at monitoring well MW-17 (Figs. 9.1 and 9.3). Individual isomer concentrations ranged from less than 0.3 µg/L for γ-BHC to 1.9 µg/L for γ-BHC (MWH 2002a, b, 2005).

BHC concentrations in groundwater at the site exceed applicable Georgia Department of Natural Resources, and Environmental Protection Division regulatory levels that range from 0.006 to 0.66 µg/L for the individual BHC isomers (Georgia Environmental Protection Division 1993). Several of the action levels for individual BHC isomers are below the detection limit for BHC in groundwater. In such cases, Georgia EPA regulations indicate that the detection limit should be used for regulatory compliance. For modeling purposes in this study, 0.044 µg/L was selected as the detection limit for BHC in groundwater. This detection limit is consistent with guidance provided by the U.S. Environmental Protection Agency (USEPA) Office of Ground Water and Drinking Water for γ-BHC in water samples (USEPA 1993).

For reference, it should be noted that $\gamma - BHC$ (lindane) is currently regulated under USEPA's national primary drinking-water standards. The USEPA maximum contaminant level (MCL) for $\gamma - BHC$ in public water-supply systems is 0.2 µg/L. MCLs represent legally enforceable contaminant concentrations in drinking water that USEPA deems attainable (using the best available treatment technology and cost considerations) and protective of public health over a lifetime (70 years) of exposure.

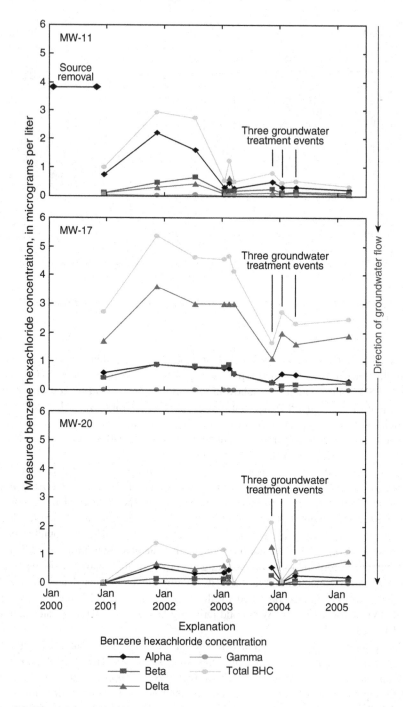

Fig. 9.3 Benzene hexachloride concentrations in groundwater measured at monitoring wells MW-18, MW-17, and MW-20 during 2000–2005 (Anderson et al. 2007)

As stated previously, the shallow groundwater in the area is not used for drinking water, and thus, exposure to contaminated shallow groundwater by ingestion is not a completed exposure pathway and was not a focus of this investigation.

9.2.2 Site Geology and Hydrogeology

Oatland Island is one of a number of islands along the eastern coast of Georgia, within the Lower Coastal Plain Province of Georgia. Layered sediments of sand, clay, and limestone in this province form a surficial aquifer (typically 20 m thick in the vicinity of Oatland Island), the Upper and Lower Brunswick aquifers (occurring at depths of 27–104 m below ground surface (bgs)), the Floridan aquifer system (occurring at depths of 34–162 m bgs), and several other, deeper aquifer systems (MWH 2003; Clarke et al. 1990). At Oatland Island and adjacent Skidaway Island (Fig. 9.1), the surficial aquifer is divided into an upper unconfined zone and a lower semi-confined zone separated by a clay layer (MWH 2003; Clarke et al. 1990).

Oatland Island is surrounded by tidal marshes, and its topography is generally flat. Soil-boring and well installation logs from the site characterize subsurface soils as fine sand with some silt from ground surface to 4.3–7.0 m bgs (MWH 2001, 2003, 2005; S&ME 1999). Grain-size analyses of samples from these shallow soils indicate greater than 90% fine sand (S&ME 1986, 1999). Some laterally discontinuous lenses of clay and silt also were documented in this unit (MWH 2003). The fine sands are underlain by about 2 m of clay followed by a layer of fine to coarse sand (MWH 2003). A representative geologic cross-section from environmental investigations at the site is shown in Fig. 9.2. During site sampling events, groundwater was typically located at 2–3 m bgs. Numerous groundwater monitoring events during 2000–2005 confirmed that groundwater flow in the study area is to the southeast (Fig. 9.1) (MWH 2003, 2005).

9.2.3 Chemical and Contaminant Properties of BHC

BHC, also known as hexachlorocyclohexane (HCH), is an organochlorine pesticide with eight chemical forms called isomers that differ from each other only in the spatial arrangement of the chlorine atoms around the benzene ring. Only four of the BHC isomers, the $\alpha-$, $\beta-$, $\delta-$ and $\gamma-$ forms, are commercially significant. Almost all of the insecticidal properties reside in the $\gamma-$ isomer, which is often distilled into a 99% pure form commercially known as lindane.

Organochlorine compounds were developed after World War II for pest eradication and control. BHC is no longer produced in the United States, and most registered agricultural uses have been banned or voluntarily cancelled (USEPA 2006a). It is still used as an active ingredient in some pharmaceutical products formulated for lice and scabies (mites) treatment in humans (ATSDR 2005b;

USEPA 2006a). Because of widespread past and current use, BHC has been detected in air, soil, sediment, surface water, groundwater, ice, fish, wildlife, and humans (USEPA 2006b).

As with most organochlorine pesticides, acute exposure to BHC affects the central nervous system in humans, causing dizziness, headaches, convulsions, and seizures. Exposure to BHC also can affect the liver and kidneys and can cause blood disorders. Animal studies indicate that all BHC isomers may cause cancer in humans (ATSDR 2005b). However, USEPA classifies only $\alpha-$ BHC and technical-grade BHC as probable carcinogens and $\beta-$ BHC as a possible carcinogen. USEPA determined that the evidence is not sufficient to assess or classify the carcinogenic potential of $\gamma-$ BHC and $\delta-$ BHC.

The aqueous solubility of each of the four BHC isomers is in the range of 5–17 mg/L (ATSDR 2005b). The density of the BHC isomers is 1.87–1.89 g/cm^3 at 20°C (ATSDR 2005b), making them denser than water (water density is 1.0 g/cm^3). Because BHC isomers are not very soluble in water and are denser than water, they are considered dense non-aqueous phase liquids (DNAPLs). When a DNAPL is released into groundwater, it can migrate downward (vertically) as a separate phase liquid, leaving residual DNAPL in soil pore spaces below the groundwater surface. Residual DNAPL can act as an ongoing source of groundwater contamination as it partitions into the groundwater slowly, over a long period of time. A DNAPL also can migrate downward in an aquifer until it encounters a confining unit such as a clay layer, where it can accumulate in pools (Environment Agency 2003; Pankow and Cherry 1996; Schwille 1988).

9.3   Method of Analysis

The analytical contaminant transport analysis system (ACTS) version 8 software used in the analysis is publicly available. ACTS contains more than 100 models and associated analytical solutions for assessing fate and transport of contaminants within four environmental transport pathways – air, soil, surface water, and ground-water (Anderson et al. 2007). Monte Carlo simulation is integrated into the ACTS software and dynamically linked to transport pathway modules so that there is no need to export simulation results into a separate software application. Several publications are available that detail the capabilities of ACTS and present case studies for its application in the context of environmental health (Anderson et al. 2007; Maslia and Aral 2004; Maslia et al. 1997; Rodenbeck and Maslia 1998).

9.3.1   Modeling Approach and Assumptions

The Oatland Island site was characterized within the ACTS software using a two-dimensional, saturated, infinite aquifer model with constant dispersion coefficients. The model domain was defined with a length of 259 m (x-axis) and a width of 152 m (y-axis). The x-axis of the model grid is aligned parallel to the groundwater flow direction, and the y-axis is centered on the former pesticide source area (Fig. 9.1). Total BHC concentrations (versus measured concentrations for individual isomers) in the groundwater were used to calibrate the model and conduct fate and transport simulations of the contaminant plume at the site. Model results for total BHC are considered representative of all four BHC isomers detected at the site. For modeling purposes, the following assumptions were made:

- Groundwater flow is steady and uniform. This assumption is supported by field measurements and analysis of the potentiometric surface (MWH 2001, 2003, 2005).Rauchen.
- The surficial aquifer is characterized as infinite with a constant dispersion coefficient. This assumption implies that actual aquifer boundaries are far enough away that they do not affect groundwater flow and contaminant transport in the area of interest.
- The time required for leakage and vertical transport of contaminants from the buried containers to the groundwater is negligible. This is consistent with the shallow depth to groundwater at the site (the surficial aquifer occurs at 2–3 m bgs) and the types of containers excavated from the site (metal containers, cardboard boxes, plastic bags) (MWH 2001; S&ME 1986, 1999).
- The time required for vertical transport of contaminants through the thickness of the surficial aquifer is negligible; the contaminants are well mixed throughout the 4.3 m depth.
- Vertical transport (sinking) of the dissolved-phase contaminant plume due to contaminant density effects is negligible. This is a valid assumption given the low aqueous solubility of the contaminants at the site (Schwille 1988).
- Tidal effects within the study area would primarily increase contaminant dispersion. This is accounted for in the probabilistic analyses by random variation of the longitudinal and transverse dispersion coefficients. Field data collected over a full tidal cycle indicated water level fluctuations of only 1 cm in monitoring wells within the study area (MWH 2001).
- Biological and chemical degradation of the contaminants are negligible. This is a conservative assumption. Various studies indicate the half-life for γ-BHC in water (groundwater and sterilized natural water) may range from 32 to more than 300 days (ATSDR 2005b; Mackay et al. 1997).

9.3.2 Model Input Parameters and Source Definition for Deterministic Simulations

In this application relevant input parameter values for groundwater flow and contaminant transport are obtained from a variety of sources, including existing site field data and previous modeling analyses, available U.S. Geological Survey (USGS) information for the area, and relevant literature articles (Table 9.1). The media, contaminant, and calibrated model parameter values used to simulate the fate and transport of total BHC in groundwater for deterministic (single-value

Table 9.1 Summary of relevant hydrogeologic and contaminant transport parameters used in Oatland Island study

Property	Value	Units	References	Notes
Hydraulic conductivity (K)	0.007, 0.029	m/day	S&ME (1999)	Calculated from site-specific rising head slug tests on 2 site monitoring wells
	2.43–2.89		MWH (2005)	Calculated from five rising head slug tests on site monitoring wells
	12		ATSDR calculated from data in S&ME (1999)	Calculated using site-specific grain-size analysis data and Hazen's approximation (Freeze and Cherry 1997)
	16		ATSDR calculated from data in MWH (2001)	Calculated using site-specific tidal and monitoring well (Data from MWH 2001)
	0.6–20		S&ME (1986), Clarke et al. (1990)	Calculated from grain-size analysis data for adjacent Skidaway Island, Georgia (S&ME 1986)
	0.017–17		Domenico and Schwartz (1998)	Typical range of K values for fine sand
Hydraulic gradient (i)	0.006	m/m	S&ME (1999)	Calculated as average gradient between monitoring wells MW-5 and K-6 at the site
	0.01		ATSDR calculated from MWH (2003), Figure 4.7	Calculated gradient between monitoring wells MW-11 and MW-23
	0.005		MWH (2005)	Calculated gradient measured in shallow groundwater across distal portion of dissolved BHC plume
Porosity of soil (θ)	0.3	—	S&ME (1999)	Based on site-specific grain-size analyses of subsurface soil indicating 90% fine sand
Recharge to surficial aquifer (q)	0.3–0.5		Freeze and Cherry (1997)	Typical range for nonindurated sands
	0.26–0.53		Morris and Johnson (1967)	Typical range for fine sands
				$q = P - Q_s - ET$
Precipitation (P)	77.4–168	cm/year	Southeast Regional Climate Center online data	Range of annual means (1948–1977) for Savannah Beach, GA (www.dnr.state.sc.us/climate/sercc/climateinfo/historical/historical.html)

	Value	Units	Reference	Comment
	127		National Weather Service data cited in Priest (2004)	Mean annual precipitation (1971–2000)
Surface runoff (Q_s)	25		Krause and Randolph (1989), Priest (2004)	
Evapotranspiration (ET)	86–89		Krause and Randolph (1989), Priest (2004)	
Stream baseflow	13.7–18.2		Priest (2004)	Mean annual stream baseflow estimates for a number of sites in coastal Georgia where the surficial aquifer is the outcropping unit
Bulk density of soil (ρ_b)	1.55	g/cm^3	Morris and Johnson (1967)	Mean value for fine sand
	1.12–1.99		Morris and Johnson (1967)	Typical range for fine sand
Longitudinal dispersivity (α_L)	7.6–15	ft	5–10% of aquifer length, from Fetter (1993), Gelhar et al. (1992); site-specific aquifer length (contaminant travel distance) estimated as 500 ft	Dispersivity is a function of the scale of the study. As the flow path increases, contaminant dispersion (mechanical mixing effects) will likewise increase (Fetter 1993)
Ratio of dispersivities (α_L/α_y)	10–100		Fetter (1993); Gelhar et al. (1992)	
Solubility in water (S)				
BHC isomers	Insoluble–69.5	mg/L	Range of six values listed in ATSDR (2005b) [Clayton and Clayton (1981), Kurihara et al. (1973), HSDB (2003), Hollifield (1979)]	
Gamma-BHC	6.6, 7, 11		USDA (2004)	
Gamma-BHC	6.8, 0.14–15.2		Mean and range of 63 values listed in Mackay et al. (1997)	
Retardation coefficient (R)				$R = 1 + (\rho_b/\theta) K_D$; $\rho_b = 1.6$ g/mL; $\theta = 0.4$; $K_D = f_{oc} \times K_{OC}$, where $f_{oc} = 0.001$, K_{OC} determined experimentally or from K_{OC}–K_{OW} relationships.
Alpha-BHC	3.5–15.9	—	Calculated using 3 K_{OC} values cited in ATSDR (2005b)	K_{OC} determined experimentally

(continued)

Table 9.1 (continued)

Property	Value	Units	References	Notes
	8.0, 5.1–12.6		Calculated using mean and range of 12 K_{OC} values from USEPA (1996)	K_{OC} determined experimentally
	12.1–22.5	–	MWH (2005); R calculated using 3 K_{OW} values	K_{OC} calculated from 3 different empirical K_{OC}–K_{OW} relationships
Beta-BHC	15.9		Calculated using K_{OC} values from Ripping (1972) [cited in ATSDR (2005b)]	K_{OC} determined experimentally
	9.6, 5.6–15.3		Calculated using mean and range of 14 K_{OC} values from USEPA (1996)	K_{OC} determined experimentally
	11.8–21.5		MWH (2005); R calculated using 3 K_{OW} values	K_{OC} calculated from 3 different empirical K_{OC}–K_{OW} relationships
Delta-BHC	26.2	–	Calculated using K_{OC} values from Weiss (1986) [cited in ATSDR (2005b)]	K_{OC} determined experimentally
	19.1–49.1		MWH (2005); R calculated using K_{OW} value	K_{OC} calculated from 3 different empirical K_{OC}–K_{OW} relationships
Gamma-BHC	3.6–15.9	–	Calculated using 6 K_{OC} values cited from various sources in ATSDR (2005b)	K_{OC} determined experimentally
	6.4, 3.9–14.0		Calculated using mean and range of 65 K_{OC} values from USEPA (1996)	K_{OC} determined experimentally
	11.1–18.8		MWH (2005); R calculated using 3 K_{OW} values	K_{OC} calculated from three different empirical K_{OC}–K_{OW} relationships

Table 9.2 Calibrated parameters values used in the Oatland Island study

Parameter	Value	Data source[a] or calculation method
Hydraulic conductivity (K), m/day	2.7	Selected from site-specific data and literature values (see Table 9.1)
Hydraulic gradient (i), m/m	6.7	Geometric mean of site-specific hydraulic gradients; (S&ME 1999; MWH 2003, 2005)
Porosity of soil (θ), unitless	0.4	Selected from site-specific data and literature values (see Table 9.1)
Bulk density of soil (ρ_b), g/cm^3	1.60	Typical for fine sand (Morris and Johnson 1967)
Specific discharge (V_d), m/year	6.7	$V_d = Ki$
Groundwater velocity (V), m/year	17	$V = Ki/\theta$
Longitudinal dispersivity (α_L), m	8	Selected as 5.5% of estimated aquifer length of 500 ft; (Fetter 1993, Gelhar et al. 1992)
Ratio of dispersivities (α_L/α_y), unitless	100	(Fetter 1993, Gelhar et al. 1992)
Longitudinal dispersion coefficient (D_x), m^2/year	140	$D_x = \alpha_L V$
Lateral dispersion coefficient (D_y), m^2/year	1.4	$D_y = D_x (\alpha_y/\alpha_L)$
Recharge to surficial aquifer (q), cm/year	15	Calculated from precipitation, surface runoff, and evapotranspiration data (see Table 9.1)
Retardation coefficient for BHC (R), unitless	14.5	Geometric mean of literature values for BHC partition coefficients (see Table 9.1)
Source concentration (C), in µg/L		
At time $= 0$ years (1970)	400	Initial concentration in groundwater assumed to be at 5% of the solubility limit for BHC. Source removal during 2000 simulated as a stepped reduction
At time $= 30.5$ years (June 2000)	200	
At time $= 31$ years (2001)	0.0	
Standard deviation (σ_W) of contaminant source width, assuming a Gaussian distribution, m	9	For $\sigma_W = 30$ ft, $\pm 2\ \sigma_W$ spans 120 ft and encompasses 95% of the source area. The actual width of the source area excavation is approximately 150 ft (MWH 2001)
Aquifer thickness, m	4.3	(MWH 2003, 2005)
X-coordinate length, m	259	ACTS computational grid geometry; x-axis aligned with direction of groundwater flow
Discretization along x-direction, m	7.6	ACTS computational grid geometry
Y-coordinate length, m	152	ACTS computational grid geometry
Discretization along x-direction, m	3	ACTS computational grid geometry
Contaminant source location (x, y), m	0.0, 76	ACTS computational grid location
Duration of simulation, years	300	ACTS simulation period
Temporal discretization, year	0.5	Selected time step for ACTS simulations

[a]Notes on parameter values are included in Table 9.1. References are included in the body of this report

input) analyses are summarized in Table 9.2. The analytical model was calibrated by iterative manual adjustment of input parameters to achieve simulation results that best matched contaminant concentrations measured in site monitoring wells. Simulation results were compared to field measurements from three monitoring wells (MW-17, MW-20, and MW-23 – Fig. 9.1) located along the centerline of the contaminant plume for longitudinal contaminant transport. To calibrate the model for lateral contaminant transport, simulation results were compared with the maximum contaminant plume width during 2003.

Contaminant source characteristics were defined in the model based upon existing contaminant and site information. The source area concentration of total BHC is modeled as a Gaussian distribution in the lateral direction (along the y-axis of the model grid). The maximum width of the source perpendicular to groundwater flow is assumed to be 46 m, based on the horizontal footprint of the source area excavation (MWH 2001). Seven different initial source concentration values of total BHC in groundwater were evaluated during model calibration: 10,000, 7,000, 1,000, 500, 400, 200, and 100 µg/L. An initial concentration of 400 µg/L total BHC in groundwater at the source offered the best calibration results and was, therefore, assumed for all model simulations presented in this report. This concentration is approximately 5% of the solubility limit of BHC in groundwater and two orders of magnitude higher than the maximum concentration of total BHC detected in the groundwater to date.

Two deterministic scenarios were simulated using the model to explore the change in total BHC concentration at the source as a result of source remediation during 2000. In the first scenario, it was assumed that the remediation, which included excavation of pesticide containers and contaminated soil down to groundwater, effectively eliminated the contaminant source. The "eliminated source scenario" is characterized in the model as a stepped reduction in source concentration from 400 to 0.0 µg/L over a 1-year period during 2000. In the second scenario, the source concentration of total BHC in groundwater was modeled as constant. The "constant source scenario" is characterized in the model as a sustained source concentration of 400 µg/L during the entire simulation period. This model scenario represents site conditions that may have resulted if source remediation activities had not taken place. The constant source scenario also could represent post-remediation site conditions where residual DNAPL is present in the source area below the water table. Residual DNAPL could act as an ongoing source to groundwater until the residual DNAPL is depleted. This is a purely hypothetical scenario; site-specific data are not available to confirm or quantify DNAPL in the source area.

9.3.3 Model Input Parameters for Probabilistic Simulations

A two-stage Monte Carlo simulation was used to conduct probabilistic analyses of contaminant transport. In stage 1, probability density functions (PDFs) were generated for eight input parameters – referred to as variants – using Monte Carlo

Table 9.3 Parameters values used to simulate Oatland Island model in probabilistic application

Parameter or variant	Mean value[a]	Range[a] used to generate PDF	PDF type[b]	Statistics for Monte Carlo simulated PDF	
Specific discharge (V_d), m/year	6.7	1.5–30	Lognormal	Minimum: 1.5 Maximum: 30	Mean: 6.7 Standard deviation: 3.3
Longitudinal dispersion coefficient (D_x), m²/year	140	30–560	Lognormal	Minimum: 28 Maximum: 550	Mean: 140 Standard deviation: 73
Lateral dispersion coefficient (D_y), m²/year	1.4	0.3–5.6	Lognormal	Minimum: 0.28 Maximum: 5.5	Mean: 1.4 Standard deviation: 0.74
Porosity of soil (θ), unitless	0.4	0.3–0.5	Uniform	Minimum: 0.3 Maximum: 0.5	Mean: 0.4 Standard deviation: 0.06
Recharge to surficial aquifer (q), cm/year	15	3–46	Normal	Minimum: 3.0 Maximum: 40	Mean: 15 Standard deviation: 6.1
Retardation coefficient for BHC (R), unitless	14.5	4–22	Triangle	Minimum: 4.2 Maximum: 21.9	Mean: 14.5 Standard deviation: 3.8
Aquifer thickness, m	4.3	4.0–4.6	Triangle	Minimum: 4.0 Maximum: 4.6	Mean: 4.3 Standard deviation: 0.12
Standard deviation (σ_W) of contaminant source width, assuming a Gaussian distribution, m	9	6–12	Normal	Minimum: 6.1 Maximum: 12	Mean: 9.1 Standard deviation: 1.1
Hydraulic conductivity (K), m/day	3	0.6–12[c]	–	PDF was not generated for this parameter.[c]	
Groundwater velocity (V), m/year	17	3.6–75[c]	–	PDF was not generated for this parameter.[c]	

[a]Notes on parameter values are included in Table 9.1

[b]10,000 Monte Carlo iterations were performed for each variant to generate a probability distribution function (PDF)

[c]Parameter range shown for reference only; hydraulic conductivity and groundwater velocity are not varied explicitly in the model simulations. These two parameters are varied implicitly: hydraulic conductivity in deriving Darcy velocity and groundwater velocity by varying Darcy velocity and porosity

simulation (Table 9.3). In stage 2, values from each PDF were randomly selected as input parameters for the fate and transport model to generate contaminant concentrations in groundwater distributed-value output for a specific time and location at the site (e.g., selected monitoring wells or the wetlands boundary – Fig. 9.1).

Each PDF generated in stage 1 consists of a range of 10,000 random values (or realizations) that characterizes the variability and uncertainty of a given input parameter. Within ACTS, the type (shape) and the descriptive statistics (mean, minimum, maximum, and variance) for each PDF were defined on the basis of the available site-specific field data and applicable literature references. The mean values specified for each PDF corresponded to the calibrated single-value parameter values used in the deterministic analysis (Table 9.2).

In stage 2, 10,000 Monte Carlo realizations of the model were run within ACTS using input values selected randomly from each of the eight PDFs developed in stage 1 (Table 9.3), thus generating a range of possible contaminant concentrations in groundwater for a specific time and location of interest. The probabilistic simulations were conducted using the eliminated source scenario. Simulation results for multiple time periods were aggregated for selected locations that are described below (Fig. 9.1):

- MW-17, located approximately 61 m downgradient from the source area, is within the contaminant plume. BHC has been detected at this monitoring well during annual sampling events since 2000.
- MW-20, located approximately 91 m downgradient from the source area, is within the contaminant plume. BHC has been detected at this monitoring well during annual sampling events since 2000.
- MW-23, located along Barley Drive, approximately 152 m downgradient from the source area, has not been affected by the contaminant plume. BHC has never been detected at this monitoring well during annual sampling events.
- The wetlands boundary, located approximately 244 m downgradient from the source area, has not been affected by the BHC contaminant plume. The wetlands extend along Richardson Creek, south and west of the Oatland Island Education Center.

9.4   Modeling Results

For the deterministic (single-value input) analysis, calibrated model parameter values were developed by manually varying input parameter values in ACTS and obtaining a best fit when compared with available field data. Comparisons of measured data and calibrated model simulation results of total BHC in groundwater at wells MW-17, MW-20, and MW-23 for 2001, 2003, and 2005 are listed in Table 9.4 and shown in Fig. 9.4. Having achieved a set of calibrated model parameter values given available field data, simulations were then conducted using single-value (deterministic) and distributed-value (probabilistic) inputs for a number of future time endpoints of interest. The results of these simulations are described in Sections 9.4.1 and 9.4.2.

Table 9.4 Deterministic results for four selected calibration targets, Oatland Island, Georgia, USA

Model calibration target	Distance from source (ft)	2001 Total BHC concentration (µg/L)		2003 Total BHC concentration (µg/L)		2005 Total BHC concentration (µg/L)	
		Measured[a]	Simulated	Measured[a]	Simulated	Measured[a]	Simulated
MW-17	200	5.4	10.2	1.6–4.7	11.0	2.5	11.7
MW-20	300	1.4	0.6	0.8–2.2	0.8	1.1	1.0
MW-23	500	<0.055	0.000041	<0.055	0.00011	<0.055	0.00026
Plume width at MW-20	300	Measured plume width at 0.02 µg/L total BHC contour = 52 m in 2003. Simulated plume width at 0.02 µg/L total BHC contour = 52 m in 2003						

[a]Site-specific data derived from consultant reports (MWH 2002a, b, 2005)

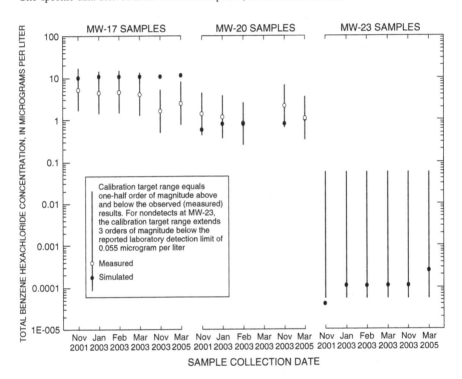

Fig. 9.4 Comparison of measured concentrations and deterministic simulation results for total benzene hexachloride in groundwater for 2001, 2003, and 2005 calibration targets (Anderson et al. 2007)

9.4.1 Deterministic Model Simulations

Deterministic results using calibrated, single-value parameter input (Table 9.2) are presented as spatial distributions of total BHC concentrations in groundwater for representative time periods (Fig. 9.5). The simulation times are for 30, 80, and 120 years from the presumed start of contamination (1970). Results are shown for

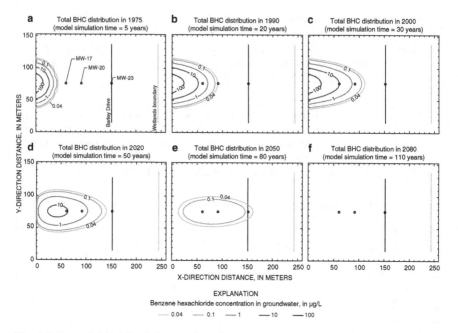

Fig. 9.5 Deterministic simulation results of total benzene hexachloride concentrations in groundwater for an eliminated source scenario during (**a**) 1975, (**b**) 1990, and (**c**) 2000, and for a constant source scenario during (**d**) 2020, (**e**) 2050, and (**f**) 2080 (Anderson et al. 2007; 1 ft equals 0.3048 m)

the eliminated source scenario (Fig. 9.5a–c) and the constant source scenario (Fig. 9.5d–f).

Deterministic results for the eliminated source and constant source scenarios are presented for comparison in Fig. 9.6 as total BHC concentration versus time. The input parameters for each scenario are identical. In the eliminated source scenario, the total BHC concentration in groundwater at the source is initially 400 µg/L and steps down to 0.0 µg/L by the end of 2000. In the constant source scenario, total BHC in groundwater at the source is 400 µg/L for all simulation times.

9.4.2 Probabilistic Model Simulations

The values generated within ACTS for the eight model variants can be expressed as frequency distributions (histograms) or PDFs. Both a histogram and a PDF are shown in Fig. 9.7 for Darcy velocity (specific discharge), one of the eight variants. The histogram (Fig. 9.7a) depicts the frequency, and the PDF (Fig. 9.7b) depicts the relative frequency (an approximation of probability) with which each value of Darcy velocity occurs in the set of 10,000 random Monte Carlo realizations. For this particular variant, a lognormal distribution was assigned as the PDF. Descriptive

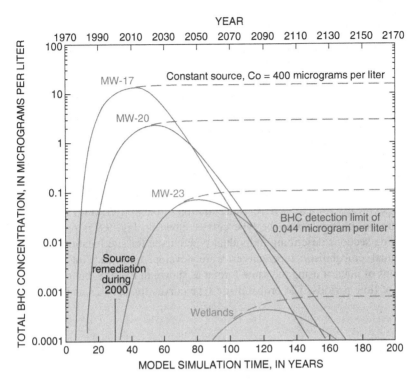

Fig. 9.6 Deterministic simulation results of total benzene hexachloride concentration in groundwater over time for an eliminated source scenario and a constant source scenario (Anderson et al. 2007)

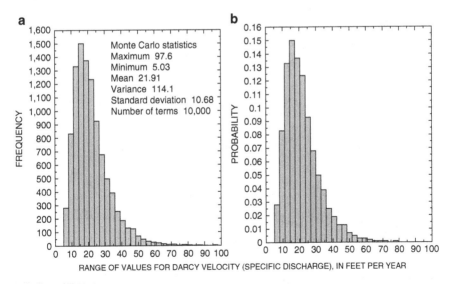

Fig. 9.7 Darcy velocity (specific discharge) values generated for probabilistic modeling: (**a**) frequency distribution (histogram) and (**b**) probability density function (PDF) (Anderson et al. 2007)

statistics for each of the eight variants used in the probabilistic fate and transport modeling are listed in Table 9.3.

Results for the probabilistic analysis of total BHC in groundwater at four different site locations (MW-17, MW-20, MW-23, and the wetlands) are depicted in Fig. 9.8 in terms of probability of exceedance versus total BHC concentration in groundwater. Here, the probability of exceedance is defined as the complementary cumulative probability function (or 1 minus the cumulative probability function).

These results are for the eliminated source scenario. Selected time periods are shown, beginning in 2005 and progressing to the time period that corresponds to the maximum probability of exceeding the detection limit of BHC (0.044 µg/L) at each site location. The maximum probability of exceeding the detection limit at each location is clearly shown in the probability of exceedance versus time plot developed specifically for a total BHC concentration of 0.044 µg/L (Fig. 9.9).

More information about the type curves shown in Fig. 9.8 is provided in the following sections describing individual type curves for site locations of interest. Individual probabilistic type curves were developed for each of the four site locations of interest using the same format as shown in Fig. 9.8, but over a broader range of time periods. The probabilistic type curves for MW-23 and the wetlands

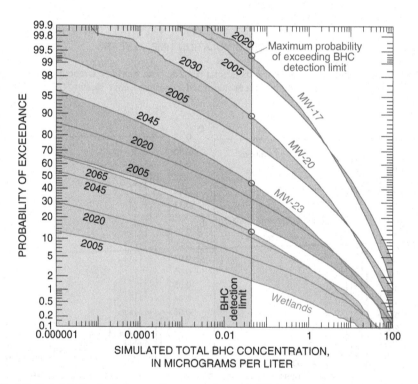

Fig. 9.8 Probabilistic simulation results expressed as probability of exceedance versus simulated total benzene hexachloride concentration for four site locations of interest (Anderson et al. 2007)

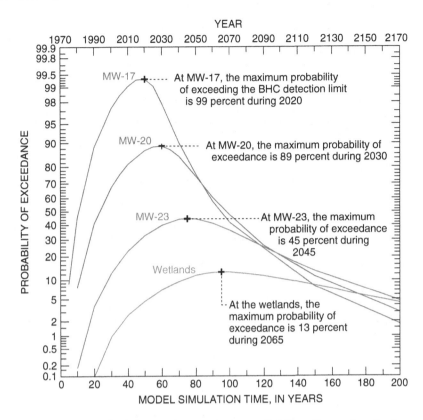

Fig. 9.9 Probabilistic simulation results expressed as probability of exceeding the benzene hexachloride detection limit versus simulation time for four site locations of interest (Anderson et al. 2007)

are discussed in the following sections and presented graphically in Figs. 9.10 and 9.11, respectively.

Probabilistic Type Curves for Monitoring Well MW-23: MW-23 is located along Barley Drive approximately 152 m downgradient from the source area. Probabilistic type curves for MW-23 were developed by aggregating results (concentration of total BHC in groundwater at MW-23) from multiple probabilistic model simulations at different time periods (Fig. 9.10).

Using Fig. 9.10, the probability of exceeding a selected concentration at a given time period is determined by selecting the concentration of interest on the x-axis (for example, 0.044 µg/L), moving vertically upward until intersecting the selected simulation time curve (for example, 2045), and then moving horizontally to the left until intersecting the y-axis. Using Fig. 9.10, at MW-23, the maximum probability of exceeding the detection limit for BHC is 45% during 2045. The results for time periods before and after 2045 are presented in separate plots for clarity:

i. The probabilistic results shown in Fig. 9.10a are presented in terms of probability of exceedance versus total BHC concentration in groundwater

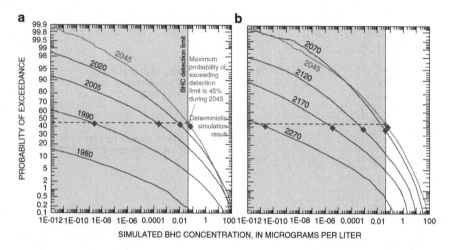

Fig. 9.10 Probabilistic simulation results at MW-23 for time periods (**a**) before and (**b**) after 2045, when the contaminant plume reached the maximum probability of exceeding the benzene hexachloride detection limit of 0.044 μg/L (Anderson et al. 2007)

for model simulations at 1980, 1990, 2005, 2020, and 2045. The probability of exceeding 0.044 μg/L at MW-23 steadily increases over time from 4% during 1990, to 18% during 2005, to 32% during 2020. The maximum probability of exceeding 0.044 μg/L total BHC at MW-23 is 45% during 2045.

ii. The probabilistic results shown in Fig. 9.10b are presented in terms of probability of exceedance versus total BHC concentration in groundwater for model simulations of 2045, 2070, 2120, 2170, and 2270. For example, the maximum probability of exceeding the BHC detection limit of 0.044 μg/L during 2170 (200 years from the start of simulation) is 5%.

Probabilistic Type Curves for Wetlands Boundary: The wetlands boundary is approximately 244 m downgradient from the source area. Probabilistic type curves for the wetlands boundary also were developed by aggregating results from multiple model simulations at different time periods (Fig. 9.11). At the wetlands boundary, the maximum probability of exceeding the detection limit for total BHC (0.44 μg/L) is 13% during 2065 (Fig. 9.11). The results for time periods before and after 2065 are presented in separate plots for clarity:

i. The probabilistic results shown in Fig. 9.11a are presented in terms of probability of exceedance versus total BHC concentration in groundwater for model simulations of 1980, 1990, 2000, 2020, 2050, and 2065. The probability of exceeding 0.044 μg/L at the wetlands steadily increases over time from 1% during 2000, to 5% during 2020, to 11% during 2050. The maximum probability of exceeding 0.044 μg/L total BHC at the wetlands is 13% during 2065.

ii. The probabilistic results shown in Fig. 9.11b are presented in terms of probability of exceedance versus total BHC concentration in groundwater for model simulations of 2065, 2170, 2220, and 2270. For example, the

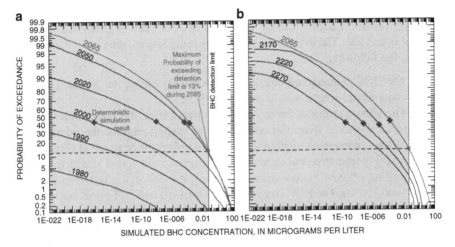

Fig. 9.11 Probabilistic simulation results at the wetlands for time periods (**a**) before and (**b**) after 2065, when the contaminant plume reached the maximum probability of exceeding the benzene hexachloride detection limit of 0.044 µg/L (Anderson et al. 2007)

maximum probability of exceeding the BHC detection limit of 0.044 µg/L during 2170 (200 years from the start of simulation) is 4%.

Deterministic model results (from single-value input parameters) also are included in Fig. 9.11 for each simulation year depicted, with the exception of 1980 and 1990 (results for these are below the range of simulated BHC concentrations shown in Fig. 9.11). Simulated results are similar to those obtained at MW-23 in that they fall in the 40–50% range of the probability of exceedance.

9.5 Discussion of Results and Conclusions

Modeling results obtained by applying ACTS in a deterministic (single-value input) analysis are in good agreement with measured groundwater contaminant concentrations at the Oatland Island site. Results from the deterministic analysis are slightly higher than measured concentrations to those at MW-17, but nearly identical to those at MW-20 and MW-23 (Table 9.4, Fig. 9.4). The reason for the greater difference between measured and simulated concentrations at MW-17 could be attributed to:

    i. Greater heterogeneity near MW-17 (Fig. 9.2) that is not accurately characterized by a single mean value (deterministic) model input;
   ii. Analytical model limitations;
  iii. Measurement error; or,
   iv. A combination of 1–3 above.

These calibration results are considered quite acceptable for fate and transport simulations using an analytical model. Both the deterministic and probabilistic simulation results obtained using two dimensional analytical fate and transport modeling provide insights into the migration of pesticides in groundwater at the Oatland Island site. The deterministic analyses using single-value input parameters indicate the following:

i.   The BHC contaminant plume develops rapidly during the 30 years that the contaminant source was in place (Fig. 9.5).
ii. The source removal during 2000 occurred before the contaminant plume reached steady state. Consequently, the leading edge of the contaminant plume continues to migrate downgradient after source removal (Fig. 9.5).
iii. In the eliminated source scenario, the maximum travel distance for the leading edge of the contaminant plume (at 0.044 µg/L) is 165 m after 80 years (2050). The BHC contaminant plume (at 0.044 µg/L) does not advance further downgradient after 80 years (Fig. 9.5).
iv. In the constant source scenario, the maximum travel distance for the leading edge of the contaminant plume (at 0.044 µg/L) is 171 m after 120 years (2090). The BHC contaminant plume does not advance further downgradient after 120 years (Fig. 9.5).
v.   Results from deterministic analysis for calibrated, single-value input parameters indicate that the contaminant plume will not affect the wetlands (Figs. 9.5 and 9.6). Simulated total BHC concentrations at the wetlands are below the BHC detection limit (0.044 µg/L) and below all relevant BHC reference values for regulatory and health guidance.
vi. Simulated total BHC concentrations in groundwater are above the detection limit (0.044 µg/L) at MW-23 during 2030–2070 (Fig. 9.6).
vii. Probabilistic analyses using 10,000 realizations for each of eight parameter variants indicate that the probability of exceeding the detection limit (0.044 µg/L) of total BHC in groundwater at the wetlands boundary increases from 1% during 2000 to a maximum of 13% during 2065 (Figs. 9.9 and 9.11).

Thus, there is at least an 87% confidence level that the downgradient wetlands will not be affected by the pesticide contaminant plume. MW-23 is located along the centerline of the plume. It is halfway between the former source area and the wetlands, it is screened across nearly the full depth of the surficial aquifer (Fig. 9.2), and it has not been affected by the contaminants (no detections of BHC in eight groundwater monitoring events during 2000–2005). Because of its location, MW-23 is particularly well-suited to use as a compliance tool for decision-making. Model simulations predict a 12–45% probability that total BHC will be above the detection limit during 2000–2045 at MW-23 (Figs. 9.9 and 9.10). Future groundwater monitoring results for MW-23 can be used both to evaluate actual contaminant plume migration and to gauge the conservatism of the model. If total BHC is consistently not detected at MW-23 during future monitoring events, then the analytical model is likely more conservative than actual site conditions, and model estimates for wetlands impact (maximum probability of 13% that total BHC will exceed the detection limit) may be too high. If BHC is detected in MW-23 during future sampling events, the MW-23 probabilistic type curves

(Fig. 9.10) provide a useful framework for interpreting the results. If sampling results at MW-23 consistently remain within model predictions, installation of an additional monitoring well to facilitate monitoring the leading edge (maximum travel distance) of the contaminant plume may be appropriate. Model simulations of the contaminant plume indicate a maximum travel distance of 12–18 m beyond Barley Drive during 2050–2090. On the basis of these results, the additional well should be located 30 m beyond Barley drive, along the centerline of the plume.

Measured site sampling results during 2000–2005 indicated detectable concentrations of BHC in MW-17 and MW-20. Future sampling results for these monitoring wells can be compared with the probabilistic type curves developed for these wells to evaluate the migration of BHC in groundwater over time. If the contaminant source was effectively eliminated during remediation during 2000, total BHC concentration in groundwater should decrease over time at MW-17 and MW-20. Given the physical and chemical properties of BHC, some contamination may be present in the source area in the form of residual DNAPL in the saturated zone (Pankow and Cherry 1996; USEPA 2003; Environment Agency 2003). If residual DNAPL is present, it will act as an ongoing source, and BHC concentrations in the groundwater at MW-17 and MW-20 could remain relatively constant, with little or no decrease, until the residual DNAPL source is depleted. Deterministic model simulations for a constant source, which could represent the upper bound for a residual DNAPL scenario, indicate that the BHC plume stabilizes at 171–18 m beyond Barley Drive – after approximately 120 years (Fig. 9.5). Future groundwater monitoring results could be compared with the eliminated source and constant source simulation results shown in Fig. 9.6 to assess the likelihood of residual DNAPL at the site.

9.6 Limitations

Analytical models such as those incorporated into the ACTS software are screening-level tools that use mathematical solutions to model groundwater and contaminant transport. Calibrating analytical models is not as rigorous or exact as calibrating more mathematically complex numerical models that are based on fewer simplifying assumptions and limitations. Additionally, as is common with many environmental analyses, the mass of contaminant released at the site is unknown. Therefore, calibration based on a contaminant mass balance is not possible. Instead, calibration of groundwater and contaminant transport is accomplished by – plume- or history-matching, which means that the model input parameters are adjusted iteratively to produce simulation results that match, as closely as possible, the contaminant concentrations measured at the site over time.

Much of the uncertainty and variability in the model parameters used to describe the hydrogeology and contaminant characteristics are captured by the PDFs generated for the eight model variants. Still, differences and heterogeneities in actual site conditions could affect groundwater and contaminant transport in ways that are not represented by the analytical model because of limitations and simplifications of the governing equations of groundwater flow and contaminant transport. The primary limitation is the assumption of a constant and uniform groundwater velocity in the longitudinal direction (along the x-axis of the model grid).

In all model simulations, it was conservatively assumed that the biological and chemical degradation of site contaminants was negligible. If BHC is (or other site contaminants are) undergoing biological or chemical transformation at the site, model results would overestimate the time required for BHC concentrations to decrease at MW-17 and MW-20. Additionally, the estimated probability of affecting the wetlands at any given compliance goal – the detection limit or other applicable state or federal standards – would likely decrease.

For the probabilistic analyses, it was assumed that remediation activities during 2000 effectively eliminated contaminants in the unsaturated zone (down to the water table) and that no residual DNAPL was present in the saturated zone below the source area. If residual DNAPL was (or is) present, model results would underestimate the time required for BHC contaminant concentrations to decrease at MW-17 and MW-20. Additionally, the presence of residual DNAPL in the saturated zone below the source area could introduce scenarios for contaminant migration that are not represented by the current model. If, for example, the residual DNAPL below the source area migrates down to the clay confining layer at 5–6 m bgs (Fig. 9.2), it could travel along the surface of the clay layer and into another area of the site. The original mass of the contaminant source is unknown and the topography of the clay layer is unknown. The potential extent, transport, and fate of residual DNAPL at the site, therefore, cannot be effectively measured or predicted.

9.7   Practical Applications

The existing site data and simulation results presented herein suggest that monitored natural attenuation of the pesticide contaminants in groundwater is a viable next step for the Oatland Island site.

  i. Source remediation (excavation of containers and contaminated soil) was completed during 2000 (MWH 2001).
 ii. In-situ treatment of source area groundwater contaminants was conducted during 2003–2004 (MWH 2001, 2005).
iii. No immediate or future public health issues are associated with the site.
 iv. Results from deterministic analysis for calibrated, single-value inputs indicate the contaminant plume will not reach the wetlands 244 m downgradient of the source area.
  v. Results from probabilistic analysis indicate that the probability of exceeding the detection limit of 0.044 µg/L total BHC in groundwater at the wetlands boundary increases from 1% during 2000 to a maximum of 13% during 2065 (Figs. 9.9 and 9.11).

vi. MW-23, located along the centerline of the plume, halfway between the former source area and the wetlands, has not yet been affected by the contaminants (no detection of BHC in eight groundwater monitoring events during 2000–2005). Model simulations predict a 12–45% probability that total BHC will affect MW-23 above the detection limit during 2000–2045 (Figs. 9.9 and 9.10). Future groundwater monitoring results for MW-23 could be used to evaluate actual contaminant plume migration and to gauge the conservatism of the model.

Based upon the actual (measured) and projected (simulated) movement of the contaminant plume, there is ample time to continue monitoring at the site without endangering the downgradient wetlands. The probabilistic type curves developed for this site should be used to evaluate future groundwater monitoring results and guide environmental and regulatory decision-making.

It should be noted that previous modeling for the site, which was completed by consultants using both analytical and computationally intensive numerical methods with single-value inputs, suggested that continued source remediation was needed to prevent the pesticide plume from reaching the wetlands. The analyses described herein utilized an analytical model linked with probabilistic techniques (publicly available in the ACTS software) to explore a wide range of parameter input values and to estimate the probability that the contaminant plume would reach the wetlands. The current analyses, in addition to being computationally efficient and easily conducted on affordable desktop computers, resulted in monetary savings for the project. Internal staff hours for analytical model calibration and probabilistic simulations are estimated at $US 50,000–60,000 during a 6-month time period compared to projected costs of $US 1,200,000 for continued remediation activities including consulting fees.

With this case study, it is demonstrated that careful application of analytical modeling techniques combined with judicious probabilistic analysis methods can be used to provide a relatively rapid, cost-effective way to explore and assess contaminant fate and transport in groundwater systems.

References

Anderson BA, Maslia ML et al (2007) Probabilistic analysis of pesticide transport in shallow groundwater at the Otland Island Education Center, Otland Island, Georgia. Agency for Toxic Substances and Disease Registry (ATSDR), Atlanta, 48 p

ATSDR [Agency for Toxic Substances and Disease Registry] (2005) Health consultation for the Oatland Island Education Center, Savannah, Chatham County, GA. U.S. Department of Health and Human Services, Agency for Toxic Substances and Disease Registry, Atlanta, GA

ATSDR (2005) Toxicological profile for hexachloro-cyclohexanes. U.S. Department of Health and Human Services, Agency for Toxic Substances and Disease Registry, Atlanta, GA

Clarke JS, Hacke CM et al (1990) Geology and ground-water resources of the coastal area of Georgia. Georgia Department of Natural Resources, Georgia Geologic Survey, Bulletin 113, Atlanta, GA

Clayton G, Clayton F (eds) (1981) Patty's industrial hygiene and toxicology, 3rd edn. New York, Wiley

Cullen AC, Frey HC (1999) Probabilistic techniques in exposure assessment: a handbook for dealing with variability and uncertainty in models and inputs. Plenum Press, New York

Domenico PA, Schwartz FW (1998) Physical and chemical hydrogeology. Wiley, New York

Environment Agency (2003) An illustrated handbook of DNAPL transport and fate in the subsurface. Environment Agency, R&D Publication 133, Bristol, UK

Fetter CW (1993) Contaminant hydrogeology. Macmillan, New York

Freeze R, Cherry J (1997) Groundwater. Prentice-Hall, Edgewood Cliffs, NJ

Gelhar LW, Welty C, Rehfeldt K (1992) A critical review of data on field-scale dispersion in aquifers. Water Resour Res 28(7):1955–1974

Georgia EPD [Georgia Environmental Protection Division] (1993) Hazardous site response. Chapter 391-3-19. In: Rules of Georgia Department of Natural Resources, Environmental Protection Division (last modified in June 2003), Atlanta, GA

Hollifield HC (1979) Rapid nephelometric estimate of water solubility of highly insoluble organic chemicals of environmental interest. Bull Environ Contam Toxicol 23:579–586

Hazardous Substances Data Bank (2003) Hexacyclochloro-hexanes. Environmental standards and regulations. National Institutes of Health, National Library of Medicine, Bethesda, MD. http://toxnet.nlm.nih.gov/cgi-bin/sis/htmlgen?HSDB February, 2010

Krause RE, Randolph RB (1989) Hydrology of the Floridan aquifer system in southeast Georgia and adjacent parts of Florida and South Carolina. U.S. Geological Survey, Professional Paper 1403-D, Denver, CO

Kurihara N, Uchida M, Fujita T et al (1973) Studies on BHC isomers and related compounds: some physiochemical properties of BHC isomers. Pestic Biochem Physiol 2:383–390

Mackay D, Shiu WY et al (1997) Illustrated handbook of physical-chemical properties and environmental fate for organic chemicals. CRC, Boca Raton, FL

Maslia ML, Aral MM (2004) ACTS – a multimedia environmental fate and transport analysis system. Practice Periodical Hazard Toxic Radioact Waste Manage – ASCE 8(3):1–15

Maslia ML, Aral MM et al (2009) Reconstructing historical exposures to volatile organic compound-contaminated drinking water at a U.S. military base. Water Qual Expo Health 1(1):49–68

Maslia ML, Aral MM et al (1997) Exposure assessment using analytical and numerical models: a case study. Practice Periodical Hazard Toxic Radioact Waste Manage-ASCE 1(2):50–60

MWH [Montgomery Watson Harza] (2001) Final remediation report, Oatland Island Education Center, Oatland Island, Georgia. U.S. Department of Health and Human Services, Centers for Disease Control and Prevention, Atlanta, GA

MWH (2002a) Draft November 2001 round of semi-annual groundwater and monitored natural attenuation report, Oatland Island Education Center. U.S. Department of Health and Human Services, Centers for Disease Control and Prevention, Atlanta, GA

MWH (2002b) Draft July 2002 round of semi-annual groundwater and monitored natural attenuation report, Oatland Island Education Center. U.S. Department of Health and Human Services, Centers for Disease Control and Prevention, Atlanta, GA

MWH (2003) Final voluntary corrective action work plan in support of UIC permit application, Oatland Island Education Center. U.S. Department of Health and Human Services, Centers for Disease Control and Prevention, Atlanta, GA

MWH (2005) Final groundwater sampling and in situ chemical oxidation summary report, Oatland Island Education Center. U.S. Department of Health and Human Services, Centers for Disease Control and Prevention, Atlanta, GA

Morris DA, Johnson AI (1967) Summary of hydrologic and physical properties of rock and soil materials as analyzed by the Hydrologic Laboratory of the U.S. Geological Survey. U.S. Geological Survey, Water-Supply Paper 1839-D

Pankow JF, Cherry JA (1996) Dense chlorinated solvents and other DNAPLs in groundwater. Waterloo, Portland, OR

Priest S (2004) Evaluation of groundwater contribution to streamflow in coastal Georgia and adjacent parts of Florida and South Carolina. U.S. Geological Survey, Scientific Investigations Report 2004-5265, Reston, VA

Ripping G (1972) Screening of the absorption behavior of new chemicals: natural spoils and model absorptions. Ecotoxicol Environ Saf 6:236–245

Rodenbeck S, Maslia ML (1998) Groundwater modeling and GIS to determine exposure to TCE at Tuscon. Practice Periodical Hazard Toxic Radioact Waste Manage 2(2):53–61

S&ME [Soil & Material Engineers, Inc] (1986). Groundwater availability of the Pliocene-recent aquifer system, Skidaway Island, Georgia. Soil and Material Engineers, Inc., Columbia, SC (unpublished report on file at the U.S. Geological Survey, Doraville, GA)

S&ME (1999) Environmental assessment report, areas "A" and "B", Oatland Island Education Center, Oatland Island, Georgia. U.S. Department of Health and Human Services, Centers for Disease Control and Prevention, Atlanta, GA

Schwille F (1988) Dense chlorinated solvents in porous and fractured media. CRC, Boca Raton, FL

USDA [U.S. Department of Agriculture] (2004) Agricultural research service pesticide properties database. http://www.ars.usda.gov/Services/docs.htm?docid=14199. Accessed April 2006

USEPA (1993) Memorandum dated December 13 to Water Management Division Directors (USEPA Regions 1–10) from James R. Elder (Director) concerning guidance and clarification on the use of detection limits in compliance monitoring. US Environmental Protection Agency, Office of Groundwater and Drinking Water, Washington DC

USEPA [U.S. Environmental Protection Agency] (1996) Soil screening guidance: technical background document. U.S. Environmental Protection Agency, Superfund Program, Washington, DC. Report No: EPA/540/R-96/018

USEPA (2003) The DNAPL remediation challenge: Is there a case for source depletion? U.S. Environmental Protection Agency expert panel on DNAPL remediation, Washington, DC. Report No.: EPA-600-R-03-143

USEPA (2006) Addendum to the 2002 Lindane Reregistration Eligibility Decision (RED). Prevention, pesticides and toxic substances. U.S. Environmental Protection Agency, Washington, DC

USEPA (2006b) Assessment of lindane and other hexachlorocyclohexane isomers. U.S. Environmental Protection Agency, Prevention, Pesticides, and Toxic Substances. Washington, DC

Weiss G (ed) (1986) Hazardous chemicals data book, 2nd edn. Noyes Data Corporation, Park Ridge, IL

Appendix 1
Definitions of Acronyms and Abbreviations

In this appendix a collection of acronyms and abbreviations that are commonly used in the environmental modeling, exposure and health risk analysis fields are given. Definitions of most of these acronyms and abbreviations are also given in various chapters of the book where appropriate.

AADI	Adjusted acceptable daily intake
ADD	Average daily dose
ADI	Acceptable daily intake
AHERA	Asbestos Hazard Emergency Response Act (1986)
ASHAA	Asbestos School Hazard Abatement Act (1984)
ASHDCA	Asbestos School Hazard Detection and Control Act (1980)
AUR	Air unit risk
BAT	Best Available Technology
BCF	Bioconcentration factor
BHP	Biodegradation, hydrolysis, and photolysis
BOD	Biochemical oxygen demand
BOD5	Biochemical oxygen demand as measured in the standard 5-day test
BPT	Best practical technology
Bw	Body weight
Bwa	Body weight (kg) for experimental animal species used in the HEC derivation of an RfC
BWh	Body weight (kg) for human used in the HEC derivation of an RfC
CAA	Clean Air Act (1970)
CDI	Chronic daily intake
CERCLA	Comprehensive Environmental Response, Compensation and Liability Act (1970)
CFR	Code of Federal Regulations
COD	Chemical oxygen demand
CRAVE	Carcinogen risk assessment verification endeavor

CSISSFRA	Chemical Safety Information, Site Security and Fuels Regulatory Relief Act (1999)
CWA	Clean Water Act (1977)
CZMA	Coastal Zone Management Act (1972)
DRA	Dose–response assessment
DRR	Dose response relationship
DW	Drinking water
DWEL	Drinking water equivalent level
EA	Exposure assessment
EC	Exposure concentration
EC_{50}	Effective concentration, 50% affected
ED	Effective dose
EED	Estimated exposure dose
EMS	Environmental management system
EMTD	Estimated maximum tolerated dose
EP	Extraction procedure
EPE	Environmental performance evaluation
EPCRA	Emergency Planning and Community Right-to-Know Act (1986)
ESA	Endangered Species Act (1973)
FEL	Frank-effect level
FFDCA	Federal Food, Drug, and Cosmetic Act (1938)
FIFRA	Federal Insecticide, Fungicide and Rodenticide Act (1947)
FMEA	Failure mode and effects analysis
FOIA	The Freedom of Information Act (1966)
FQPA	Food Quality Protection Act (1996)
FR	Federal Register
HA	Health Advisory
HAPPS	Hazardous air pollution prioritization system
HAPs	Hazardous air pollutants
HAS	Health assessment summary
HAZOP	Hazards and Operability Study
HDT	Highest dose tested
HEAST	Health effects assessment summary tables
HEC	Human equivalent concentration
HEEP	Health and environmental effects profile
HI	Hazard index. Defined in terms of the ratio [intake/RfD], is used to compare relative harmful noncarcinogenic effects
HMTA	Hazardous Materials Transportation Act (1975)
HON	Hazardous organic NESHAPs
HSDB	Hazardous Substance Data Base
ICRP	International Commission on Radiological Protection
IRAA	Indoor Radon Abatement Act (1988)
IRIS	Integrated risk information system
ISO	International Organization for Standardization

LADD	Lifetime average daily dose
LBPPPA	Lead-Based Paint Poisoning Prevention Act (1971)
LCA	Life cycle assessment
LCCA	Lead Contamination Control Act (1988)
LC_{50}	Lethal concentration, 50% affected
LCLO	Lethal concentration Low; the lowest concentration at which death occurs
LC_{50}	Lethal concentration 50; concentration hazard lethal to 50% of the animals
LD_{50}	Lethal dose 50, dose lethal to 50% of the animals
LDL	Lethal dose low – the lowest dose at which death occurs
LDT	Lowest dose tested
LEL	Lower explosive limit
LEL	Lowest-effect level
LEPC	Local Emergency Planning Committee
LOAEL	Lowest observed adverse-effect level. A result of toxicological studies which identify chemical concentration levels with critical toxic effect
LOAEL(ADJ)	LOAEL adjusted to continuous exposure duration from an intermittent regimen by hour/day and days/7 days
LOAEL(HEC)	LOAEL adjusted for dosimetric differences across species to a human equivalent concentration
LOC	Limiting oxygen concentration
LOEL	Lowest observed effect level
MACT	Maximum Achievable Control Technology
MATC	Maximum allowable toxicant concentration
MCL	Maximum contaminant level
MCLG	Maximum contaminant level goals
MEI	Maximum exposed individual
MF	Modifying factor
MOC	Management of change
MOE	Margin of exposure
MOS	Margin of safety
MPRSA	Marine Protection, Research, and Sanctuaries Act (1972)
MSDS	Material Safety Data Sheet
MTBF	Mean time between failure
MTD	Maximum tolerated dose
MTL	Median threshold limit
MTTF	Mean time to failure
MWTA	Medical Waste Tracking Act (1988)
NAAQS	National Ambient Air Quality Standards
NEEA	National Environmental Education Act (1990)
NEPA	National Environmental Policy Act of 1969
NESHAP	National Emission Standards for Hazardous Pollutants

NOAEL	No observed adverse-effect level. A result of toxicological studies which identify chemical concentration levels with no observed toxic effect
NOAEL(ADJ)	NOAEL adjusted to continuous exposure duration from an intermittent regimen by hour/day and days/7 days
NOAEL(HEC)	NOAEL adjusted for dosimetric differences across species to a human equivalent concentration
NOEC	No observed effects concentration
NOEL	No observed effect level
NPDES	National Pollutant Discharge Elimination System
NSPS	New-Source Performance Standards
NTP	National Toxicology Program
NWPA	Nuclear Waste Policy Act (1982)
ODA	Ocean Dumping Act (1972)
ODBA	Ocean Dumping Ban Act (1988)
OHM/TADS	Oil and Hazardous Materials Technical Assistance Data Systems
OPA	The Oil Pollution Act of (1990)
OSHA	The Occupational Safety and Health Act (1970)
P	Probit dose extrapolation model
PBPK	Physiologically based pharmacokinetic
PCi	Picocurie
PD	Position Document
PEL	Permissible exposure limit
PELs	Permissible exposure limits
PFD	Process Flow Diagram
PHA	Process hazards analysis
PMR	Proportionate mortality ratio
PPA	The Pollution Prevention Act (1990)
PPPA	Pollution Prevention Packaging Act (1970)
PPE	Personal protective equipment
PSA	Process safety analysis
PSI	Process safety information
PSM	Process Safety Management
RA	Risk assessment
RAFS	Risk Assessment and Feasibility Study
RCRA	Resource Conservation and Recovery Act
RDD	Regional deposited dose
RDDR	Regional deposited dose ratio used in derivation of an HEC for particles
RDDR(ER)	Regional deposited dose ratio used in the HEC derivation of an RfC for an observed extrarespiratory effect of particles
RDDR(ET)	Regional deposited dose ratio used in the HEC derivation of an RfC for an observed effect of particles in the extrathoracic region of the respiratory tract

RDDR(PU)	Regional deposited dose ratio used in the HEC derivation of an RfC for an observed effect of particles in the pulmonary region of the respiratory tract
RDDR(TB)	Regional deposited dose ratio used in the HEC derivation of an RfC for an observed effect of particles in the tracheobronchial region of the respiratory tract
RDDR(TH)	Regional deposited dose ratio used in the HEC derivation of an RfC for an observed effect of particles in the thoracic region of the respiratory tract
RDDR(TOTAL)	Regional deposited dose ratio used in the HEC derivation of an RfC for an observed effect of particles in the total respiratory tract
RECRA	Resource Conservation and Recovery Act
RGD	Regional gas dose
RGDR	Regional gas dose ratio used in the derivation of an HEC for gases
RGDR(ET)	Regional gas dose ratio used in the HEC derivation of an RfC for an observed effect of a gas in the extrathoracic region of the respiratory tract
RGDR(PU)	Regional gas dose ratio used in the HEC derivation of an RfC for an observed effect of a gas in the pulmonary region of the respiratory tract
RGDR(TB)	Regional gas dose ratio used in the HEC derivation of an RfC for an observed effect of a gas in the tracheobronchial region of the respiratory tract
RGDR(TH)	Regional gas dose ratio used in the HEC derivation of an RfC for an observed effect of a gas in the thoracic region of the respiratory tract
RGDR(TOTAL)	Regional gas dose ratio used in the HEC derivation of an RfC for an observed effect of a gas in the total respiratory tract
RfD	Reference dose
RfC	Reference concentration
RgD	Regulatory dose
RM	Risk management
RME	Reasonably maximally exposed individual
RMP	Risk management plan
RQ	Recordable quantity, reportable quantity
RRA	Resource Recovery Act (1970)
RTECS	Registry of Toxic Effects of Chemical Substances
RV	Residual volume
Sa	Surface area (in cm^2) of respiratory tract region for experimental animal species used in the HEC derivation of an RfC
Sa(ET)	Surface area (in cm^2) of extrathoracic region for experimental animal species used in the HEC derivation of an RfC

Sa(TB)	Surface area (in cm^2) of tracheobronchial region for experimental animal species used in the HEC derivation of an RfC
Sa(TH)	Surface area (in cm^2) of thoracic region for experimental animal species used in the HEC derivation of an RfC
Sa(PU)	Surface area (in cm^2) of pulmonary region for experimental animal species used in the HEC derivation of an RfC
Sa(TOTAL)	Surface area (in cm^2) of total respiratory system for experimental animal species used in the HEC derivation of an RfC
SAB	Science Advisory Board
SANSS	Structure and Nomenclature Search System
SAR	Structure activity relationship
SARA	Superfund Amendments and Reauthorization Act of 1986
SC	Subcutaneous
SCE	Sister-chromatid exchange
SDWA	Safe Drinking Water Act (1974)
SECDA	Shoreline Erosion Control Demonstration Act (1974)
SEPA	Shoreline Erosion Protection Act (1965)
SF	Slope factor, safety factor
Sh	Surface area (in cm^2) of respiratory tract for humans, used in the HEC derivation of an RfC
Sh(ET)	Surface area (in cm^2) of extrathoracic region for humans, used in the HEC derivation of an RfC
Sh(TB)	Surface area (in cm^2) of tracheobronchial region for humans, used in the HEC derivation of an RfC
Sh(TH)	Surface area (in cm^2) of thoracic region for humans, used in the HEC derivation of an RfC
Sh(PU)	Surface area (in cm^2) of pulmonary region for humans, used in the HEC derivation of an RfC
Sh(TOTAL)	Surface area (in cm^2) of total respiratory system for humans, used in the HEC derivation of an RfC
SIP	State Implementation Plan
SMCL	Secondary maximum contaminant level
SMCRA	Surface Mining Control and Reclamation Act (1977)
SMR	Standard mortality ratio
SPA	Shore Protection Act (1988)
SPCC	Spill Prevention Control and Countermeasures
STEL	Short-term exposure limit
SWDA	Solid Waste Disposal Act (1965)
TAG	Technical Advisory Group
TD	Toxic dose
TDB	Toxicology data base
TEA	Total exposure analysis
TEC	Total environmental characterization
TLV	Threshold limit value

TQ	Threshold quantity
TOC	Total organic carbon
TOTAL	Total respiratory tract
TRE	Total resource effectiveness
TSCA	Toxic Substances Control Act (1976)
TSS	Total suspended solids
TWA	Time weighed average
UCL	Upper confidence limit
UEL	Upper explosive limit
UF	Uncertainty factor
UMTRCA	Uranium Mill-Tailings Radiation Control Act (1978)
Vaa	Alveolar ventilation rate (m^3/day) for experimental animal species used in HEC derivation of an RfC
Vah	Alveolar ventilation rate (m^3/day) for human used in HEC derivation of an RfC
VOC	Volatile organic compound
v/v	Volume for volume
WBC	White blood cell(s)
WQC	Water quality criteria

Appendix 2
Environmental Modeling and Exposure Analysis Terms

In this appendix a list of terms which are commonly used in the environmental modeling, exposure and health risk analysis fields are given, along with their definitions.

Environmental Pathways and Processes

Advection
: Transport by an imposed or ambient velocity field in any of the three coordinate directions (longitudinal, lateral and transverse), as in river, canal or coastal water velocity fields, or as in a Darcy velocity field in an aquifer or a wind velocity in the atmosphere.

Absorption
: Diffusion of a chemical into solid particles by a sorption process through the boundary of the solid particles. Due to this process, the transport of chemicals may be slowed (retarded) or the solid particles may behave as long term (secondary) sources for these chemicals due to potential dissolution of the chemical from the solid particles back into the medium.

Adsorption
: Attachment of chemicals to a particle surface during a sorption process. Due to this process, the transport of chemicals may be slowed (retarded) or the solid particles may behave as long term (secondary) sources for these chemicals due to potential desorption of the chemical from the solid particles back into the medium.

Aquifer
: A geologic formation which contains water and transmits water under normal drainage conditions.

Aquiclude	A geologic formation which contains water but does not transmit water.
Aquifuge	A geologic formation that neither contains water nor transmits water.
Aquitard	A geologic formation which contains water and transmits water at low rates relative to the transmission rate of aquifers.
Conduction	Transfer of energy by the jostling motion of atoms through direct contact between atoms.
Convection	Transport induced by hydrostatic instability, such as density or temperature differences in a lake, reservoir or atmosphere.
Chemical reaction	By definition, chemical reactions involve the formation or breakage of chemical bonds between atoms of a chemical. Chemical reactions may be described in terms of chemical kinetics, which describe the rate at which a reaction takes place or of chemical equilibrium, which describe the final expected chemical composition in a control volume.
Confined aquifer	An aquifer which is bounded by impervious geologic formations both from above and below.
Decay	Radioactive chemicals or chemicals in general may undergo decay, which will reduce the concentration levels of both dissolved and sorbed phases. The rate of disappearance (decay) of a chemical is usually expressed in terms of the half life of the chemical.
Decomposition	The decomposition process is a chemical recycling process, in which organic compounds are broken down into their inorganic components (mineralization or remineralization). The agents which affect this process are bacteria or fungi, often in association with worms, insects and other organisms that aid the process by breaking up the organic matter both chemically and mechanically.
Deposition	The removal of airborne substances to available surfaces that occurs as a result of gravitational settling and diffusion, as well as electrophoresis and thermophoresis.
Detection limit	The lowest concentration of a chemical that can reliably be distinguished from a zero concentration.
Diffusion	Random walk of an ensemble of particles from regions of high concentration to regions of low concentration.
Dispersion	Refers to the mixing process by which the pollutants or natural substances are mixed within a water or air column. Four processes may contribute to dispersion (mixing) at different levels of contribution to the overall mixing: molecular diffusion, turbulent diffusion, shear

mixing and mechanical mixing. The importance or the level of contribution of these mixing processes to the overall dispersion is media dependent. For example, turbulent diffusion may not be important in subsurface transport, or mechanical mixing may not be considered in surface or air transport.

Flocculation	In surface waters, although individual small particles may remain suspended and be transported, they may also aggregate into clumps. This aggregation process is called flocculation.
Half life	Half life of a chemical is defined in terms of the time it takes to reduce its chemical activity (concentration) by half.
Hydrodynamic mixing	A process that describes the combined effects of molecular diffusion and mechanical mixing in groundwater transport.
Infiltration	The process of water moving into the soil either from the ground surface or from a surface water body.
Molecular diffusion	Mixing associated with random molecular motions based on gradients of concentration. This process may be described by Fick's law. This classical diffusion equation relates the diffusion flux to concentration gradients in terms of a "molecular diffusion" coefficient which is chemical specific.
Mechanical mixing	Mixing associated with random scattering of particles associated with heterogeneity and tortuosity of particle pathways in subsurface environments. The mathematical definition is analogous to molecular diffusion, as it contains a dynamic dispersivity coefficient, which is usually expressed in terms of dispersion coefficient times the average linear velocity, replacing the molecular diffusion coefficients. Although it is media dependent, mechanical mixing effects are usually much larger than molecular diffusion effects in defining dispersion.
Percolation	Water movement through the soil.
Retardation	As a consequence of adsorptive and absorptive processes, the transport of chemicals is slowed or retarded. In subsurface fate and transport modeling, the effect of this process is introduced to the mathematical models in terms of a retardation coefficient which is a function of the ratio of the sorbed chemical concentration to the mobile chemical concentration.
Runoff	Water which is not absorbed by the soil or infiltrated to the soil and flows to a lower ground, eventually draining into a stream, river or other surface water bodies.

Sedimentation (particle settling)	The sinking or rising of particles having densities different from the ambient fluid, such as sand grains or organic matter.
Semi-confined aquifier	An aquifer bounded by a leaky aquifer from above or below or both.
Solvent	A liquid capable of dissolving or dispersing another substance (for example, acetone or mineral spirits).
Sorption	Chemicals in the environment may "sorb" onto the grains, or onto iron oxyhydroxide, or organic coatings of grains of aquifers or sediments in surface waters. The term "sorb" includes both adsorptive and absorptive processes.
Suspension (particle entrainment)	The picking up of particles, such as sand or organic matter, from the bed of a water body by turbulent flow past the bed.
Tributary	A stream or river whose water flows into a larger stream or river.
Turbulent diffusion	The random scattering of particles by turbulent motion, considered mainly in air and surface water pathways. Its mathematical definition is analogous to molecular diffusion, with "eddy" diffusion coefficient replacing the molecular diffusion coefficients, which are much larger than molecular diffusion coefficients.
Volatilization	Transport mechanism associated with the transfer of molecules from the liquid phase to the gas phase.
Volatile organic compounds, VOCs	Organic compounds that evaporate readily into the air. VOCs include substances such as benzene, toluene, methylene chloride, and methyl chloroform.
Watershed	The sum total all of the land and smaller surface water bodies which drain into a stream or river at a point.
Water table aquifer	An aquifer which is not bounded by another geologic formation from above and the upper boundary is defined in terms of a water level where the pressure condition is atmospheric.

Environmental Modeling

| Analytic solution | A formula for the solution of a state variable in a mathematical model that uses continuous and closed form mathematical functions. |
| Analytic uncertainty | Uncertainty associated with the individual parameters used in defining a physical process in a model. |

Analytic uncertainty propagation	A procedure that examines how uncertainty in individual parameters affects the overall uncertainty in model predictions.
Boundary conditions	Conditions on the primary unknowns (state variables) of the problem posed, which must be satisfied at the boundaries of the solution domain.
Calibration	A process that yields a model simulation output as close as possible to the simulated event. The measure of calibration is statistical comparisons between model results and observations on the event. In a calibration process, adjustment of model parameters within the range of experimentally determined values reported in the literature or obtained through a site specific study is the key operation. Calibration should not be confused with the methods of parameter estimation.
Cauchy boundary condition	A boundary condition in which the gradients of the primary unknown variables (state variables) are defined at the boundary of the solution domain, as a function of the unknown variable. This boundary condition is also referred to as the "third type" boundary condition.
Conservation of energy	The energy associated with matter entering any system plus the net energy added to the system is equal to the energy leaving the system.
Conservation of mass	Mass can neither be created nor destroyed, but can be transferred or transformed. This law forms the basis of most mechanistic models.
Conservation of momentum	Momentum can neither be created nor destroyed, but can be transferred or transformed. This law forms the basis of Newton's first law of motion.
Control volume	Any closed reference volume across whose boundaries one accounts for all transport fluxes, and within whose boundaries one accounts for all processes that produce or consume matter.
Descriptive statistics	A branch of statistics that deals with the organization, summarization, and presentation of data.
Difference equation	A representation of a differential equation in terms of discrete difference ratios that represents the derivatives of the function and the value of the function at discrete points.
Differential equation	An equation involving a function and its derivatives.
Dirichlet boundary condition	A boundary condition in which the value of the primary unknown variables are defined at the boundary of the solution domain. This boundary condition is also referred to as the "first type" boundary condition.

Distribution	A set of values derived from a specific population or set of measurements that represents the range and array of data for the factor being studied.
Initial condition	A condition which defines the value and distribution of the state variables at time zero or at the staring time of the analysis.
Mass balance (mass conservation)	An accounting process of mass inputs, outputs, reactions and accumulation within a control volume.
Model	A mathematical function with parameters which can be adjusted so that the function closely describes a set of empirical data. A "mathematical" or "mechanistic" model is usually based on biological or physical mechanisms, and has model parameters that have real world interpretation. In contrast, "statistical" or "empirical" models curve-fit to data in which the mathematical function used is selected for its numerical properties. Extrapolation from mechanistic models (e.g., pharmacokinetic equations) usually carries higher confidence than extrapolation using empirical models.
Mathematical model	Representation of a chemical, physical or biological process using mathematical principles.
Model inputs	Forcing functions such as boundary or initial conditions, or other parameters that characterize the process modeled and are required to run a model.
Model parameters	Coefficients used in a model to describe the process that is modeled (e.g., rate coefficients, equilibrium coefficients).
Monte Carlo method	A repeated random sampling from the distribution of values for each of the parameters in a generic equation (a model) representing a process, whose purpose is to derive and estimate from the distribution of predictions which is based on the generic equation.
Neuman boundary condition	A boundary condition in which the gradients of the primary unknown variables are defined at the boundary of the solution domain. This boundary condition is also referred to as the "second type" boundary condition.
Non-point source	Distributed sources of contamination that may originate from runoff or percolation from land as a result of different land use activities. In this case, while the flux or load from a small area of the watershed may be low, the overall loading to the watershed may be large if the total area contributing to contamination is large.

Numerical solution	Use of a discrete (approximate) representation for the solution of a state variable in a mathematical model.
Parameter estimation	Usually identified as inverse solution in which the independent variables of the models are treated as unknowns and the dependent variables are treated as known variables. This method yields the best estimates for the parameters of the model based on data from dependent variables.
Physiologically based pharmacokinetic model	Physiologically based compartmental model that is used to describe (quantitatively) the pharmacokinetic behavior of toxicants.
Point source	Contaminants discharged to air, groundwater or surface water from a defined source.
Population	In statistical terms, it is used to define the collection of measurements on all elements of a universe about which one wishes to draw conclusions or make decisions.
Post audit	Comparison of model results with field data that is collected after the model is developed.
Probability analysis	A field devoted to the study of random variation in systems.
Probabilistic uncertainty analysis	A technique that assigns a probability density function to each input parameter associated with a generic equation (a model) representing a process, then randomly selects values from each of the distributions and inserts them into the generic equation. Repeated calculations produce a distribution of predicted values by the generic equation, which reflect the combined impact of the variability in each input to the prediction. Monte Carlo analysis is a common type of probabilistic uncertainty analysis.
Random samples	Samples selected from a statistical population such that each sample has an equal probability of being selected.
Robust	A statistical term which indicates that an association between two variables remains significant when other variables are taken into account.
Robustness	Confidence in model results and applicability of the model to represent the process that is modeled. Robustness is established after repeated applications of the model simulate an event under different circumstances.
Sample	A small part of something designed to show the nature or quality of the whole.
Sensitivity analysis	Process of changing one variable while leaving the other variables in a model constant to determine the effect of the changed variable on the output. In this procedure, the

uncertain quantity is selected at its credible lower and upper bounds, while holding all others at their nominal values, and the results are calculated at each combination of values. The results may identify the variables that have the greatest effect on predictions based on the generic equation (model) and may help focus further attention on specific aspects of data collection efforts.

Simulation	The use of a model to produce information on state variable(s) with any input data.
State variable	The dependent variable of the model.
Statistical model	A model based on random variables.
Steady state models	Models in which the simulated value of the state variable does not change with time.
Stochastic models	Models that seek to identify and predict the probability of occurrence of a given outcome in a process being modeled.
Time dependent models	Models in which the simulated value of the state variable changes with time.
Uncertainty	Uncertainty represents a lack of knowledge about factors that affect the development of a representation of a process or that could yield biased or inaccurate estimates of the predicted value based on this representation. Thus types of uncertainty may include uncertainty in a scenario, in the parameters used in a model or in the model itself.
Uncertainty analysis	Determination of the statistical uncertainty associated with a state variable, due to uncertainty in model parameters, inputs, or the initial state through statistical or stochastic methods.
Upper bound	An estimate of the plausible upper limit to the true value of the quantity. This is usually not a statistical confidence limit.
Validation	Scientific acceptance that: (i) the model includes all major and salient processes of the system modeled (ii) the processes are formulated correctly; and (iii) the model properly describes the observed phenomena within the physical, temporal and conceptual boundaries of its definition.
Variability	Statistically represented range in the values of the parameters of a model or the values of the state variable.
Verification	A statistically acceptable comparison between model simulation results and a second set of independent data on the process that is modeled. In the verification stage, all calibrated model parameters are kept fixed at their calibration values.

Environmental Toxicology

Acute exposure	Exposure over a short period of time, usually in the range of 24 h. may be used to define a single exposure or multiple exposures during that period of time.
Acute toxicity	Adverse effect that an organism experiences due to an acute exposure. The older term used to describe immediate toxicty. Its former use was associated with toxic effects that were severe (e.g., mortality) in contrast to the term "subacute toxicity" that was associated with toxic effects that were less severe. The term "acute toxicity" is often confused with that of acute exposure. The examples of acute toxicity may range from a reaction to an insect bite for very short term exposure to chemical asphyxiation from exposure to high concentrations of carbon monoxide (CO) for somewhat longer terms of exposure.
Allergen	A substance such as pollen or another protein to which a subject becomes sensitized.
Allergic reaction	A reaction to a toxicant caused by an altered state of the normal immune response. The reaction can be defined in terms of an immediate reaction such as the case of anaphylaxis or delayed reaction such as the case in cell-mediated reactions.
Anthropogenic	An event which is a results of human activity, as opposed to biogenic, which is a result of biologic activity.
Chronic exposure	Exposure that is repeated over a longer period of time (months).
Chronic toxicity	Permanent adverse effect that an organism experiences after exposure to a toxicant.
Dose	The total amount of toxicant administered to an organism within a specific time period. Units used to define this term may depend on the application method, such as quantity per unit body weight or quantity per unit body surface area. Usually dose is a part of the total exposure amount. Absorbed dose is the amount of a toxicant that passes into a tissue or an organ.
Delayed or latent toxicity	The adverse effect an organism experiences long after the initiation of exposure to a toxicant.
Exposure	The process by which an organism comes in contact with a substance.
Exposure path	The route of exposure for an organism to a substance, such as air, water, soil, food, medication. The exposure path usually resides in the ambient environment.

Idiosyncratic reaction	A response to a toxicant which may occur at exposure levels much lower than those that are required to cause the same effect in the majority of individuals within the same population.
Internal/absorbed dose	The amount of toxicant that is absorbed into the organism and distributed through systemic pathways.
In-vitro	In an artificial environment outside a living organism or body. For example, some toxicity testing is done on cell cultures or slices of tissue grown in the laboratory, rather than on a living animal
In-vivo	Within a living organism or body. For example, some toxicity testing is done on whole animals, such as rats or mice
Local toxicity	An adverse effect that is experienced at the toxicant's site of contact with an organism.
Mechanism of toxicity	The biologic or biochemical processes through which a toxicant's effect on an organism is manifested.
Maximum tolerated dose	The highest dose just below the level at which toxic effects other than cancer can occur. Acute toxicity studies yield this dose level.
Reversible toxicity	A reversible adverse effect that is experienced by an organism to a toxic substance when the exposure is terminated.
Safety	A probabilistic measure of a specific exposure or dose not producing a toxic effect.
Subacute exposure	Similar to acute exposure, except that the exposure period is longer. Usually exposure periods that ranges from days to a month fall in this category.
Subchronic exposure	Similar to chronic exposure in which the exposure period is within several months.
Systemic toxicity	An adverse effect that manifests itself at another point of vulnerability distant from the point of entry of the toxicant. The transfer processes from point of entry to vulnerability point may include absorption, diffusion and mechanical transfer. Examples would include the adverse effects on the central nervous system resulting from chronic ingestion of mercury.
Target organ dose	The integral concentration of a toxicant in a target organ that may cause health effects, over a time interval.
Teratogen	A substance that causes defects in development between conception and birth. A teratogen is a substance that causes a structural or functional birth defect.
Toxic agent	A substance that can have the characteristic of producing an undesirable or an adverse health effect.

Toxin	Any natural toxicant produced by an organism
Toxicant	Any substance that causes adverse effects when in contact with a living organism at a sufficiently high concentration.
Toxicity	The adverse effect that a chemical or physical substance might produce within a living organism.
Toxicology	The science that is involved with the qualitative and quantitative study of adverse health effects of chemicals or physical substances which can be produced in living organisms under specific conditions of exposure.

Exposure Analysis

Absorption fraction	The relative amount of a substance that penetrates through a surface into a body (unitless ratio).
Acceptable daily intake	An estimate of the daily exposure dose that is likely to be without deleterious effect even if continued exposure occurs over a lifetime.
Acute exposure	One dose or multiple doses occurring within a short time (24 h or less).
Anecdotal data	Data based on descriptions of individual cases rather than on controlled studies.
Ambient	The conditions surrounding a person, or the sampling location.
Average daily dose	Dose rate averaged over a pathway-specific period of exposure expressed as a daily dose on a per-unit-body-weight basis. The ADD, expressed in mass/mass-time units, is used for exposure to chemicals with non-carcinogenic non-chronic effects.
Best tracer method	Method for estimating soil ingestion that allows for the selection of the most recoverable tracer for a particular subject or group of subjects. Selection of the best tracer is made on the basis of the food/soil ratio.
Bioassay	The determination of the potency (bioactivity) or concentration of a test substance by noting its effects in live animals or in isolated organ preparations, as compared with the effect of a standard preparation.
Bioavailability	The degree to which a drug or other substance becomes available to the target tissue after administration or exposure.
Carcinogen	An agent capable of inducing a cancer response.
Carcinogenesis	The origin or production of cancer, very likely a series of steps. The carcinogenic event so modifies the genome and/or other molecular control mechanisms in the target cells that they can give rise to a population of altered cells.

Chronic effect	An effect that is manifest after some time has elapsed from initial exposure. See also Health hazard.
Chronic exposure	Multiple exposures occurring over an extended period of time, or a significant fraction of the animal's or the individual's life-time.
Chronic intake	The long term period over which a substance crosses the outer boundary of an organism without passing an absorption barrier.
Chronic study	A toxicity study designed to measure the (toxic) effects of chronic exposure to a chemical.
Chronic toxicity	The older term used to describe delayed toxicity. However, the term "chronic toxicity" also refers to effects that persist over a long period of time whether or not they occur immediately or are delayed. The term "chronic toxicity" is often confused with that of chronic exposure.
Confounder	A condition or variable that may be a factor in producing the same response as the agent under study. The effects of such factors may be discerned through careful design and analysis.
Core grade(s)	Quality ratings, based on standard evaluation criteria established by the Office of Pesticide Programs, given to toxicological studies after submission by registrants.
Critical effect	The first adverse effect, or its known precursor, that occurs as the dose rate increases.
Developmental toxicity	The study of adverse effects on the developing organism (including death, structural abnormality, altered growth, or functional deficiency) resulting from exposure prior to conception (in either parent), during prenatal development, or postnatally up to the time of sexual maturation.
Exposure	Contact of a chemical, physical or biological agent with the outer boundary of an organism. Exposure is quantified as the concentration of the agent in the medium in contact integrated over the time duration of the contact.
Exposure assessment	A process in which potential recipients (individual or ecosystem) of a hazardous source are identified. This includes, the magnitude of concentration levels, the duration, the frequency and the route of exposure.
Exposure concentration	The concentration of a hazardous substance in its transport or carrier medium at the point of contact.
Exposure duration	The total time an individual or an ecosystem is exposed to an hazardous substance.
Exposure pathway	The physical path a chemical takes from the source of the hazard to the exposure point.

Exposure registry	A system of ongoing follow-up of people who have had documented environmental exposures.
Exposure route	The way a chemical pollutant enters an organism after contact, e.g., by ingestion, inhalation, or dermal absorption.
Exposure scenario	A set of facts, assumptions, and inferences about how exposure takes place that aids the exposure assessor in evaluating, estimating, or quantifying exposures.
Endpoint	A response measure in a toxicity study.
Estimated exposure dose (EED)	The measured or calculated dose to which humans are likely to be exposed considering exposure by all sources and routes.
Frank-effect level (FEL)	Exposure level which produces unmistakable adverse effects, such as irreversible functional impairment or mortality, at a statistically or biologically significant increase in frequency or severity between an exposed population and its appropriate control.
Health advisory	An estimate of acceptable drinking water levels for a chemical substance based on health effects information; a Health Advisory is not a legally enforceable Federal standard, but serves as technical guidance to assist Federal, state, and local officials.
Immediate versus delayed toxicity	Immediate effects occur or develop rapidly after a single administration of a substance, while delayed effects are those that occur after the lapse of some time. These effects have also been referred to as acute and chronic, respectively.
Intake	The process by which a substance crosses the outer boundary of an organism without passing an absorption barrier (e.g., through ingestion or inhalation).
Intake rate	Rate of inhalation, ingestion, and dermal contact depending on the route of exposure.
Initiation	The ability of an agent to induce a change in a tissue which leads to the induction of tumors after a second agent, called a promoter, is administered to the tissue repeatedly.
Latency period	The time between the initial induction of a health effect and the manifestation (or detection) of the health effect; crudely estimated as the time (or some fraction of the time) from first exposure to detection of the effect.
Local versus systemic toxicity	Local effects refer to those that occur at the site of first contact between the biological system and the toxicant; systemic effects are those that are elicited after absorption and distribution of the toxicant from its entry point to a distant site.

Lowest-observed-adverse-effect level (LOAEL)	The lowest exposure level at which there are statistically or biologically significant increases in frequency or severity of adverse effects between the exposed population and its appropriate control group.
Malignant	Tending to become progressively worse and to result in death if not treated; having the properties of anaplasia, invasiveness, and metastasis.
Margin of exposure (MOE)	The ratio of the no observed adverse effect level (NOAEL) to the estimated exposure dose (EED).
Margin of safety (MOS)	The older term used to describe the margin of exposure.
Metastasis	The transfer of disease from one organ or part to another not directly connected with it adj., metastatic.
Microenvironment	The combination of activities and locations that yield potential exposure.
Modifying factor (MF)	An uncertainty factor which is greater than zero and less than or equal to 10; the magnitude of the MF depends upon the professional assessment of scientific uncertainties of the study and database not explicitly treated with the standard uncertainty factors (e.g., the completeness of the overall database and the number of species tested); the default value for the MF is 1.
Morbidity	The state of being ill or diseased. Morbidity is the occurrence of a disease or a condition that alters health and quality of life.
Mortality	Death. Usually the cause (a specific disease, a condition, or an injury) is stated.
No evidence of carcinogenicity	According to the U.S. EPA Guidelines for Carcinogen Risk Assessment, a situation in which there is no increased incidence of neoplasms in at least two well-designed and well-conducted animal studies of adequate power and dose in different species.
No-observed-adverse-effect level (NOAEL)	An exposure level at which there are no statistically or biologically significant increases in the frequency or severity of adverse effects between the exposed population and its appropriate control; effects may be produced at this level, but they are not considered as adverse, nor precursors to adverse effects. In an experiment with several NOAELs, the regulatory focus is primarily on the highest one, leading to the common usage of the term NOAEL as the highest exposure without adverse effect.
No-observed-effect level (NOEL)	An exposure level at which there are no statistically or biologically significant increases in the frequency or

	severity of any effect between the exposed population and its appropriate control.
Occupational mobility	An indicator of the frequency at which workers change from one occupation to another.
Occupational tenure	The cumulative number of years a person has worked in his or her current occupation, regardless of the number of employers, interruption in employment, or time spent in other occupations.
Organoleptic	Affecting or involving a sense organ as of taste, smell, or sight.
Pathway	The physical course that a chemical or pollutant takes from the source to the organism exposed.
Per capita intake rate	The average quantity of food consumed per person in a population composed of both individuals who ate the food during a specified period and those that did not.
Population mobility	An indicator of the frequency at which individuals move from one residential location to another.
Promoter	In studies of skin cancer in mice, an agent which results in an increase in cancer induction when administered after the animal has been exposed to an initiator, which is generally given at a dose which would not result in tumor induction if given alone. A cocarcinogen differs from a promoter in that it is administered at the same time as the initiator. Cocarcinogens and promoters do not usually induce tumors when administered separately. Complete carcinogens act as both initiator and promoter. Some known promoters also have weak tumorigenic activity, and some also are initiators. Carcinogens may act as promoters in some tissue sites and as initiators in others.
Proportionate mortality ratio (PMR)	The number of deaths from a specific cause and in a specific period of time per 100 deaths in the same time period.
Residential volume	The volume of the structure in which an individual resides and may be exposed to contaminants.
Residential occupancy period	The time between when a person moves into a residence and when the person moves out or dies.
Reportable quantity	The quantity of a hazardous substance that is considered reportable under CERCLA. Reportable quantities are: (1) 1 lb, or (2) for selected substances, an amount established by regulation either under CERCLA or under Section 311 of the Clean Water Act. Quantities are measured over a 24-h period.
Reversible versus irreversible toxicity	Reversible toxic effects are those that can be repaired, usually by a specific tissue's ability to regenerate or

	mend itself after chemical exposure, while irreversible toxic effects are those that cannot be repaired.
Route	The way a chemical or pollutant enters an organism after contact, e.g., by ingestion, inhalation, or dermal absorption.
Safety factor	See Uncertainty factor. Short-term exposure – multiple or continuous exposures occurring over a week or so.
Screening level assessment	Examination of exposures that would fall on or beyond the high end of the expected exposure distribution.
Standardized mortality ratio (SMR)	The ratio of observed deaths to expected deaths.
Subchronic exposure	Multiple or continuous exposures occurring usually over 3 months.
Subchronic study	A toxicity study designed to measure effects from subchronic exposure to a chemical.
Sufficient evidence	According to the U.S. EPA's Guidelines for Carcinogen Risk Assessment, sufficient evidence is a collection of facts and scientific references which is definitive enough to establish that the adverse effect is caused by the agent in question.
Superfund	Federal authority that was established by the Comprehensive Environmental Response, Compensation, and Liability Act (CERCLA) in 1980 to respond directly to releases or threatened releases of hazardous substances that may endanger health or welfare.
Supporting studies	Those studies that contain information that is useful for providing insight and support for conclusions.
Systemic effects	Effects that require absorption and distribution of the toxicant to a site distant from its entry point, at which point effects are produced. Most chemicals that produce systemic toxicity do not cause a similar degree of toxicity in all organs, but usually demonstrate major toxicity to one or two organs. These are referred to as the target organs of toxicity for that chemical.
Systemic toxicity	See Systemic effects.
Target organ of toxicity	See Systemic effects.
Threshold	The dose or exposure below which a significant adverse effect is not expected. Carcinogens are thought to be non-threshold chemicals, to which no exposure can be presumed to be without some risk of adverse effect.
Threshold limit values (TLVs)	Recommended guidelines for occupational exposure to airborne contaminants.

Total tap water	Water consumed directly from the tap as a beverage or used in the preparation of foods and beverages (i.e., coffee, tea, soups, etc.).
Total fluid intake	Consumption of all types of fluids including tap water, milk, soft drinks, alcoholic beverages and water intrinsic to purchased foods.
Uncertainty factor	One of several, generally tenfold factors, used to derive operationally the reference dose (RfD) from experimental data. UFs are intended to account for (1) the variation in sensitivity among the members of the human population (2) the uncertainty in extrapolating animal data to the case of humans (3) the uncertainty in extrapolating from data obtained in a study that is of less-than-lifetime exposure and (4) the uncertainty in using LOAEL data rather than NOAEL data.
Uptake	The process by which a substance crosses an absorption barrier and is absorbed into the body.
Weight-of-evidence for carcinogenicity	The extent to which the available biomedical data support the hypothesis that a substance causes cancer in humans.

Environmental Risk Analysis

Added risk	The difference between the cancer incidence under the exposure condition and the background incidence in the absence of exposure.
Attributable risk	The difference between the risk of exhibiting a certain adverse effect in the presence of a toxic substance and that risk in the absence of the substance.
Contaminant concentration	The concentration of the contaminant in the medium (air, water, soil, etc.). It has the units of mass/volume or mass/mass.
Dose	The amount of a substance available for interaction with metabolic processes or biologically significant receptors after crossing the outer boundary of an organism.
Dose–response assessment	A process in which the quantity of a hazard that may reach organs or tissues and the percentage of the exposed populations are identified.
Dose–response relationship	A relationship between the amount of an agent (either administered, absorbed, or believed to be effective) and changes in certain aspects of the biological system (usually toxic effects), apparently in response to that agent.

Excess lifetime risk	The additional or extra risk incurred over the lifetime of an individual by exposure to a toxic substance.
Extra risk	The added risk to that portion of the population that is not included in measurement of the background disease rate.
Hazard	Hazard refers to the source of risk but is not synonymous with risk. A chemical which is hazardous to human health may not be considered to be a risk unless there is an exposure pathway which links that chemical to humans or to the environment.
Hazard control	The process, through which one recognizes, evaluates and eliminates the potential hazards of a hazardous source.
Hazard identification	Hazard identification is the first step in risk analysis, which is the process of identifying all the hazards with the potential to harm an individual or the environment.
Human equivalent concentration	Exposure concentration for humans that has been adjusted for dosimetric differences between experimental animal species and humans to be equivalent to the exposure concentration associated with observed effects in the experimental animal species. If occupational human exposures are used for extrapolation, the human equivalent concentration represents the equivalent human exposure concentration adjusted to a continuous basis.
Human equivalent dose	The human dose of an agent that is believed to induce the same magnitude of toxic effect as that which the known animal dose has induced.
Incidence	The number of new cases of a disease within a specified period of time.
Incidence rate	The ratio of the number of new cases over a period of time to the population at risk.
Individual risk	The probability that an individual person will experience an adverse effect. This is identical to population risk unless specific population subgroups can be identified that have different (higher or lower) risks.
Inhaled dose	The amount of an inhaled substance that is available for interaction with metabolic processes or biologically significant receptors after crossing the outer boundary of an organism.
Internal dose (absorbed dose)	The amount of substance penetrating across absorption barriers of an organism, via either physical or biological processes.
Interspecies dose conversion	The process of extrapolating from animal doses to equivalent human doses.

Lifetime average daily dose	Dose averaged over a lifetime. LADD is used for compounds with carcinogenic or chronic effects and is expressed in terms of mg/kg-day or other mass/mass-time units.
Limited evidence	According to the U.S. EPA's Guidelines for Carcinogen Risk Assessment, limited evidence is a collection of facts and accepted scientific inferences which suggest that the agent may be causing an effect, but this suggestion is not strong enough to be considered an established fact.
No data	According to the U.S. EPA Guidelines for Carcinogen Risk Assessment, "no data" describes a category of human and animal evidence in which no studies are available to permit one to draw conclusions as to the induction of a carcinogenic effect.
Potential dose	The amount of a chemical contained in material ingested, air breathed, or bulk material applied to the skin.
Principal study	The study that contributes most significantly to the qualitative and quantitative risk assessment.
Process hazard analysis	An analysis conducted to identify, evaluate, eliminate or control potential hazards within a plant of facility.
Prospective study	A study in which subjects are followed forward in time from initiation of the study. This is often called a longitudinal or cohort study.
Reasonable risk	Risk levels which may be considered tolerable. This level may be determined through consensus building, by comparing costs, benefits and alternative risks that have previously been accepted as tolerable.
Reference concentration (RfC)	An estimate (with uncertainty spanning perhaps an order of magnitude) of a continuous inhalation exposure to the human population (including sensitive subgroups) that is likely to be without an appreciable risk of deleterious noncancer effects during a lifetime.
Reference dose (RfD)	An estimate (with uncertainty spanning perhaps an order of magnitude) of a daily exposure to the human population (including sensitive subgroups) that is likely to be without an appreciable risk of deleterious effects during a lifetime.
Regional deposited dose (RDD)	The deposited dose of particles calculated for the region of interest as related to the observed effect. For respiratory effects of particles, the deposited dose is adjusted for ventilatory volumes and the surface area of the respiratory region effected (mg/min-cm^2). For extrarespiratory effects of particles, the deposited dose in the total respiratory system is adjusted for ventilatory volumes and body weight (mg/min-kg).

Regional deposited dose ratio (RDDR)

The ratio of the regional deposited dose calculated for a given exposure in the animal species of interest to the regional deposited dose of the same exposure in a human. This ratio is used to adjust the exposure effect level for interspecies dosimetric differences to derive a human equivalent concentration for particles.

Regional gas dose (RGD)

The gas dose calculated for the region of interest as related to the observed effect for respiratory effects. The deposited dose is adjusted for ventilatory volumes and the surface area of the respiratory region effected ($mg/min\text{-}cm^2$).

Regional gas dose ratio (RGDR)

The ratio of the regional gas dose calculated for a given exposure in the animal species of interest to the regional gas dose of the same exposure in humans. This ratio is used to adjust the exposure effect level for interspecies dosimetric differences to derive a human equivalent concentration for gases with respiratory effects.

Regulatory dose (RgD)

The daily exposure to the human population reflected in the final risk management decision; it is entirely possible and appropriate that a chemical with a specific RfD may be regulated under different statutes and situations through the use of different RgDs.

Relative risk (sometimes referred to as risk ratio)

The ratio of incidence or risk among exposed individuals to incidence or risk among unexposed individuals.

Risk

Risk is the probability of injury, disease, death or exposure of individuals, populations or ecosystems which creates a hazardous or adverse condition. In quantitative terms, risk is expressed in values ranging from 0 (representing the certainty that harm will not occur) to 1 (representing the certainty that harm will occur). The following are examples showing the manner in which risk is expressed in IRIS: E-4 = a risk of 1/10,000; E-5 = a risk of 1/100,000; E-6 = a risk of 1/1,000,000. Similarly, 1.3E-3 = a risk of 1.3/1000 = 1/770; 8E-3 = a risk of 1/125; and 1.2E-5 = a risk of 1/83,000.

Risk assessment

Risk assessment is a process in which the severity of adverse effects imposed by a hazardous condition on a population or an ecosystem is estimated.

Risk characterization

A process in which a numerical value is associated with the risk.

Risk communication

The results of risk assessment and risk management must be communicated to the technical and public communities as the best alternative, based on the assumption and

Appendix 3
Definitions and Operations of the ACTS/RISK Software

The Analytical Contaminant Transport Analysis System (ACTS) software is a computational platform designed to provide environmental engineers and health scientists with a user friendly interface to access commonly used environmental multimedia transformation and transport models for the four environmental pathways, i.e. air, soil, surface water, and groundwater pathways. The complementary health risk analysis software (RISK) provides a computational platform to analyze the health risk of exposure to contaminants that may be linked to the environmental pathway analysis provided in ACTS, although it can also be used as an independent application. ACTS and RISK contain more than 300 models that are available in the public domain literature. The analytical solutions to these models are provided under a unified computational platform and are accessed through an easy to learn and easy to use graphical user interface. A powerful feature of both of these computational platforms is their ability to conduct probabilistic analyses using one- and two-stage Monte Carlo simulations that are dynamically linked to all environmental pathway and risk analysis modules. Thus, third-party software is not required to conduct Monte Carlo simulations in the ACTS and RISK applications. Publication-quality graphical output of simulation results can be developed using the graphic utilities included in the software.

Contributors to the ACTS/RISK Software

An application software that covers a wide range of environmental pathways and exposure models that is developed in a user friendly interface necessarily requires the contribution of numerous participants to the effort. In this case the participants to the project are the former students of the author of this book. The author is indebted to these contributors and would like to acknowledge their contributions by including their names here in an alphabetical order: Dr. Jiabao Guan, Dr. Elcin Kentel, Dr. Orhan Gunduz, Dr. Wasim Khan, Dr. Boshu Liao, Mr. Morris Maslia, Ms. Jenny Morgan, Mr. William Morgan, Mr. Babar Sani.

The design of the ACTS/RISK software architecture is provided by Dr. Jiabao Guan and Dr. Wasim Khan.

ACTS and RISK Software Download

The ACTS and RISK software has been updated several times over the years since 1993. These updates have been made to provide error free software as much as possible. As this book goes into publication, we are now at Version 9 for ACTS and Version 1 for RISK. The latest versions of these two software tools can be downloaded from the web addresses given below, free of charge.

http://mesl.ce.gatech.edu/SHARE/Share.html or http://extras.springer.com

Information on this can also be found on the home page of this volume on http://www.springer.com

Disclaimer

Analytical models used in the ACTS and RISK software are available in the public domain literature and have been used by the U. S. Geological Survey (USGS), U. S. Environmental Protection Agency (EPA), U. S. Department of Health and Human Services (DHHS) and other federal agencies, as well as private consulting companies and educational institutions. No warranty is expressed or implied by the author, as to the accuracy and functioning of the programs and related material included in the software tools that accompany this book. Nor shall the fact of distribution of the ACTS and RISK software through the web page given above constitute any such warranty, and no responsibility is assumed by the author for the use of this software and the book in connection herewith. Any use of trade products, or company names in this book is for descriptive purposes only and does not imply endorsement by the author.

Background Information

In the development of the two computational platforms, a decision was made to focus on four objectives:

- Developing a computational platform that would be easy to learn and easy to use with a user-friendly graphical interface
- Inclusion of analytical models that focus on key parameters and their relationship to exposure thresholds, which would allow users to make timely public health decisions

- Developing an analysis system with the capability to conduct probabilistic analyses to address issues of uncertainty and variability for all applications included in the two software tools
- Ability to produce publication quality graphical output internally and also provide text or spread sheet type output for use with external and third-party applications if needed

To meet these objectives, the analytical contaminant transport analysis system ACTS and health risk analysis system RISK was developed with partial funding provided by the Agency for Toxic Substances and Disease Registry (ATSDR). The resulting computational platform is in public domain and can be downloaded from the Multimedia Environmental Simulations Laboratory (MESL) web page at the Georgia Institute of Technology: http://mesl.ce.gatech.edu/SHARE/Share.html or Springer web page: http://extras.springer.com

Overview of the Computational Platform

The ACTS and RISK software have been developed to provide professionals in the fields of hydrogeology, environmental engineering, and environmental health with compact analytical tools to evaluate the transformation and transport of contaminants in multimedia environments such as air, soil, surface water, and groundwater. ACTS and RISK were been initially released in 1993 and the most current version is available over the Internet as described above. The transformation and transport models included in the software are dynamic models that can be used to assess steady state and time dependent contaminant concentrations introduced to soil layers or contaminants released into air or water (Fig. A3.1). Transformation processes that can be simulated in ACTS include sorption, decay, first-order biodegradation and multispecies by-product generation. Analysis of the soil/air pathway within ACTS includes a chemical database, several emission models are included for simulating the emission rates from a contaminated soil layer to the land surface, and Box and Gaussian dispersion models are included for evaluating migration of contaminants through air. The air pathway transport analysis module also includes indoor air and landfill emission models proposed by the USEPA. The surface water pathway module includes near-field mixing, far-field mixing, and sediment transport models. The groundwater pathway module includes unsaturated and saturated zone models with a capability of multi-species analysis within the saturated flow zone. Again, within the saturated flow zone, constant hydrodynamic dispersivity and variable hydrodynamic dispersivity conditions may also be evaluated. To analyze cases involving uncertainty and variability of the input parameters, the Monte Carlo simulation computational platform is dynamically linked to all pathway models (Fig. A3.1). Finally, results of the analyses can be retrieved and viewed in either graphical or text formats.

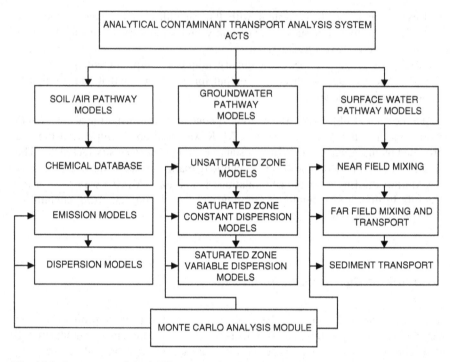

Fig. A3.1 Components of the ACTS computational platform

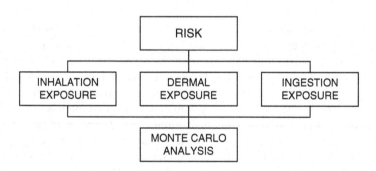

Fig. A3.2 Components of the RISK computational platform

The main components of the RISK platform, as shown in Fig. A3.2, include standard exposure models due to inhalation, ingestion and dermal intake. In RISK, Monte Carlo analysis is also linked to all models provided through the same computational platform that is used in the ACTS software.

Preferences Menu

It is important for the user to start each project with appropriate selections made on the preferences menu. The preferences menu can be accessed from the opening window of the ACTS or RISK software (Fig. A3.3).

In this window several options are available to the user at the start of a new project. As expected, the software platform generates numerous text and graphics output files during an analysis sequence. It is important for the user to collect these output files in certain folders of the user's choice in order not to mix and match the output files of different projects. Thus it is recommended that the user should first create these folders on his or her computers and then link them using the "Browse..." command to the initiated project task. Other options that are available to the user also include the selection of the text editor to be used with the software to view the text output files, and the selection of the chemical database to be used in the project. The default selection for these options can be seen in Fig. A3.3. It is recommended that these options be changed for each project to keep the output of each project stored in separate folders, or to provide access to different interfaces as described below.

The text editor selection is the first choice that needs to be made. Any text editor that is available on one's computer, i.e. NOTEPAD, WORDPAD or MS WORD can be used for this purpose. The default text editor is set to NOTEPAD as seen in Fig. A3.3. This default selection can be changed by pointing and selecting the executable file of the text editor in the Windows^{TM} systems folder using the "Browse..." button.

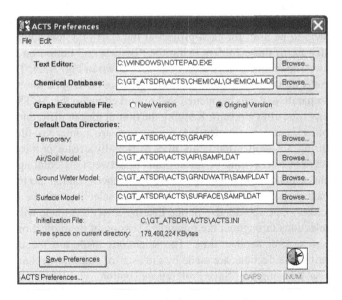

Fig. A3.3 Preferences menu window of the ACTS and RISK software

The next selection is the chemical database selection option. When the ACTS software is installed three chemical database files, CHEMICAL.MDB, CHEM1. MDB and CHEM2.MDB are automatically installed on the computer. Initially all three of these files contain the same chemical list and database. The CHEMICAL. MDB file is a file that cannot be edited and the two other files are the editable versions of the CHEMICAL.MDB data file. These two files provide the user with the option of upgrading the database based on the specific needs of the project under study. These files are stored under the file path C:\GT_ATSDR\ACTS\ CHEMICAL which is created during the installation process. The selection of any of these databases can be made using the "Browse…" button on the preferences window. As can be seen in Fig. A3.3, the default option is the CHEMICAL.MDB database. The editing procedures of the editable databases CHEM1. MDB and CHEM2.MDB will be described in the following sections of this appendix.

The next selection to be made is the choice of the graphics software module (Fig. A3.3). Here there are two options available to the user: the "New Version" and the "Original Version" of the graphics module. The "New Version" provides more options to the user to display and print the output of the ACTS and RISK software, such as three dimensional plots. The "Original Version" is more stable but does not provide the three dimensional plotting option. The use of the "Original Version" is recommended initially since it will simplify the choices of the interface for a novice user. The choice between the two can be made by clicking on the radio button for either case. As can be seen in Fig. A3.3 the default option is the original version. This selection can be changed by the user during an application anytime before selecting the graphical output menus of the models.

The next set of selections refers to the folders that will be used to store the temporary and final output files of a project. It is recommended that the user create separate folders to be used for each project at the start of a new application and point the software to those folders using the "Browse…" button on the preferences window. The default folder selections are shown in Fig. A3.3. Through the appropriate selection of these folders, the files that are developed for each project can be kept separate and can be opened and reworked using the ACTS and RISK software at any time after a restart. Keeping the output files of different projects in the same folder is not recommended since there are numerous files that are generated during an application. This may eventually create problems for the user or inadvertent overwriting of certain files in between different applications.

Once all the selections are made, the "Save Preferences" button has to be clicked to save the selections made. Otherwise the default selection will be kept as the choice of the user. One should also notice that the selections made will remain as the default selections throughout the remaining session of the project or during the use of the software at another time. The selections made are the new default selections until new selections are made by the user for a new project at a different time and are saved using the "Save Preferences" button.

Menu Definitions and Operations of the ACTS Software

The menu options and operations for all modules of the ACTS software are based on standard Windows^{TM} operating system functions and menus. In this appendix a list of these menu options is provided with a brief description of their function. The user may need to familiarize himself or herself with these operations, either by preparing and running test databases or by using the sample data provided for each ACTS and RISK module. The ACTS software includes numerous checks for potential data input errors that may trigger an error during the execution of the programs. Although all modules have been rigorously tested over several years, it is almost impossible to develop a completely error free and crash resistant software especially for a complex computational domain such as ACTS or RISK. If the reader notices an error during the operation of the system, and if he or she thinks it is not a user-generated error, please send an e-mail message to the author including the data file used. We will try to correct that error in our next revision of the ACTS and RISK systems and inform the reader of the outcome. Also, if the reader is an active user of the software and would like to receive periodic information on upgrades or other ACTS/RISK news material or short courses that are offered by the author, he or she should register his or her name using the list server information on the download web page or send the information to the author at (maral@ce.gatech.edu).

Standard Menu Options and Operations

File Menu

Open: Opens an existing ACTS/RISK data file. ACTS and RISK software use an extension which is specific to models included in the software. When a user database is saved, this specific extension is automatically added to the data file stored. This data file can be found in the default data path chosen by the user under the preferences menu. If the user has not modified this path, the data files will be saved under the default ACTS and RISK folder paths. It is recommended that at the start of a new project, specific folders be prepared and the default path be modified to point to these folders using the preferences menu option. This process will allow the user generated files to be sorted and saved in specific files related to the project. This choice of specific folders to store the data in a specific session can be accessed under the "Options→Preferences" pull down menu at the start of the ACTS/RISK software as described above (Fig. A3.3).

Save: Saves the active data file with the name and location previously set in the "Save As" dialogue box. When the user saves a data set for the first time ACTS displays the "Save As" dialogue box. If the user prefers to change the name and location of an existing data set, one should choose the "Save As" command from the File menu.

Save As: Displays the "Save As" dialogue box. In this dialogue box one may specify the name and location of the data file to be saved.

Print: Prints the active data file on the selected printer available on the computer. Before one uses this command, one should install a printer. See the WINDOWS™ documentation on how to install a printer on a computer. To select a printer and adjust the parameter setting of a printer, choose "Printer Setup" from the file menu.

Printer Setup: Provides a list of installed printers, allows choosing the default printer and setting other printing options.

Exit: Exits the model window. ACTS and RISK at this step prompts the user to save any unsaved changes made to the data file. The data remains saved in the memory unless one terminates the program by selecting "RETURN TO THE MAIN SHELL" from the model window. The user may also choose the Edit model option to return to the data saved from the main menu.

Calculate Menu

Executes the model equations based on the input data prepared. Any computational error encountered during the execution will be displayed, and recommended changes that are required to avoid the error will be displayed. Results are stored in the memory to be displayed later on. At this stage, only a specific output requested by the user at a certain spatial point and time is displayed in the analytical calculation results grid in the input data window. The specification of this spatial point and time is entered into the input data file by the user as will be described below. If an input variable is using a Monte Carlo analysis generated set of values, a "Monte Carlo" heading is displayed in the results grid. To view the complete results, the user should choose the "View" or "Graph" menu option from the "Results" menu. For the Air dispersion models the computation is only performed using the "Current Chemical" data when multiple chemicals are selected. The current chemical can be changed by a mouse click on the name of the chemical in the analytical calculation results grid if more than one chemical is selected while using the "Chemical" menu. Pressing function key F7 may also perform this operation.

Results Menu

If the calculation performed is not a single output as displayed on the results grid, or if at least one variable of the model is computed as a "Monte Carlo" variable, this menu option allows viewing of the results in a bar chart or in other two- or three-dimensional graphical formats.

Graph: Uses the built in graphics software to display the results. This option may only be used if the "Monte Carlo" heading is displayed in the analytical result grid for the current chemical, or the results are one-, two- and three-dimensional time dependent or steady state results. Only these types of output can be displayed in graphical format. If the model used generates only one output for the problem that is not spatially variable or time independent, then the graph option cannot be used. This is the case for some of the surface water pathway models. There are two graphical packages that are available for the user. These are identified as: (i) New Version; and, (ii) Original Version. The choice of using either of these graphic packages needs to be made by the user at the start of the ACTS and RISK software. This choice can be accessed under the "Options→Preferences" pull down menu at the start of the ACTS/RISK software. The difference between these two graphical packages is that the "New version" provides the user with more options to plot the results of a computation. The use of the "Original Version" is initially recommended since it will simplify the choices and the interface. Once a selection is made, the "Save Preferences" button must be clicked to save the option that is selected in the preferences window. Throughout the remaining session, and in future sessions, the selected set of preferences will remain as the default selection until a new selection is made during another session.

    View: Displays numerical output generated for a simulation in a WINDOWS^{TM} Editor. The editor to be used for this purpose can be selected under the "Options→Preferences" menu at the start of the project as discussed earlier.

Special Menu (Only in AIR Pathway Model)

Use Same Soil and Physical Data: Allows the user to use the same input data for all of the selected chemicals under the air dispersion models. Whenever the user selects a different chemical, the data for soil and physical parameters of the problem are copied from the old chemical input window to the newly selected chemical input window. In this case the Monte Carlo simulation values are also copied. The chemicals selected for an application remain unchanged when the user saves the data file. The status of this option is also saved with the data file. When the user opens the file during another time the same selection of chemicals must be used, or one may restart a new project with a new set of chemical selections if changes are necessary.

    Note: Selection of this option will overwrite data from the old chemical to the new selected chemical. The data for the newly selected chemical would be lost.

    Chemical Constants Usage: Displays a dialogue box that indicates which properties of the selected chemicals are being used in the analytical computation.

Help Menu

Displays help for the current model. Function key F1 can also be used to display related information. When selected, sections of help menu information will appear as context sensitive information on the model used.

Chemical Menu Options

OK button: Choosing the OK button will allow the use of the chemicals listed in the selected chemicals list box. Selecting Exit from the File menu is the same as choosing OK.

Cancel button: Choosing this option will discard any changes made in the chemical data editing process and will exit the chemical selection module with the original selection intact.

Clear button: Unselects all of the selected chemicals. A loss of data in the emission and concentration models may result if the OK or EXIT option are selected after unselecting all chemicals. Choose CANCEL to undo the selection process if the original files are to be kept.

Printing chemical properties: To print properties of the selected chemicals, choose the Print option from the File menu. The software uses the default printer defined in the print manager to print the text using the default editor. To select a different printer choose Printer Setup from the file menu of the opening window of the air model and make the appropriate selection at the beginning of the project.

Selection of chemicals: A chemical can be selected by moving the mouse cursor to the left margin of the grid. When the left margin is reached the mouse cursor becomes an arrow mark. Click the left mouse button on the row of the chemical while pressing the "CTRL" button on the keyboard. The row color of the selected chemical changes to red, which indicates the selection of the chemical. All selected chemicals are also displayed in a list in the lower right corner of the "Selected Chemicals" window. After the desired selections are made one may choose the OK button to use the selected chemicals in the contaminant transport and/or emission models of the air pathway. All chemical properties that are necessary to execute the model in use will automatically be transferred from the chemical database to the model data input grid. This process simplifies the chemical properties data entry for the user. The user needs to be careful with the units of the other parameters that are entered to execute a specific model. These units are displayed on the data input window grid.

Unselecting the chemicals: To unselect a chemical move the mouse cursor to the left margin of the grid. The mouse cursor changes to an arrow mark. Click the left mouse button while the "CTRL" button is depressed. The color of the row changes from red to white and the chemical is unselected. If the user leaves this window by selecting the OK button at this stage, the unselected chemical will not be available for the next step of calculations. To undo this change, one either has to re-select the chemical or choose the "Cancel" option to disregard the changes made in the chemical selection operation. In this case, the first selection made will be available for the next step.

Modifying Chemical Database: The ACTS software does not allow modification of the CHEMICAL.MDB database, which is the master data file. There are two other databases included in the software, CHEM1.MDB and CHEM2.MDB. These two databases are duplicates of the CHEMICAL.MDB database and they are editable. If either of these two databases is selected as the default database from

"Options → Preferences" menu, then the "Chemicals" pull down menu in the "Select Chemicals" window becomes active and editing of the database can proceed. All air model files save chemical properties data in addition to other input and output data that are entered by the user. When the user opens a file, the chemical properties are read and compared with the database. If the chemical properties in a data file do not match the data in a default chemical database, a message box is displayed indicating this discrepancy. This discrepancy occurs by changing the chemical database after preparing an air pathway input data file or by changing the chemical's data in the input box grid, which is not usually recommended. In any case, the user is given the choice to use either chemical properties from the default chemical database selected or from those saved in the file. If the data saved in the model option are selected the chemical database will be unaffected and the data entered manually will be used. The user needs to be very careful in selecting this option regarding the accuracy of the data used for a specific chemical.

Adding a new chemical: New chemicals can be added by selecting ADD NEW from the CHEMICAL menu on the form. All chemicals must have different names and symbols cannot be used as names. This operation is not available if the Chemical database selected is the master dataset, which is not editable. When adding a new chemical to a database, all of the chemical properties listed in the table must be entered. Otherwise errors will result during the calculation steps since multiple parameters are accessed by the models and this process is transparent to the user. Leaving a parameter blank for a new chemical may initiate an error during calculation if the model uses that parameter. When a new chemical is entered the units of the data should match the units of the property as indicated in the title bar of the chemical database. If this rule is not followed errors will occur.

Editing existing chemical: To change properties of the chemical, first select ENABLE EDIT from the CHEMICAL menu. Then place the cursor on the box of the chemical property to be changed and type in the new value. Moving to a different row would result in updating the chemical property entered. The change would be permanent and the old value would be lost. After editing is complete, select the DISABLE EDIT option from the CHEMICAL menu to avoid any accidental changes on the database during another session.

Deleting a chemical from the database: A chemical can be deleted by selecting this option.

Note: For any given input file prepared and saved for the air pathway models, the user can only use the same set of chemicals in all the emission and air dispersion models. If this is not what is desired, that is if a new chemical needs to be introduced a new data base must be prepared.

Monte Carlo Module of the ACTS/RISK Software

The transformation and transport of contaminants depends on media-specific parameters. Typically many of these parameters exhibit spatial and temporal variability as well as variability due to measurement errors. ACTS/RISK software provides the capability to analyze the impact of the uncertainty and variability in the model inputs on the outputs, using the Monte Carlo simulation technique.

The Monte Carlo module allows the user to generate random variables for most of the model parameters included in the ACTS/RISK software. Based on these random variables, a one-stage or two-stage Monte Carlo simulation may be performed in emission models, groundwater contaminant transport models, surface water mixing models and air pathway models. Figure A3.4 shows a typical data entry window for this operation. In this window, the upper workspace grid is the input area and the lower grid is the output area.

In order to produce a Monte Carlo simulation, the user will first click an empty box under the variant column of the input grid, which will lead to a list of all the available variants of the model used, and will appear in a pull down window. From this window, the user should select an appropriate variable for which a probability density function will be generated. Once a selection is made, the value of the parameter, as defined in the input database prepared by the user in the previous data entry window, will automatically appear in the column identified as "Mean." Thus in the ACTS/RISK software, the input value for a parameter entered in the input data window is always considered to be the mean value for that parameter during the Monte Carlo analysis phase. However, the user has the option of changing this value at this stage if desired. It is important to note that the intended change will also alter the original value of the parameter in the parent model input window. Next, the user should enter the minimum, the maximum, the variance and the number of random parameters to be generated in the corresponding columns to the right of the "Mean" column. The distribution type can be selected by placing the

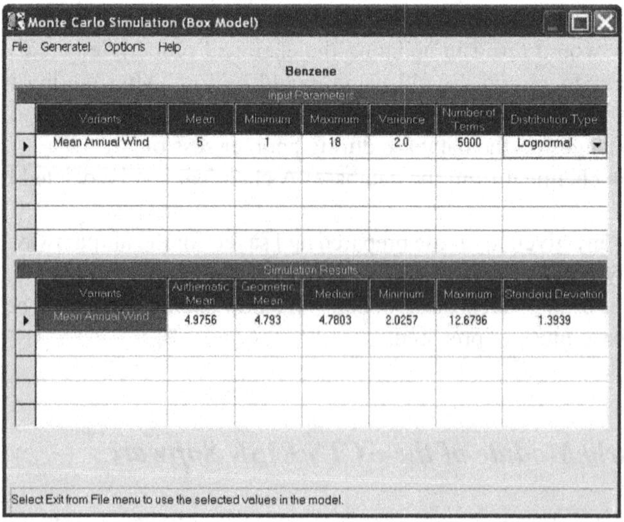

Fig. A3.4 Monte Carlo input window

mouse cursor in the "Distribution Type" column and by double clicking the selection that will appear in a pull down menu. The selection of one of the distributions will place the desired distribution name into the "Distribution Type" column. This completes the input data preparation phase of a Monte Carlo analysis.

After completing the input data entry for all desired variants that will be generated randomly, the user may press the F7 button on the keyboard or click on the "Generate" menu on the menu bar to calculate the distributions for all of the selected variables. Output values will be displayed in the "Simulation Results" grid. At this point several options are available to the user. To select a computed output "Arithmetic Mean" or "Geometric Mean" or other representation of the data generated as an input value to the parent model, the user should click on the value box. The background color of this value box changes to red indicating the selected value. This process allows the user, for example, to use the mean of the randomly generated parameters as an input value in the parent model. When the user exits the Monte Carlo window after this selection, the selected value will appear in the input data box of the parameter in the model input data window. The other option is to select all of the values generated for the specific parameter. To initiate this option, the user should click on the variant name, first column on the left, in the "Simulation Results" grid. Clicking there will change the background color of the variant name selected to red. After completing either of these selections, the user will exit the Monte Carlo module by selecting the "EXIT" option from the "FILE" menu. The selected parameter value(s) will automatically be transferred to the parent model for use as simulation as Monte Carlo values. The user may also unselect a selected value by clicking in the corresponding cell if the choice was made in error, before exiting the Monte Carlo module. If no selection is made and if the Monte Carlo module is exited, then the value returned to the parent model will be the original value entered in the parent model. If this value was modified in the "Mean" input box as described above, the returned value will be the new value entered in the Monte Carlo input box. Whenever the model file is saved, all of the associated Monte Carlo simulation results are saved in the same file. In all ACTS/RISK simulations, Monte Carlo analysis may be performed for multiple variants that appear in the first column of the input area for the model selected in the previous step.

Display of Results in the GRAPHIX Module

Numerical results of parameter distributions computed in this window can be displayed in graphical format on the screen, or hard copy printouts of these distributions can be made if desired. For this choice "Options" menu will be used.

The options available in the graphics option include the choice of a display of the results in "histogram", "frequency distribution", "cumulative frequency distribution", "probability distribution", "cumulative probability distribution" and "complementary cumulative probability function" formats.

Description of Menu Options in the Monte Carlo Simulation Window

File Menu: Same as Before

Print: Prints the active graph on the selected printer. Before using this command, a printer must be installed. See the Windows documentation on how to install a printer. To select a printer choose PRINTER SETUP from the file menu.

Exit: Ends the Monte Carlo simulation and returns to the parent model data entry window. Any output values selected for a variant in the "Simulation Results" grid are displayed in the parent model input box for the parameters. The data remains saved in the memory unless the user terminates the program by selecting RETURN TO THE MAIN SHELL from the initially selected pathway model window.

Generate: Generates a Monte Carlo probability density function for the pre-selected parameters in the model. After selecting values from the output of the Monte Carlo simulation window, as described above, the user may "Calculate" the Monte Carlo simulation based analysis from the input window of the project by closing the Monte Carlo window and returning to the input window. Any computational error encountered during this analytical calculation will also be displayed. Output results of the Monte Carlo generation will be displayed in the lower grid.

For the air pathway, the computation is only performed on the "Current Chemical." The current chemical can be changed by selecting CHEMICAL/MODEL from the SELECT menu in the air pathway and unsaturated groundwater pathway models. Simulation may also be performed by pressing the function key F7.

Graph: Displays the Monte Carlo simulation results in graphical format.

View: Displays the Monte Carlo simulation result on parameters in the text editor.

Help: Displays help for the active model. Function key F1 can also be used to display the help menu.

Graphics Module of the ACTS/RISK Software

The graphics module of the ACTS and RISK software uses the two different graphics module software options identified as: (i) New Version; and, (ii) Original Version, as outlined earlier. The selection for either of these options can be made in the "Preferences" window as described above. The interface window for the two graphics options is the same (Figs. A3.5–A3.7) with the only difference being that the "New Version" selection provides an option to plot three-dimensional surface plots of the output data as shown in Fig. A3.8. The use of the "Original Version" is recommended initially since it will simplify the choices on the interface for a novice user.

Fig. A3.5 Breakthrough
curve plot input window for
two-dimensional output data

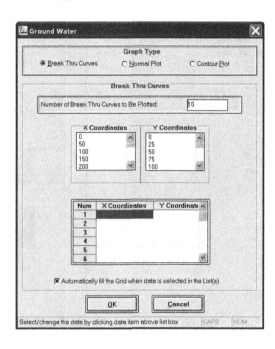

Fig. A3.6 Normal curve plot
input window for two-
dimensional output data

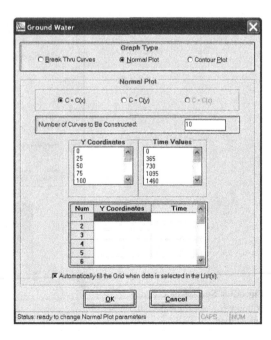

Fig. A3.7 Contour plot input window for two-dimensional output data

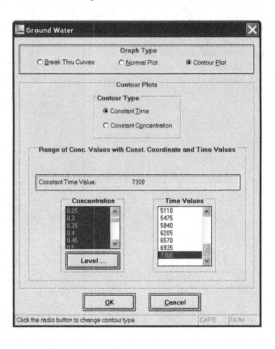

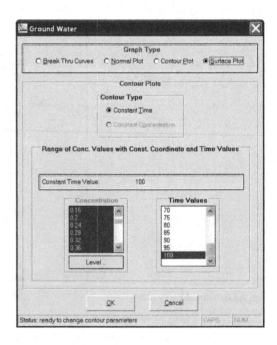

Fig. A3.8 Surface plot input window for two-dimensional output data

The first option to display the output is the plot of the breakthrough curves of the output concentrations at one or more points of the solution domain. In Fig. A3.5 the selection options are shown for a two-dimensional model. In this case, first the number of breakthrough curves that one would like to display will be entered (here the default is 10) and then the (x, y) coordinates of the point where the output will be displayed will be selected by clicking on the coordinate values in the x- and y-coordinates window. All selections made will appear in the grid below the x- and y-coordinates columns. If the model used generates one-dimensional or three-dimensional output the data entry window will automatically change, allowing only for x-coordinate selection for one dimensional models or x-, y-, z-coordinate selection for three dimensional models at this step. Once these selections are made the user may click on the "OK" button to display the output in the desired format. In the window where the graphical plot of the ACTS output is displayed there are several other options which allow the user to modify the output presented, such as the selection of the display colors, the type of lines or symbols used, changing the title of the plot or the axis titles of the plot, and finally adding footnotes to provide information about the plot prepared. All of these options can be added to the plot generated using the pull down menus at the top bar of the plot window or by clicking in the various locations of the plot area where a change or modification is desired. This process is the same in all figures that are generated in the ACTS and RISK software. One should also notice that the user may generate standard templates during this process, save templates generated or use an existing template generated earlier in the current plot. All of these options can be accessed from the pull down menu on the menu bar. It is recommended that the user get familiar with this process by using the interface provided in the plot window while experimenting with test databases.

The second plot option is identified as the "Normal" plot, which implies that the numerical results obtained from the ACTS software will be plotted as $C(x)$, $C(y)$ or $C(z)$ results at various times. The input data entry window for this case is shown in Fig. A3.6 where the input window for a two-dimensional model output is displayed. As can be seen in Fig. A3.6, the option to select $C(z)$ is inactive since the output data is generated using a two-dimensional model. If the model used is a one-dimensional model, the $C(y)$ option will also be inactive. After selecting which option to plot using the radio buttons, the user will again select the number of plots to be displayed. This selection is followed by the selection of the y-coordinates and time values of the desired $C(x)$ plot. At this step, if one selects to plot the $C(y)$ plot, than the x-coordinates and time values of the desired $C(y)$ plot will be entered. As before, the selected coordinates will appear in the grid below. Once the user is satisfied with the selections made, the "OK" button can be clicked to display the graph. Similar to the case above the title, the axis titles and footnotes can be added to the figure to finalize the plot. The templates option is available in this case as well. It is recommended that the user get familiar with this process by using the interface provided in the plot window while experimenting with sample databases.

The third option that is available to the user is the possibility of displaying the results of the computation in a contour plot format. This window is shown in Fig. A3.7 for the case of a two-dimensional analysis. The contour plot option is not available for one dimensional analysis since contour plots cannot be generated for one-dimensional analysis. As such, this option is only available for two- or three-dimensional analysis. As can be seen in Fig. A3.7 the choice for this case is the plot of various concentration contours at a fixed time or the plot of the spatial variation of constant concentrations at selected times. These selections can be made using the radio buttons and the grid data that are available for the output generated during an application. Once appropriate selections are made one may click on the "OK" button to display the contour plot desired. As before the title, the axis titles and footnotes can be added to the figure to finalize the plot in the plot window. The templates option is also available in this case as well. It is recommended that the user get familiar with this process by using the interface provided in the plot window while using sample data files provided.

The fourth option that is available to the user is the possibility of displaying the results of the computation in a surface plot format. This option is only available if the "New Version" of the graphics module is selected as the default graphics module in the preferences window as discussed above. This window is shown in Fig. A3.8 for the case of a two-dimensional analysis. The surface contour plot option is not available for one dimensional analysis since surface contour plots cannot be generated for one-dimensional analysis. As such, this option is only available for two- or three-dimensional analysis results. As can be seen in Fig. A3.8 the choice for this case is the plot of various concentration contours at a fixed time, or the plot of the constant concentration at selected times as described above for the contour plot option. These selections can be made using the radio buttons and the grid data that are available for the output generated during an application. Once appropriate selections are made, one may click on the "OK" button to display the surface plot desired. As before the title, the axis titles and footnotes can be added to the figure to finalize the plot in the plot window. The templates option is available in this case as well. It is recommended that the user get familiar with this process by using the interface provided in the plot window and the sample data files provided.

The operational characteristics of these plotting routines are standardized such that all modules in the ACTS and RISK software use the same interface. One should also notice that the graphics package that displays the statistical analysis is the same as the one discussed in this appendix. However, in the case of a RISK model, only the bar charts, the statistical output or the probability density functions of the statistical output can be displayed. The user, through practice, may immediately recognize that the procedures that are used to change the titles, colors and displayed lines follow the same procedures described in this appendix. Familiarity with these procedures may only be achieved through practicing the different options that are available in each window.

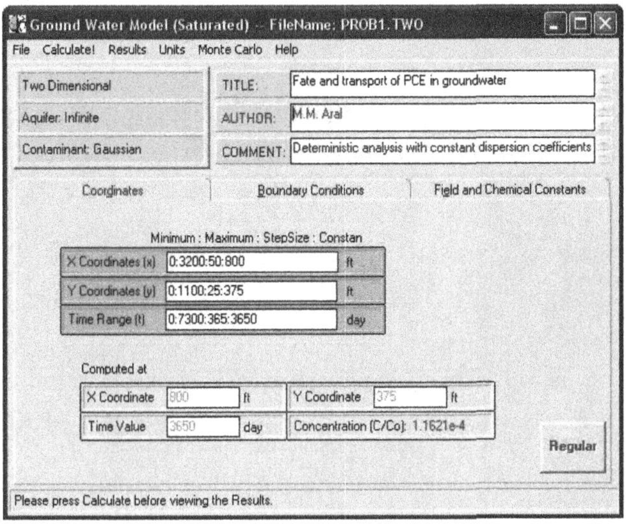

Fig. A3.9 Standard data entry window for groundwater module

Rules for Data Entry in a Typical Input Window

The data entry procedure for all models follows a very simple rule. The discretization of the selected coordinate axis will follow the following format described below as shown in Fig. A3.9. To discretize the x-coordinate, the beginning value (minimum), the final value (maximum), the discretization interval (step size) and the constant value are entered while separating the data entry values with (:) as shown below and in Fig. A3.9.

0:3200:50:800

The last data entry need not be entered, in which case the maximum value will be selected as the constant value from which the concentration will be computed and displayed in the output grid below. As the data are entered, the user will recognize the reflection of the input in the output grid below. The other coordinate discretization data follows the same format.

This is the only data entry procedure that is standardized according to a format. The other data entry options in the "Boundary Conditions" and "Field and Chemical Constants" folders will follow a simple numerical data entry action.

Appendix 4
MCL Levels on Contaminants

Table A4.1 USEPA Drinking-Water Standards on health goals

Chemical	MCLG (μg/L)	MCL (μg/L)	SMCL (μg/L)
Synthetic organic chemicals			
Acrylamide (1)	0[a]	Treatment technique[a]	Treatment technique[a]
Adipates (di(ethylhexyl)adipate)	400[b]	400[b]	
Alachor	0[a]	2[a]	
Aldicarb	1[c]	3[c]	
Aldicarb sulfoxide	1[c]	4[c]	
Aldicarb sulfone	1[c]	2[c]	
Atrazine	3[a]	3[a]	
Benzene	0[d]	5[e]	
Benzo[a]pyrene	0[b]	0.2[b]	
Carbofuran	40[a]	40[a]	
Carbontetrachloride	0[d]	5[e]	
Chlorinated Benzenes	0[b]	75–100[b]	
Chlorodane	0[a]	2[a]	
Chloroform	0[b]	100[b]	
Chlorophenoxy (herbicides 2,4,5,-TP)	0[b]	50[b]	
Chlorophenoxy (herbicides 2,4,-D)	0[b]	70[b]	
Dalapon	200[b]	200[b]	
Dibromochloropropane	0[a]	0.2[a]	
o-Dichlorobenzene (5)	600[a]	600[a]	10
p-Dichlorobenzene (5)	75[e]	75[e]	5
1,2-Dichloroethylene	0[d]	5[e]	
1,1-Dichloroethylene	7[d]	7[e]	
cis-1,2-Dichloroethylene	70[d]	70[e]	
Trans-1,2-Dichloroethylene	100[a]	100[a]	
1,2-Dichloropropane	0[a]	5[a]	
2,4-Dichlorophenoxyacetic acid (2,4-D)	70[a]	70[a]	
Di(ethylhexyl)phthalate	0[b]	6[b]	
Diguat	20[b]	20[b]	

(*continued*)

Table A4.1 (continued)

Chemical	MCLG (μg/L)	MCL (μg/L)	SMCL (μg/L)
Dinoseb	7[b]	7[b]	
Endothall	100[b]	100[b]	
Endrin	2[b]	2[b]	
Epichlorohydrin (1)	0[a]	Treatment technique[a]	
Ethylbenzene (5)	700[a]	700[a]	30
Ethylene dibromide	0[a]	0.05[a]	
Glyphosate	700[b]	700[b]	
Heptachlor	0[a]	0.4[a]	
Heptachlor epoxide	0[a]	0.2[a]	
Hexachlorobenzene	0[b]	1[b]	
Hexachlorocyclopentadiene [HEX] (5)	50[b]	50[b]	8[b]
Lindane	0[b]	0.4[b]	
Methoxychlor	40[a]	40[a]	
Methylene chloride	0[b]	5[b]	
Monochlorobenzene	100[a]	100[a]	
Oxamyl (vydate)	200[b]	200[b]	
PCBs as decachlorobiphenol	0[a]	0.5[a]	
Pentachlorophenol	0[a]	1[a]	
Picloram	500[b]	500[b]	
Simaze	4[b]	4[b]	
Styrene (5)	100[a]	100[a]	10
2,3,7,8-TCDD (dioxin)	0[b]	3.0×10^{-5}[b]	
Tetrachloroethylene	0[a]	5[a]	
1,2,4-Trichlorobenzene	70[b]	70[b]	
1,1,2-Trichloroethane	3[b]	5[b]	
Trichloroethylene (TCE)	0[d]	5[d]	
1,1,1-Trichloroethane	200[d]	200[d]	
Toluene (5)	1,000[a]	1,000[a]	40
Toxaphene	0[a]	3[a]	
2-(2,4,5-Trichlorophenoxy)-propionic acid (2,4,5-TP, or Silvex)	50[a]	50[a]	
Vinyl chloride	0[d]	2[d]	
Xylenes (total) (5)	10,000[a]	10,000[a]	20
Inorganic chemicals			
Antimony	6[b]	6[b]	
Arsenic		50[i]	
Asbestos (fibers per liter)	7×10^{6}[a]	7×10^{6}[a]	
Barium	2,000[c]	2,000[c]	
Berylium	0[b]	1[b]	
Cadmium	5[a]	5[a]	
Chromium (hex)	100[a]	100[a]	
Chromium (tri)	100[a]	100[a]	
Copper (4)	1,300[h]	1,300[h]	
Cyanide	200[b]	200[b]	
Fluoride	4,000[d]	4,000[d]	2,000[d]
Lead (4)	0[h]	15[h]	
Mercury	0.2[h]	0.2[a]	
Nickel	100[b]	100[b]	

(*continued*)

Table A4.1 (continued)

Chemical	MCLG (μg/L)	MCL (μg/L)	SMCL (μg/L)
Nitrate (as N) (2)	10,000^a	10,000^a	
Nitrite (as N)	1,000^a	1,000^a	
Selenium	50^a	50^a	
Silver			100
Sulfate	5×10^{5b}	5×10^{5b}	
Thallium	0.5^b	2^b	
Microbiological parameters			
Bacteria		<1/100 ml	
Giardic lamblia	0 Organisms^f		
Legionella	0 Organisms^f		
Heterotropic bacteria	0 Organisms^f		
Viruses	0 Organisms^f		
Radionuclides			
Radium 226 (3)	0^g	20 pCi/L^g	
Radium 228 (3)	0^g	20 pCi/L^g	
Radon 222	0^g	300 pCi/L^g	
Uranium	0^g	20 μg/L (30 pCi/L)^g	
Beta and Photon emitters (excluding radium 228)	0^g	4 mrem ede/year^g	
Adjusted gross alpha emitters (excluding radium 226, uranium and radon 222)	0^g	15 pCi/L^g	

A pCi (picocorrie) is a measure of the rate of radioactive disintegration. Mrem ede/year is a measure of the dose of radiation received by either the whole body or a single organ

1. This is a chemical used in treatment of drinking water supplies. The U.S. EPA specifies how much may be used in the treatment process. It would be unlikely to find this chemical in contaminated water

2. The total nitrate plus nitrite cannot exceed 10 mg/L

3. This MCL would replace the current MCL of 5 pCi/L for combined 226 Rsa and 228 Ra. The radionuclide rules were under review as of Spring, 1997

4. There is no MCL for copper and lead. The U.S. EPA requires the treatment of water before it enters a distribution system to reduce the corrosiveness so that these chemicals do not leach from the distribution system back into the water supply

5. SMCL is a suggested value only. Concentrations above this level may cause adverse taste in water. See Federal Register, January 30, 1991

6. The MCL for arsenic is under review as of Spring, 1997

^{a}Final value. Published in Federal Register, January 30, 1991
^{b}Final value. Published in Federal Register, July 17, 1992
^{c}Final value. Published in Federal Register, July 1, 1991
^{d}Final value. Published in Federal Register, April 2, 1986
^{e}Final value. Published in Federal Register, July 7, 1987
^{f}Final value. Published in Federal Register, June 29, 1989
^{g}Proposed value. Published in Federal Register, July 18, 1991
^{h}Final value. Published in Federal Register, July 7, 1991
^{i}Proposed 1 value. Published in Federal Register, November 13, 1985
^{j}Proposed value. Published in Federal Register, February 12, 1978

Appendix 5
Conversion Tables and Properties of Water

Table A5.1 Physical properties of water

Units	Specific weight γ	Density ρ	Viscosity μ	Kinematic viscosity ν	Surface tension σ	Vapor pressure
At normal conditions (20.2°C–68.4°F and 760 mm Hg–14.7 lb/in.^{2})						
SI	9,790 N/m^3	998 kg/m^3	1.0×10^{-3} N s/m^2	1.0×10^{-6} m^2/s	7.13×10^{-2} N/m	2.37×10^3 N/m^2
BU	62.3 lb/ft^3	1.94 slugs/ft^3	2.09×10^{-5} lb s/ft^2	1.08×10^{-5} ft^2/s	4.89×10^{-3} lb/ft	3.44×10^{-1} lb/in.^{2}
At standard conditions (14°C–39.2°F and 760 mm Hg–14.7 lb/in.^{2})						
SI	9,810 N/m^3	1,000 kg/m^3	1.57×10^{-3} N s/m^2	1.57×10^{-6} m^2/s	7.36×10^{-2} N/m	8.21×10^2 N/m^2
BU	62.4 lb/ft^3	1.94 slugs/ft^3	3.28×10^{-5} lb s/ft^2	1.69×10^{-5} ft^2/s	5.04×10^{-3} lb/ft	1.19×10^{-1} lb/in.^{2}

SI: Standard International Units; BU: British Units

Table A5.2 Common constants

Constants	SI	BU
Standard atmospheric pressure	1.014×10^5 N/m^2 (pascals)	14.7 lb/in.^{2}
	760 mm Hg	29.9 in. Hg
	10.3 m H$_2$O	33.8 ft H$_2$O
Gravitational constant	9.81 m/s^2	32.2 ft/s^2

Table A5.3 Conversion table

A	B	C
Length		
Inch	Meter	2.54E-2
Feet	Meter	0.3048
Yard	Meter	0.9144
Mile	Kilometer	1.609
Inch	Centimeter	2.54

(continued)

Table A5.3 (continued)

A	B	C
Area		
Square inch	Square centimeter	6.452
Square feet	Square meter	9.29E-2
Square yard	Square meter	0.8361
Square mile	Square kilometer	2.59
Acre	Square kilometer	4.047E-3
Acre	Hectare	0.4047
Volume		
Cubic feet	Cubic meter	2.832E-2
Cubic yard	Cubic meter	0.7646
Cubic inch	Cubic centimeter	1.639E1
Quart	Liter	0.9464
Gallon	Liter	3.785
Gallon (UK)	Liter	4.546
Acre-feet	Cubic meter	1.234E3
Million Gallon	Cubic meter	3.785E3
Gallon (UK)	Gallon (US)	1.2
Mass		
Pound (lb)	Kilogram	0.4536
Ounce	Gram	2.835E1
Ton, short	Tone metric	0.9072
Ton, long	Tone	1.016
Velocity		
Feet/second	Meter/second	0.3048
Mile/hour	Meter/second	0.447
Flow rate		
Gallons/minute	Liter/second	6.309E-2
Gallons/minute	Cubic meter/day	5.3
Acre-feet/day	Liter/second	1.458E-1
Hyd. conductivity		
Feet/year	centimeter/second	9.665E-7
Feet/year	Meter/day	8.351E-4
Darcy	feet/day	2.433
Darcy	Meter/day	0.7416
Transmissivity		
Square feet/second	Square meter/day	8.027E3
Square feet/day	Square meter/day	9.290E-2
Force		
Pound	Newton	4.448
Pound/square feet	Pascal	4.788E1
Pound/square inch	Kilogram/square centimeter	7.031E-2
Temperature		
Degree Fahrenheit	Degree Celsius	$5(^\circ F - 32)/9$
Degree Celsius	Degree Fahrenheit	$1.8(^\circ C) + 32$
Kelvin	Degree Celsius	Kelvin $- 273.2$

To convert A to B, multiply A by C; To convert B to A, divide B by C or perform the operation indicated in the box

Index